Handbook of Experimental Pharmacology

Volume 279

The *Handbook of Experimental Pharmacology* is one of the most authoritative and influential book series in pharmacology. It provides critical and comprehensive discussions of the most significant areas of pharmacological research, written by leading international authorities. Each volume in the series represents the most informative and contemporary account of its subject available, making it an unrivalled reference source.

HEP is indexed in PubMed and Scopus.

Jörg Striessnig
Editor

Voltage-gated Ca^{2+} Channels: Pharmacology, Modulation and their Role in Human Disease

Springer

Editor
Jörg Striessnig
Center for Chemistry and Biomedicine
UI, Pharmacology and Toxicology
Innsbruck, Austria

ISSN 0171-2004 ISSN 1865-0325 (electronic)
Handbook of Experimental Pharmacology
ISBN 978-3-031-38436-3 ISBN 978-3-031-38437-0 (eBook)
https://doi.org/10.1007/978-3-031-38437-0

This Springer imprint is published by the registered company Springer Nature Switzerland AG
The registered company address is: Gewerbestrasse 11, 6330 Cham, Switzerland

Preface

More than 50 years after the first approval of Ca^{2+}-channel blockers for the treatment of hypertension and myocardial ischemias, recent progress in voltage-gated Ca^{2+}-channel (VGCC) research now brings us much closer to discover next-generation Ca^{2+}-channel modulators for the treatment of human diseases also outside the cardiovascular system. Some key findings along this progress are outlined by leading experts in the field in this book: it brings together novel insight into human pathophysiology from channelopathies, newly discovered modulatory mechanisms, and advanced technology therapeutic strategies toward the isoform- and tissue-selective modulation of VGCCs, including gene therapy.

1. A Bumpy Road Toward Drug Therapies in Ca^{2+} Channelopathies

With next-generation DNA sequencing and advanced bioinformatic pipelines, we witnessed a dramatic increase in the identification of disease-causing genetic variants in all ten VGCC types. Functional expression of disease variants and characterization of their gating defects provided us with a better, although still incomplete, understanding of how altered channel function leads to human disease, often including neurodevelopmental and neurological symptoms. It also reveals the underlying molecular pathophysiology as an essential basis for the design of personalized therapies, including the repurposing of existing medicines. Together with long-term clinical data, this can improve care for affected individuals.

In this regard, research on *CACNA1A* channelopathies has taken a leading role in the Ca^{2+}-channel field. In *CACNA1A* channelopathies, disease-causing variants alter the function or expression of the pore-forming α1-subunit of Cav2.1 channels. *CACNA1A* channelopathies were the first Ca^{2+}-channel channelopathies (Episodic Ataxia, Familial Hemiplegic Migraine Type I and Spinocerebellar Ataxia Type 2) described in 1996–1998. Today, we know that *CACNA1A* genetic variants cause neurological and neurodevelopmental disorders much more frequently than variants in any of the other nine α1-subunit genes. In this book, Indelicato and Boesch have nicely summarized the large body of clinical data, which in recent years have expanded the clinical phenotype of *CACNA1A* channelopathies also to developmental, cognitive, and behavioral symptoms. Although specific and personalized therapies are not yet available, the authors outline the potential and limitations of current therapeutic options (such as the use of acetazolamide, aminopyridine, and antiepileptic and antimigraine drugs). As in other Ca^{2+}-channelopathies, this evidence mostly accumulates from case reports or small, non-controlled studies. This reflects the difficulty for controlled randomized trials with rare channelopathies and emphasizes the importance of data exchange between centers experienced in the treatment of such patients often in close exchange with families of affected individuals. The *CACNA1A* foundation (cacna1a.org) has already emerged as a platform to find specific treatment options by providing a collaborative network of patients, families, clinicians, and scientists. It could serve as a role model to coordinate research and therapy with other Ca^{2+}-channelopathies. Along these lines, Herold and colleagues and Nadine Ortner provide excellent overviews of the disease spectrums caused by gain- or loss-of-function mutations in the α1-subunits of Cav1.2 (*CACNA1C*) and Cav1.3 (*CACNA1D*) Ca^{2+}-channels, respectively. They discuss genotype–phenotype correlations emerging from the constantly rising number of disease variants reported in these genes. Both channels are members of the L-type channel family with known high sensitivity to clinically already used Ca^{2+}-channel blockers (such as nifedipine or verapamil). This raises the intriguing question if symptoms arising from the characteristic gain-of-channel function induced by these variants may respond to Ca^{2+}-channel blocker treatment. There is an unmet medical need to treat associated neuropsychiatric and neurological (seizures) symptoms (both channelopathies) as well as additional severe cardiac (*CACNA1C*, LQT8) or endocrine symptoms (*CACNA1D*, hyperaldosteronism, hyperinsulinism). Both articles explain why repurposing these drugs may or may not be efficacious with certain variants in these patients. Patient-derived induced pluripotent stem cells are also mentioned as an excellent system to test drug effects and can help to guide personalized treatments.

The articles by Nadine Ortner and Dinh et al. also emphasize how pathogenic variants in different Ca^{2+}-channels provide converging new insight into the role of Ca^{2+}-channels in a given tissue – in this case aldosterone-producing adrenal zona glomerulosa (ZG) cells. Gain-of-function variants both in *CACNA1D* (L-type Cav1.3) and in *CACNA1H* (T-type Cav3.2) cause hyperaldosteronism, either as somatic variants in tumors (aldosterone-producing adenomas, APAs) or as germline mutations. Intracellular Ca^{2+} is required for the activity of aldosterone synthase (*CYP11B2*) and other steps of aldosterone synthesis. Both channels have negative activation thresholds, can therefore be active close to the negative resting membrane potential (−80 mV) of ZG cells and could thus contribute to their electrical activity and intracellular Ca^{2+} oscillations. The pathologic hyperactivity of these channels therefore drives excess aldosterone production. However, in healthy ZG cells, the individual inhibition of Cav1.3 or Cav3.2 does not seem to affect aldosterone secretion, suggesting that they can compensate for each other. At present, there is also no systematic evidence for clinical benefit of treatment with currently available blockers of T-type or L-type Ca^{2+}-channels in patients with APAs. As discussed in these chapters, Cav1.3-selective blockers or combined blockers of T- and L-type channels (dual blockers inhibiting both L-type and T-type channels at therapeutic concentrations are not yet available) would be more suitable candidates.

Ghaleb and Flucher also provide an excellent overview over a recently discovered channelopathy resulting from *de novo* gain-of-function variants in *CACNA1I* (T-type Cav3.3 channels). Affected individuals present with variable degrees of a congenital neurodevelopmental disorder and epileptic seizures. So far only a small number of affected individuals have been reported and therefore it remains unclear if especially their seizures can be preferentially controlled by existing antiepileptic drugs with T-type channel blocking activity or even by new potent and highly selective T-type channel inhibitors currently in early clinical development (such as Z944/PRAX-944 for tremor or epilepsy).

Finally, Hessenberger et al. review the channel-dependent and more recently discovered channel-independent functions of the four α2δ-subunit isoforms. Thereby, they help us to better understand how genetic variants in their genes (*CACNA2D1-CACNA2D4*) may affect normal cellular signaling resulting in a number of human genetic diseases.

2. Toward Advanced Technology Medicinal Products Targeting Ca^{2+}-Channels

Mouse knockout and knockin models and more recently, as outlined in this book, in particular also Ca^{2+}-channelopathies have provided us with detailed insight into the physiological and pathophysiological role of the different Ca^{2+}-channels. More than 50 years after the first approval of (L-type) Ca^{2+}-channel blockers, this raises new interest in subtype-selective Ca^{2+}-channel modulators for the treatment of human disease. One example is the selective inhibition of Cav1.3 channel blockers for various indications (see chapter by Ortner in this book) including neuroprotection in early Parkinson disease but also the prevention of levodopa-induced dyskinesias (LID). In their article, Caulfield et al. nicely illustrate the role of Cav1.3 channels in the striatum and in substantia nigra dopamine neurons, providing a rationale for the specific inhibition of Cav1.3 as a therapeutic approach. No satisfactory therapies are available so far for most patients for the treatment of LID, justifying more advanced therapeutic strategies. The authors summarize compelling preclinical evidence that gene-therapy based knockdown of striatal Cav1.3 in parkinsonian rats can ameliorate LID in a preventive setting. It also reduced its severity once LID is established during high-dose levodopa treatment without obvious adverse effects. This approach highlights the potential of not only isoform- but also tissue- or even cell-specific Ca^{2+}-channel modulation. Likewise, Gavin and Colecraft also describe a novel approach, designer genetically-encoded Ca^{2+}-channel inhibitors (GECCIs) to achieve such maximal tissue- or organ-specific inhibition of a particular VGCC isoform for maximal on-target but minimal off-tissue adverse effects. They guide us through their continuous efforts to modify endogenous proteins known to constitutively inhibit many VGCCs in a way that enables the selective modulation of certain Ca^{2+}-channel isoforms even allowing their acute regulation. As an important delivery of these proof-of-concept studies, they now describe "Cav-aβlator," a nanobody-based gene-therapy approach, to silence Cav1 and Cav2 channels by selective degradation of their functionally relevant accessory β-subunits. They provide experimental evidence for its usefulness as a potential novel approach for treating pain in a preclinical neuropathic pain model.

3. Ca^{2+}-Channel Modulation: New Discoveries Answer (Very) Old Questions

Despite intensive research over more than 40 years, the research community was unable to answer two questions regarding key physiological processes known to be tightly controlled by Ca^{2+}-channels: (1) how are cardiac Cav1.2 channels upregulated by protein kinase A (PK-A) to contribute to the adrenergic stimulation of the fight-or-flight response in the heart, and (2) how do skeletal muscle Cav1.1 channels trigger the opening of ryanodine receptors and calcium release from the sarcoplasmic reticulum. In this book, Del Rivero Morfin et al. address the first question. They review the many unsuccessful attempts over the years in which researchers mainly focused on the stimulation of cardiac Cav1.2 channels by the *direct* PK-A-dependent phosphorylation of its α1- and β-subunits. The authors then explain their exciting discovery of the unexpected *indirect disinhibition* of channel activity induced by the PK-A phosphorylation of the inhibitory RGK-protein Rad. Rad phosphorylation dislodges it from the auxiliary β-subunit and thereby relieves constitutive channel inhibition. However, unlike in the heart, in neurons and in arterial smooth muscle cells PK-A activates Cav1.2 channels by direct phosphorylation of the α1-subunit at serine-1928. Baudel et al. review findings showing that this phosphorylation is required for the enhanced Cav1.2 activity in arterial smooth muscle cells induced by diabetic hyperglycemia. It involves the upstream activation of purinergic P2Y11 receptors, which promotes adenylate cyclase 5 activity. This enables the localized cAMP production within the channel's nanodomain also requiring PK-A anchoring to the α1-subunit by A-kinase anchoring protein 5 (AKAP5).

Campiglio et al. address the second "old question" by showing that the Cav1.1-associated protein STAC3 was an important missing link, transmitting fast voltage-dependent conformational changes of the plasma membrane Cav1.1 α1-subunit to the intracellular ryanodine receptors. STAC3 loss-of-function variants were discovered as the cause of a human congenital myopathy, which – together with other congenital myopathies resulting from mutations from proteins involved in excitation-contraction coupling – are summarized in this chapter.

Many of the findings described in the above chapters emphasize the important role of protein interactions with the channel complexes to mediate specific physiological functions. Loh et al. add another excellent example of how the Cav1.2-associated proteins Galectin-1 and Cereblon control the activity of these channels in the cardiovascular system and discuss their roles in hypertension and heart failure.

Taken together, this tremendous progress made during the past years in VGCC research holds promise that our advanced understanding of VGCC involvement in rare and common human disease together with rapidly emerging novel concepts of their isoform-selective and even tissue-specific modulation can pave the way toward next-generation of Ca^{2+}-channel modulators for the treatment of conditions with unmet clinical need.

Innsbruck, Austria Jörg Striessnig

Contents

Part I

Modulation

Ca_V1.1 Calcium Channel Signaling Complexes in Excitation–Contraction Coupling: Insights from Channelopathies

Marta Campiglio, Agnieszka Dyrda, Wietske E. Tuinte, and Enikő Török

Contents

Abstract

In skeletal muscle, excitation–contraction (EC) coupling relies on the mechanical coupling between two ion channels: the L-type voltage-gated calcium channel (Ca_V1.1), located in the sarcolemma and functioning as the voltage sensor of EC

M. Campiglio (✉) · A. Dyrda · W. E. Tuinte · E. Török
Institute of Physiology, Medical University Innsbruck, Innsbruck, Austria
e-mail: marta.campiglio@i-med.ac.at

J. Striessnig (ed.), *Voltage-gated Ca^{2+} Channels: Pharmacology, Modulation and their Role in Human Disease*, Handbook of Experimental Pharmacology 279,
https://doi.org/10.1007/164_2022_627

coupling, and the ryanodine receptor 1 (RyR1), located on the sarcoplasmic reticulum serving as the calcium release channel. To this day, the molecular mechanism by which these two ion channels are linked remains elusive. However, recently, skeletal muscle EC coupling could be reconstituted in heterologous cells, revealing that only four proteins are essential for this process: $Ca_V1.1$, RyR1, and the cytosolic proteins $Ca_V\beta_{1a}$ and STAC3. Due to the crucial role of these proteins in skeletal muscle EC coupling, any mutation that affects any one of these proteins can have devastating consequences, resulting in congenital myopathies and other pathologies.

Here, we summarize the current knowledge concerning these four essential proteins and discuss the pathophysiology of the $Ca_V1.1$, RyR1, and STAC3-related skeletal muscle diseases with an emphasis on the molecular mechanisms. Being part of the same signalosome, mutations in different proteins often result in congenital myopathies with similar symptoms or even in the same disease.

Keywords

Calcium signaling · Congenital myopathies · Malignant hyperthermia · STAC3

1 Skeletal Muscle Excitation–Contraction (EC) Coupling

Excitation–contraction (EC) coupling is a fundamental process in muscle physiology, in which an electric signal is converted into a mechanical response. In skeletal muscle, the voltage-gated calcium channel $Ca_V1.1$ functions as the voltage sensor of EC coupling (Beam et al. 1986; Rios and Brum 1987). When an action potential is generated at the neuromuscular junction, it travels along the sarcolemma and activates $Ca_V1.1$ in the transverse (T-) tubules of muscle cells. Upon activation, $Ca_V1.1$ undergoes a conformational change, which in turn activates the opening of the calcium release channel, the ryanodine receptor (RyR1) in the sarcoplasmic reticulum (SR) (Schneider and Chandler 1973). The increase of intracellular calcium concentration allows calcium to bind to the myofilament protein troponin C, which switches on the contractile machinery (Fig. 1).

Unlike cardiac muscle EC coupling, in which RyR2 is activated by the calcium entering through the voltage-sensor $Ca_V1.2$, skeletal muscle EC coupling is independent of calcium influx. In fact, the RyR1 receptor is mechanically activated by $Ca_V1.1$ in skeletal muscle in a process also called conformational coupling (Armstrong et al. 1972; Rios and Brum 1987). In teleost fish (e.g., zebrafish), $Ca_V1.1$ does not conduct any calcium current (Schredelseker et al. 2010) and functions merely as a voltage sensor. The signal transduction between the voltage sensor and the calcium release channel requires a highly specialized membrane system, called the triad. Triads consist of transverse tubules (T-tubules), i.e. invaginations of the sarcolemma where $Ca_V1.1$ channels are located, flanked on both sides by cisternae of the SR, the calcium stores where the RyR1 are positioned. The conformational coupling between $Ca_V1.1$ and RyR1 in skeletal

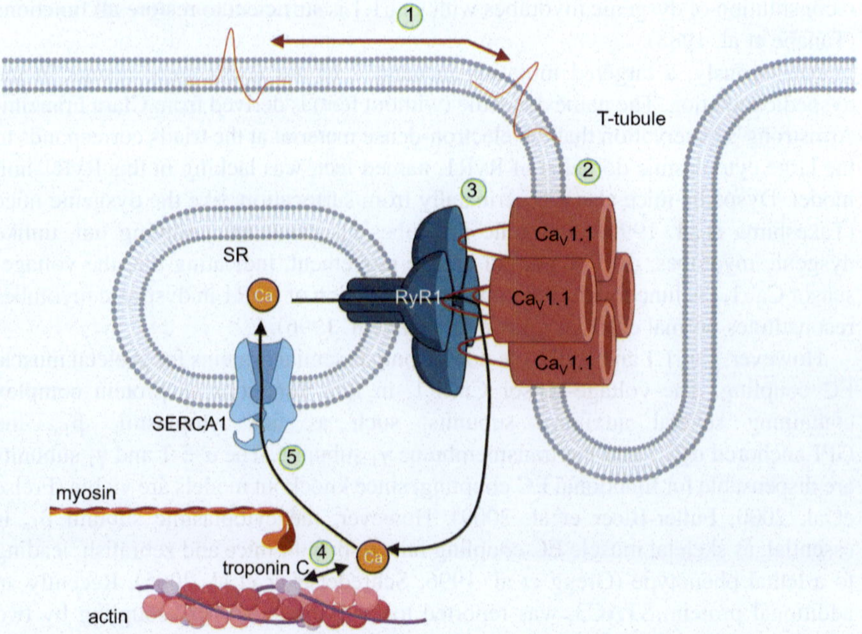

Fig. 1 Sequence of events in skeletal muscle EC coupling. (1) An action potential is generated at the neuromuscular junction (not shown) and travels down the T-tubule. (2) The action potential is sensed by the voltage-sensor $Ca_V1.1$, which through a conformational change activates the calcium release channel RyR1 on the sarcoplasmic reticulum (SR). (3) RyR1 releases calcium into the cytosol. (4) Calcium binds to troponin C, enabling repetitive cross-bridge formation between myosin heads and actin molecules. (5) Calcium is transported back into the SR by SERCA1. Created with BioRender.com

muscle requires not only their precise colocalization in the triads of muscle cells, but also the physical interaction of the two proteins. As a result, $Ca_V1.1$ forms ordered arrays of tetrads opposite the RyR1 (four $Ca_V1.1$ opposite every other RyR1) (Fig. 3c) (Franzini-Armstrong et al. 1998).

2 Essential Proteins for Skeletal Muscle Excitation–Contraction Coupling

The voltage-sensor $Ca_V1.1$ and the calcium release channel RyR1 are the two key players of EC coupling and, consequently, loss-of-function mutations in either protein is lethal. In mice, a single nucleotide deletion at position 4010 of the $Ca_V1.1$ transcript (Chaudhari and Beam 1989; Tanabe et al. 1988) results in a lethal dysgenic phenotype (Pai 1965). Dysgenic mice are paralyzed and die at birth from respiratory failure (Pai 1965). In fact, they lack EC coupling (Powell and Fambrough 1973), L-type channel conductance (Beam et al. 1986), and charge movement associated with voltage sensing (Adams and Beam 1990). However, recombinant reconstitution of dysgenic myotubes with $Ca_V1.1$ is sufficient to restore all functions (Tanabe et al. 1988).

Analogously, a targeted mutation in exon 2 of the RyR1 gene results in the dyspedic mutation. The name dyspedic (without feet) is derived from Clara Franzini-Armstrong's observation that the electron-dense material at the triads corresponds to the large cytoplasmic domains of RyR1, named feet, was lacking in this RyR1-null model. Dyspedic mice also die perinatally from suffocation, like the dysgenic mice (Takeshima et al. 1994). Dyspedic myotubes also lack EC coupling but, unlike dysgenic myotubes, display normal charge movement, indicating that the voltage-sensor $Ca_V1.1$ is functional. Recombinant expression of RyR1 in dyspedic myotubes reconstitutes normal calcium release (Nakai et al. 1996).

However, $Ca_V1.1$ and RyR1 are not the only essential proteins for skeletal muscle EC coupling. The voltage-sensor $Ca_V1.1$, in fact, functions as protein complex containing several auxiliary subunits, such as the cytoplasmic β_{1a}, the GPI-anchored $\alpha_2\delta$-1 and the transmembrane γ_1 subunits. The $\alpha_2\delta$-1 and γ_1 subunits are dispensable for functional EC coupling, since knockout models are viable (Freise et al. 2000; Fuller-Bicer et al. 2009). However, the cytoplasmic subunit β_{1a} is essential, as skeletal muscle EC coupling fails in β_1-null mice and zebrafish, leading to a lethal phenotype (Gregg et al. 1996; Schredelseker et al. 2005). Recently an additional protein, STAC3, was reported to be essential for EC coupling by two studies, which analyzed the effects of STAC3 knockout in mice and fish, respectively (Horstick et al. 2013; Nelson et al. 2013). Lack of STAC3 resulted in a paralyzed phenotype in zebrafish and caused rapid death at birth from suffocation in mice, similarly to the other EC coupling null models.

Following the discovery of STAC3, skeletal muscle EC coupling could be reconstituted in heterologous cells (Perni et al. 2017), defining that for this transduction process only four proteins ($Ca_V1.1$, RyR1, β_{1a}, and STAC3) are necessary and sufficient (Fig. 2).

2.1 The Voltage-Sensor $Ca_V1.1$

Voltage-gated calcium channels play a crucial role in many physiological processes in excitable cells. They open upon depolarization of the plasma membrane, resulting in influx of extracellular calcium into the cytosol, leading to various cellular responses. Voltage-gated calcium channels can be grouped into three categories: Ca_V1 (L-type), Ca_V2 (P/Q-, R-, N-), and Ca_V3 (T-type). The voltage sensor of EC coupling $Ca_V1.1$ belongs to the family of L-type calcium channels. L-type calcium channels are high-voltage activated channels and, therefore, require a strong depolarization for activation, with $Ca_V1.1$ requiring the strongest one, and display long-lasting currents. L-type currents can be blocked by three classes of antagonists, i.e. 1,4-dihydropyridines (DHP), phenylalkylamines, and benzothiazepines (Catterall 2011). Because skeletal muscle is the largest organ and because $Ca_V1.1$ binds with high affinity to the DHP nitrendipine, $Ca_V1.1$ was the first channel to be purified, cloned, and sequenced from rabbit T-tubule membranes of skeletal muscle

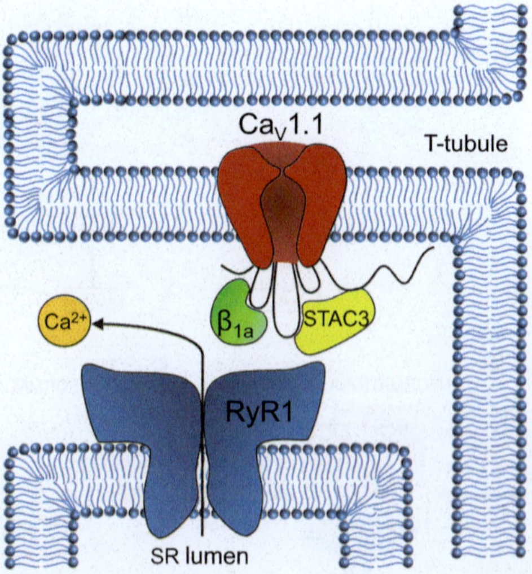

Fig. 2 Schematic view of the essential proteins in EC coupling. The voltage-sensor $Ca_V1.1$ is located in the T-tubule membrane. The $β_{1a}$ subunit interacts with the intracellular I-II loop of $Ca_V1.1$, while the adaptor STAC3 protein interacts with the II-III loop and the proximal carboxyl-terminus

(Curtis and Catterall 1984; Tanabe et al. 1987). Interestingly, the primary sequence of $Ca_V1.1$ revealed a basic structure very similar to that of the previously cloned $Na_V1.1$ (M. Noda et al. 1986; Tanabe et al. 1987). $Ca_V1.1$, as well as other Ca_V channels, consists of four non-identical homologous repeats, each comprised of six transmembrane α-helices (S1-S6). The S1-S4 of each repeat function as voltage-sensor domains (VSDs), a conserved structural and functional module capable of sensing a change in the cell membrane electrical potential and transducing it into a conformation change. On the other hand, the S5-S6 and their connecting loop form the pore and selectivity filter of the channel (Catterall 2011). The cytoplasmic amino- and carboxyl-termini, as well as the large loops connecting the four repeats (I-II, II-III, and III-IV), establish important interactions with proteins modulating the expression, targeting and gating of the channel (Fig. 3a) (Flucher et al. 2000; Kugler et al. 2004; Linsley et al. 2017; Nakada et al. 2018; Niu et al. 2018b; Pragnell et al. 1994).

Among the L-type calcium channels, $Ca_V1.1$ is unique, as it performs its main function as voltage sensor of EC coupling independently of calcium influx (Armstrong et al. 1972; Dulhunty and Gage 1988). The trigger for calcium release from the RyR1 is instead the outward movement of gating particles (Schneider and Chandler 1973), which were demonstrated to be the charged residues of the voltage-sensing S4 segments of $Ca_V1.1$ (Beam et al. 1986; Rios and Brum 1987) (Fig. 3b). Supporting the expendability of $Ca_V1.1$ calcium permeation for skeletal muscle EC coupling, in teleost fish $Ca_V1.1$ merely functions as the voltage sensor of EC coupling, as the two $Ca_V1.1$ isoforms have distinct residue substitutions that block calcium permeation (Dayal et al. 2021; Schredelseker et al. 2010). Accordingly, a mouse model with one of these pore substitutions in $Ca_V1.1$ (N617D), blocking

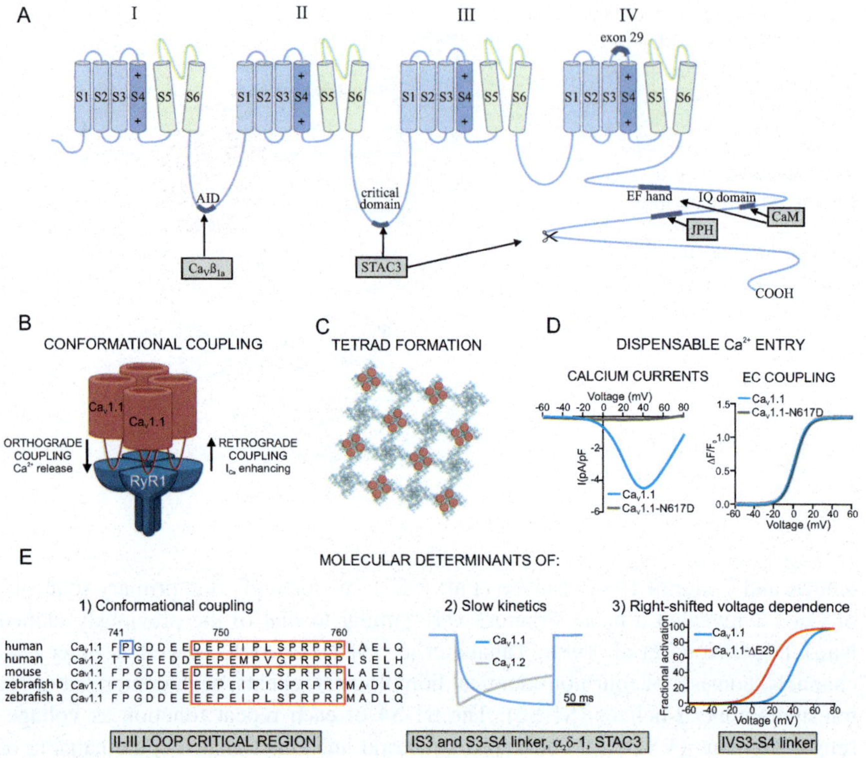

Fig. 3 Domain structure and functional properties of $Ca_V1.1$. (**a**) Cartoon depicting the domain structure of $Ca_V1.1$. The four homologous transmembrane domains (I–IV) consist each of six transmembrane domains. Within each domain, S1-S4 form the voltage-sensing domain (VSD, in blue) and S5 and S6 form the pore (in light green). The large intracellular loops provide binding sites for the interaction of auxiliary subunits and modulators. The I–II loop contains the AID (alpha interacting domain), responsible for the interaction with $Ca_V\beta_{1a}$, the II–III loop contains the critical region for EC coupling, responsible for the interaction with STAC3, while the carboxyl-terminus contains the regions responsible for the interaction with junctophilin (JPH), with calmodulin (CaM) and STAC3 (IQ domain and EF-hands). (**b**) The signaling with the RyR1 is bidirectional. The orthograde coupling from $Ca_V1.1$ to RyR1 activates the calcium release, while the retrograde coupling, from RyR1 to $Ca_V1.1$ enhances $Ca_V1.1$ currents. (**c**) As a result of the conformational coupling, $Ca_V1.1$ is arranged in tetrads opposite every other RyR1 tetramer. (**d**) The N617D mutation abolishes $Ca_V1.1$ currents, but does not affect EC coupling, demonstrating the expendability of $Ca_V1.1$ currents for EC coupling. (**e**) The molecular determinant of the conformational coupling with the RyR1 (1) is the critical region within the II–III loop of $Ca_V1.1$. The minimal peptide necessary for the binding to STAC3 is highlighted in red, while the P742, the mutation of which results in congenital myopathy is highlighted in blue. The molecular determinants of the "poor" channel characteristics of $Ca_V1.1$ (2) are: the IS3, S3–S4 linker, as well as the $\alpha_2\delta$-1 and STAC3 subunits for the slow kinetics of activation, and for the right-shifted voltage-dependence (3) the long IVS3–S4 linker, which is excluded in the ΔE29 variant. Panels B and C were created with BioRender.com

calcium permeation but allowing voltage sensing, displayed no aberrant phenotype (Dayal et al. 2017) (Fig. 3d).

Not only is calcium entry through $Ca_V1.1$ not required for skeletal muscle EC coupling, but it is also questionable whether $Ca_V1.1$ calcium currents are activated in the short duration of a T-tubular action potential (1–2 ms). In fact, $Ca_V1.1$ displays extremely slow kinetics of activation, right-shifted voltage-dependence of activation, small current amplitude, and low open probability compared to other calcium channels. Several mechanisms contribute to the slow kinetics of activation of $Ca_V1.1$. First, the S3 segment and the S3–S4 linker of the first repeat confer the slow kinetics of channel activation, which can be transferred to the fast-activating $Ca_V1.2$ when these segments are inserted (Nakai et al. 1994). Similarly, insertion of any heterologous sequence in the VSDI disrupts $Ca_V1.1$ slow kinetics (Tuluc et al. 2016). Secondly, auxiliary subunits, such as $\alpha_2\delta$-1 and STAC3 further decelerate the kinetics of activation of $Ca_V1.1$ (Obermair et al. 2005; Tuinte et al. 2022). The right-shifted voltage-dependence of activation and the low open probability are instead encoded in the long S3–S4 linker of the VSDIV (Tuluc et al. 2009, 2016) (Fig. 3e). Altogether, the presence of several mechanisms aimed at curtailing $Ca_V1.1$ currents in skeletal muscle suggests that calcium influx through $Ca_V1.1$ may not only be dispensable for muscle function, but also disadvantageous.

An exception to the "poor" channel characteristics of $Ca_V1.1$ is the embryonic $Ca_V1.1e$ splice variant. In fact, $Ca_V1.1e$ can support substantial calcium entry, as it displays enhanced kinetics of activation, a voltage-dependence of activation shifted by 30 mV toward more negative potentials and 5-folds-increased current amplitude due to a higher open probability compared to the adult $Ca_V1.1a$ variant (Tuluc et al. 2009). On a molecular level, the embryonic isoform lacks exon 29, encoding 19 residues in the S3–S4 linker of the fourth repeat of $Ca_V1.1$. Expression of $Ca_V1.1e$ is restricted to fetal and early post-natal development and completely absent in adult healthy skeletal muscle. Supporting the notion that $Ca_V1.1$ currents may be disadvantageous in adult skeletal muscle, a mouse model in which exon 29 is deleted, resulting in the continuous expression of the embryonic $Ca_V1.1e$ isoform, had mild effects on muscle differentiation and strength. Specifically, the elevated calcium influx through $Ca_V1.1e$ resulted in a shift of the fiber type composition toward the slower myosin isoform, and, in older mice, dramatic mitochondrial damage and loss have been observed (Sultana et al. 2016). The notion, that the increased calcium influx through an ectopically expressed embryonic $Ca_V1.1e$ isoform is harmful to skeletal muscles, is substantiated by the observation that in type 1 myotonic dystrophy (DM1) patients the severity of the muscle dysfunction correlates with the amount of re-expression of the embryonic isoform (Tang et al. 2012).

Ultimately, the main role of $Ca_V1.1$ is not calcium conduction, but sensing the action potential, and, in response, activating the RyR1 by a unique conformational coupling process. As discussed above this is supported by the formation of tetrads, i.e. four $Ca_V1.1$ opposite every other RyR1 (Fig. 3c). This is not the case for $Ca_V1.2$, the L-type calcium channel expressed in heart, which functions as the voltage sensor of EC coupling in cardiac muscle. $Ca_V1.2$ activates the RyR2 through the conduction of calcium, in a process called calcium-induced calcium release (CICR) (Franzini-Armstrong et al. 1998). Functionally the $Ca_V1.1$/RyR1 interaction depends on a specific sequence in the II-III cytoplasmic loop of $Ca_V1.1$ (Fig. 3e), which was identified using an elegant chimeric approach. Channel chimeras in which the intracellular loops and the amino- and carboxyl-termini of $Ca_V1.2$ were substituted with the corresponding regions of $Ca_V1.1$ were expressed and analyzed in dysgenic myotubes ($Ca_V1.1^{-/-}$) (Tanabe et al. 1990). Further mutagenesis analysis pinned down the essential residues of $Ca_V1.1$ involved in the linkage to RyR1 to eight amino acids within the II-III loop (Kugler et al. 2004).

Another way in which $Ca_V1.1$ differs from $Ca_V1.2$ is that the signaling between $Ca_V1.1$ and RyR1 is bidirectional (Fig. 3b). In addition to the orthograde signal from $Ca_V1.1$ activating RyR1 (i.e., EC coupling), the RyR1 mediates a retrograde signal, enhancing $Ca_V1.1$ currents. Consequently, dyspedic myotubes (RyR1 knockout) displayed $Ca_V1.1$ currents characterized by a very small amplitude. This is not due to a reduced functional expression of $Ca_V1.1$, as charge movement measurements were comparable to those from control myotubes. Instead, reconstitution of the RyR1 increased the $Ca_V1.1$ current amplitude, suggesting that the interaction with RyR1 increases $Ca_V1.1$ open probability (Nakai et al. 1996). Functionally, the interaction with the RyR1 seems to alleviate the inhibition of the auxiliary γ_1 subunit (Polster et al. 2016). Interestingly, the embryonic $Ca_V1.1e$ subunit is neither subjected to retrograde coupling by the RyR1, nor to inhibition by the γ_1 subunit, but still supports orthograde coupling (Benedetti et al. 2015; El Ghaleb et al. 2022; Tuluc et al. 2009). Not only the retrograde coupling modulates the current amplitude by increasing $Ca_V1.1$ open probability, but it also slows down the activation kinetics (Avila and Dirksen 2000). Interestingly, the $Ca_V1.1$ region critical for retrograde coupling is located in the II-III loop and coincides with that essential for the orthograde coupling (Grabner et al. 1999).

2.2 The Calcium Release Channel RyR1

Ryanodine receptors are the largest known ion channels (>2.5 MDa) and are named after the alkaloid with insecticidal properties found in the South American plant *Ryania speciosa* that binds and modulates gating of RyRs (Rogers et al. 1948). In mammals, three RyR genes (RyR1, RyR2, RyR3) encode subunits of ~5,000 amino acids, which assemble as homotetramers. The three RyR isoforms are differentially expressed. Whereas RyR1 is the skeletal muscle isoform, RyR2 is known as the cardiac isoform, but is also widely expressed in cells of the nervous system and other tissues. RyR3, while originally isolated from brain tissue, is ubiquitously expressed, and its physiological role is less defined (Hakamata et al. 1992; McPherson and Campbell 1993; Nakai et al. 1990; Otsu et al. 1990). Within the 5,000 amino acids monomer, the 1,000 carboxyl-terminal amino acids form the calcium conductive pore of the channel, which similarly to voltage-gated ion channels consists of six (or possibly eight) SR transmembrane helices. The first 4,000 amino acids of each subunit form the cytoplasmic domains (the observed "feet" in electron microscopy),

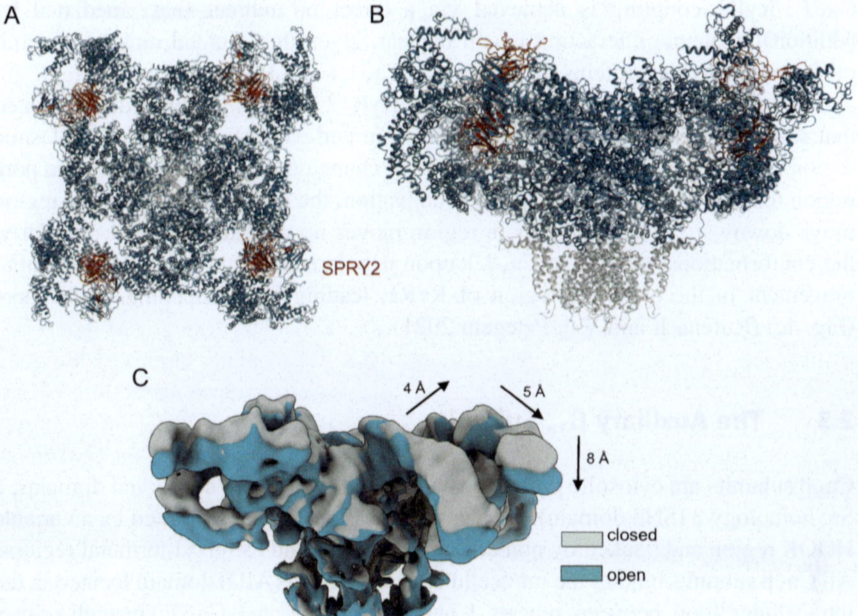

Fig. 4 RyR1 structures and opening. (**a**, **b**) Top and side view of a rabbit RyR1 tetramer (PDB 5TB0). The SPRY2 domain, which was suggested to interact with the II-III loop of $Ca_V1.1$, is colored in bromide red and mostly not exposed. The transmembrane regions are in gray. (**c**) Superposition of the density maps of RyR1 in the closed (EMD-8391, containing EGTA only) and open (EMD-8376, containing caffeine, ATP, and calcium). The arrows indicate the allosteric movements of the cytosolic cap, with the regions closer to the symmetry axis moving upward, and the peripheral parts moving downward

which contain several binding sites for various modulators (Fig. 4a, b) (Efremov et al. 2015; Yan et al. 2015; Zalk et al. 2015). Among these are calcium, ATP, magnesium, calsequestrin, calmodulin, S100A1, FKBP, triadin, SRP27, and calumenin (Bleunven et al. 2008; Buratti et al. 1995; Jones et al. 2005; Jung et al. 2006; Lai et al. 1988; Meissner 1984; Prosser et al. 2008; Tripathy et al. 1995; Wei et al. 2006; Wium et al. 2012). However, opening of the skeletal muscle isoform RyR1 is mainly triggered by the voltage-sensor $Ca_V1.1$. RyR1 is still sensitive to modulation by calcium and other factors, however its activity is less stringently regulated by the intracellular calcium concentration compared to the cardiac/neuronal RyR2 isoform (Laver and Lamb 1998).

The region on the RyR1 involved in the interaction with $Ca_V1.1$ remains elusive. Two studies have suggested an interaction between the SPRY2 domain of RyR1 and the II-III loop of $Ca_V1.1$ (Leong and MacLennan 1998; Tae et al. 2009). However, the crystal structure of the SPRY2 domain revealed that it is mostly not accessible and isothermal calorimetry experiments failed to detect any interaction with the II-III loop of $Ca_V1.1$ (Fig. 4a, b) (Lau and Van Petegem 2014). Altogether, whether the

$Ca_V1.1$/RyR1 coupling is achieved via a direct or indirect (e.g., mediated by additional proteins) interaction is still unclear, given that skeletal muscle EC coupling requires other proteins, in addition to $Ca_V1.1$ and RyR1.

How can $Ca_V1.1$ induce the opening of RyR1? Cryo-EM studies demonstrated that some allosteric activators, such as calcium and caffeine, bind the cytoplasmic region of RyR1 and induce a conformational change that is transmitted to the pore region (des Georges et al. 2016). Upon activation, the edges of the cytosolic region move downward, whereas the pore region moves upward and outward. Similarly, the conformational change in $Ca_V1.1$ upon depolarization may stimulate a similar movement in the cytosolic region of RyR1, leading to the opening of the pore (Fig. 4c) (Rufenach and Van Petegem 2021).

2.3 The Auxiliary β_{1a} Subunit

$Ca_V\beta$ subunits are cytosolic proteins consisting of two highly conserved domains, a Src homology 3 (SH3 domain) and a guanylate kinase (GK), separated by a variable HOOK region and flanked by non-conserved amino- and carboxyl-terminal regions. All $Ca_V\beta$ subunits bind to the intracellular α-interaction (AID) domain located in the intracellular loop between repeats I and II of Ca_V1 and Ca_V2 channels, via a hydrophobic groove in the GK domain (Fig. 5). The AID/$Ca_V\beta$ interaction plays an essential role in regulating the membrane expression and gating properties. In fact, in the absence of a $Ca_V\beta$, Ca_V1 and Ca_V2 channels show little or no functional expression. In addition, $Ca_V\beta$ subunits shift the voltage-dependence of activation toward more hyperpolarized potentials, modulate inactivation, and mediate the interaction of the channel with other regulatory proteins (e.g., RGKs, G-proteins, AHNAK).

There are four mammalian genes encoding for $Ca_V\beta$ (β_{1-4}) subunits, and each undergoes alternative splicing, generating multiple isoforms. Four different splice isoforms of the β_1 gene have been reported. The β_{1a} isoform, the first to be purified and cloned, is expressed only in skeletal muscle and it is the exclusive partner of $Ca_V1.1$. In accordance with the crucial role of $Ca_V\beta$ in targeting and modulating the

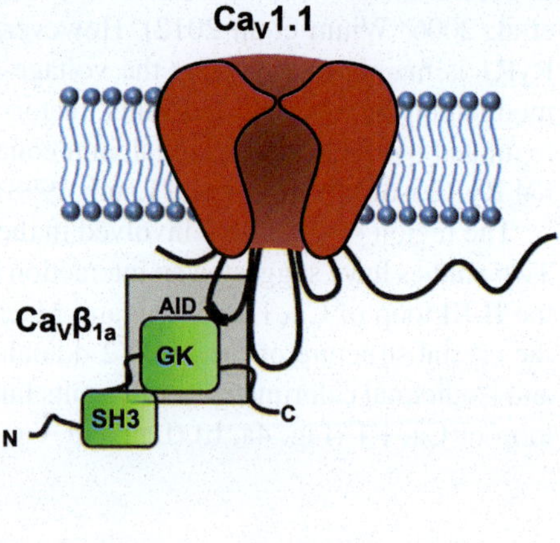

Fig. 5 Model showing the interaction between $Ca_V\beta_{1a}$ and $Ca_V1.1$. The $Ca_V\beta_{1a}$ subunit interacts with the AID region in I-II loop through the GK domain. The GK/AID interaction is conserved among channel and $Ca_V\beta$ isoforms. However, the SH3 domain and the carboxyl (C-) terminus of specifically $Ca_V\beta_{1a}$ have been shown to be essential for EC coupling

function of Ca_V1 channels, β_1-null models are lethal, similarly to $Ca_V1.1$ knockout models (Gregg et al. 1996; Schredelseker et al. 2005). In fact, β_1-null myotubes showed reduced $Ca_V1.1$ membrane expression and charge movement. Surprisingly, lack of the β_1 subunit results also in no tetrad formation, suggesting that the physical link to RyR1 is missing which is a prerequisite for skeletal muscle EC coupling (Schredelseker et al. 2005). Reconstitution of β_1-null zebrafish myotubes with non-muscle β paralogs rescued triad targeting of $Ca_V1.1$ and charge movement, indicating that these functions are common to all β subunits. However, tetrad formation and consequently skeletal muscle EC coupling is only promoted by the skeletal muscle β_{1a} isoform (Schredelseker et al. 2009). A chimeric approach similar to that taken to identify the critical $Ca_V1.1$ region for EC coupling revealed that the SH3 and carboxyl-terminal domains of β_{1a} are essential for restoring skeletal muscle EC coupling (Dayal et al. 2013). The lack of tetrad formation in the absence of the β_1 subunit suggests that β_1 is crucial for the $Ca_V1.1$/RyR1 interaction either by binding $Ca_V1.1$ and allosterically changing its conformation, or by directly binding the RyR1. Exploring the latter hypothesis, it has been reported that the last 35 amino acids of the β_{1a} subunit bind and activate purified RyR1 in vitro (Rebbeck et al. 2011). Particularly, within these 35 amino acid residues, a unique hydrophobic heptad motif is essential for skeletal muscle EC coupling, as a β_{1a} subunit with alanine substitutions of three of these residues failed to restore EC coupling in β_1-null mouse myotubes (Sheridan et al. 2004). However, reconstitution of the same triple mutant in β_1-null zebrafish myotubes resulted in normal EC coupling and tetrad formation (Dayal et al. 2010), challenging the notion that the carboxyl-terminus of β_{1a} binds to the RyR1. Altogether, while the importance of the carboxyl-terminus of the β_{1a} subunit for skeletal muscle EC coupling is established, which exact region is involved and its interacting partner are still elusive.

2.4 The Adaptor Protein STAC3

STAC proteins (SH3 and cysteine-rich containing proteins) form a family of three members (STAC1, STAC2, and STAC3) of adaptor proteins, i.e. proteins which facilitate interactions between protein-binding partners and thus the formation of macromolecular signaling complexes (Bower et al. 2012; Nelson et al. 2013). STACs are small proteins (364–403 amino acids) and consist of a variable amino-terminus, a protein kinase C (PKC) C1 domain, a conserved linker region, and two Src homology 3 (SH3) protein interaction domains. STAC1 and STAC2 are expressed in brain and a variety of other tissues, while expression of STAC3 is restricted to skeletal muscle (Nelson et al. 2013; Suzuki et al. 1996).

The importance of STAC3 in skeletal muscle EC coupling was first demonstrated by two studies, which analyzed the effects of STAC3 knockout in mouse and zebrafish (Horstick et al. 2013; Nelson et al. 2013). In both animal models, lack of STAC3 resulted in a EC coupling deficient phenotype, causing rapid death at birth from suffocation in mouse and in paralysis in fish.

STAC3 was reported to have multiple functions in skeletal muscle. First, STAC3 is important for $Ca_V1.1$ membrane expression. Indeed, $Ca_V1.1$ expression levels are reduced both in zebrafish and mouse STAC3 knockout myotubes (Linsley et al. 2017; Polster et al. 2016). Second, studies performed in zebrafish and mouse STAC3 KO myotubes demonstrated that STAC3 is also essential for the voltage-sensing function of $Ca_V1.1$. In fact, deletion of STAC3 causes reduced $Ca_V1.1$ charge movement in both models, more than the reduction in $Ca_V1.1$ levels would predict (Linsley et al. 2017; Polster et al. 2016). Additionally, in the mouse model, which unlike zebrafish expresses a conducting $Ca_V1.1$ channel, the currents are dramatically decreased (Polster et al. 2016). Third, analysis of zebrafish myotubes demonstrated that the formation of tetrads was essentially abolished in the absence of STAC3 (STAC3 KO and morpholino knock-down of the maternal RNA), suggesting that STAC3, similarly to β_{1a}, plays a crucial role in the physical arrangement of $Ca_V1.1$ opposite the RyR1 (Linsley et al. 2017). Additionally, reconstitution of KO mouse myotubes with a STAC3 mutant (W284S) linked to a severe myopathy (Native American myopathy or NAM) revealed that, beyond its role in supporting $Ca_V1.1$ function, STAC3 is essential for the conformational coupling between $Ca_V1.1$ and RyR1 (Linsley et al. 2017; Polster et al. 2016).

This three-fold function of STAC3 in trafficking, supporting charge movement, and EC coupling is reminiscent of that of the calcium channel β_{1a} subunit (Schredelseker et al. 2005). However, while a single crucial interaction site between β_{1a} and $Ca_V1.1$ has been identified (between the AID and the GK domain) (Chen et al. 2004; Opatowsky et al. 2004; Van Petegem et al. 2004), two distinct interactions between STAC3 and $Ca_V1.1$ have been reported (Fig. 6). The first interaction to be characterized was between the SH3 domains of STAC3 and the II-III loop of $Ca_V1.1$. Using an isothermal calorimetry assay, binding between the II-III loop of $Ca_V1.1$ and the SH3 domains of all three STAC isoforms was detected. Additionally, the crystal structure of the SH3 domains of the STAC2 isoform with the $Ca_V1.1$ minimal peptide, i.e. the 14 amino acids that are sufficient to detect binding between the two domains, revealed that, of the two SH3 domains, SH3-1 is mainly involved in the interaction with the II-III loop of $Ca_V1.1$ (Wong King Yuen et al. 2017). In an independent study, the group of Kurt Beam reached a similar conclusion. Using a colocalization assay in tsA201 cells, they demonstrated that the SH3-1 domain of STAC3 binds the II-III loop region of $Ca_V1.1$, which was previously found to be essential for EC coupling (Polster et al. 2018b). Together these two studies demonstrate that the SH3-1 domain of STAC3 associates with the II-III loop of $Ca_V1.1$ and that this interaction is crucial for the conformational coupling of $Ca_V1.1$ and the RyR1.

The second interaction between $Ca_V1.1$ and STAC3 is established between the C1/linker region of STAC3 and the proximal carboxyl-terminus of $Ca_V1.1$. First, we reported that the C1 domain of STAC3 is crucial for its stable incorporation in the calcium channel complex in skeletal muscle myotubes (Campiglio and Flucher 2017). Additionally, STAC3 increases the membrane incorporation and the gating of $Ca_V1.1$ in tsA201 cells (Polster et al. 2015), and, using a FRET approach, it was shown that STAC3 does so by interacting with the proximal carboxyl-terminus of

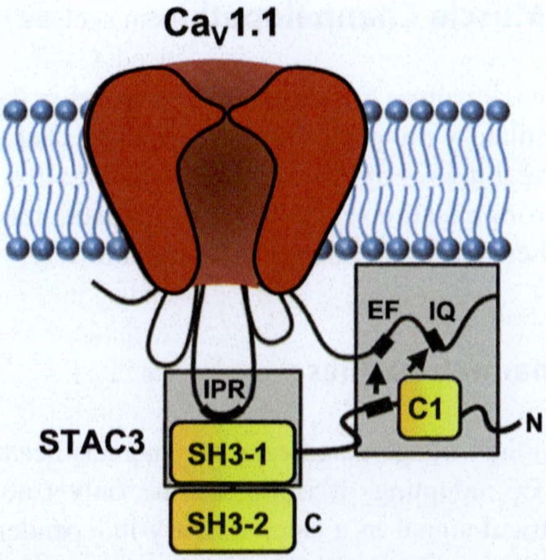

Fig. 6 Model showing two distinct interaction sites between STAC3 and $Ca_V1.1$. The first interaction is established between the $Ca_V1.1$ II–III loop and the SH3-1 domain of STAC3. The integrity of the IPR motif as well as the tryptophan mutated in Native American myopathy is crucial for this first interaction. The second interaction is established between the carboxyl-terminus of Ca_V1 channels and the linker region of STAC proteins. The integrity of the EF-hands and IQ domain of Ca_V1 channels as well as the STAC linker and C1 domain is important for this second interaction

$Ca_V1.1$, and more specifically with the EF-hands region (Niu et al. 2018b). This second STAC3/$Ca_V1.1$ interaction, conserved among STAC isoforms and L-type calcium channels, inhibits a negative feedback mechanism called calcium-dependent inactivation (CDI) and has been particularly studied in $Ca_V1.2$ and $Ca_V1.3$. The STAC3 binding site on the channel side was mapped on the proximal carboxyl-terminus, between the EF-hands and the IQ domain (Campiglio et al. 2018a; Niu et al. 2018a; Polster et al. 2018a), while the $Ca_V1.1$ binding site was identified in the linker connecting the C1 and the SH3 domains (Niu et al. 2018a; Polster et al. 2018a). Recently, this interaction was reported to contribute also to the slow kinetics of activation and inactivation of $Ca_V1.1$, by reconstituting a linker mutant (STAC3-ETLAAA) in STAC3 knockout myotubes. Interestingly, STAC3 does not alter CDI, which $Ca_V1.1$ lacks, but instead inhibits the other voltage-dependent component of inactivation (VDI) of $Ca_V1.1$ currents (Tuinte et al. 2022).

Finally, as STAC3 is an adaptor protein, and it interacts with the $Ca_V1.1$ critical region for EC coupling, it is tempting to speculate that STAC3 may mediate the coupling with RyR1. Indeed, it was reported that endogenous STAC3 localizes in the triads of myotubes even in the absence of $Ca_V1.1$, suggesting the presence of an additional interaction partner for this otherwise cytosolic protein (Campiglio et al. 2018b). However, to date, there is no further evidence that STAC3 may interact with the RyR1, leaving the question of how $Ca_V1.1$ and RyR1 interact still unanswered.

3 Skeletal Muscle Channelopathies

Skeletal muscle channelopathies are caused by mutations in the genes encoding ion channels or their regulatory proteins. They are typically characterized by recurring episodes of muscle weakness or by stiffness (myotonia). In this review, we will focus on diseases arising from mutations of the proteins essential for EC coupling. For the β_1-subunit, to date there are no disease-causing mutations reported.

3.1 $Ca_V1.1$ Channelopathies

$Ca_V1.1$ is unique among voltage-gated calcium channels because of its pivotal role in skeletal muscle EC coupling. It is indeed the only Ca_V isoform capable of transducing an electrical signal in a calcium entry-independent manner, through a conformational coupling with the RyR1, besides conducting calcium through its pore. Accordingly, mutations in $Ca_V1.1$ can cause abnormal ionic currents, as in the case of hypokalemic periodic paralysis (HypoPP) and normokalemic periodic paralysis (NormoPP), or result in a defective coupling with RyR1, as in the case of malignant hyperthermia susceptibility (discussed in the paragraph of diseases caused by mutations in multiple EC coupling proteins) (Fig. 7, Table 1).

3.1.1 Hypokalemic Periodic Paralysis (HypoPP)

Hypokalemic periodic paralysis is a dominantly inherited disorder characterized by recurrent episodes of muscle weakness or paralysis, with concomitant low serum K^+ (<3 mM) (Cannon 2015; Weber and Lehmann-Horn 1993). Attacks are usually triggered by rest after intense exercise, meals rich in carbohydrates, and cold. During an episode, the membrane potential of the affected fibers shifts to more positive values, causing the sodium channels to be inactivated. This, in turn, causes a reduction in fiber excitability and therefore muscle weakness and paralysis (Cannon 2017; Jurkat-Rott et al. 2012).

Most HypoPP cases origin from missense mutations in two voltage-gated ion channels, $Ca_V1.1$ (~60% of the cases) and $Na_V1.4$ (~20%) (Cannon 2015; Sternberg et al. 2001). Remarkably, 22 out of 23 HypoPP currently known missense mutations in $Ca_V1.1$ and $Na_V1.4$ neutralize gating charges R1 or R2 in the S4 segments (Cannon 2017; Matthews et al. 2009), leading to altered activation, current densities, and calcium release (Cannon 2015; DiFranco and Cannon 2022; Wu et al. 2012). However, the reduced activation and current density are not the cause of the disorder. In fact, the positive gating charges in the S4 segments are not only important for the voltage-sensing function, but also for sealing the hydrophilic gating pore through which the S4 segments move in response to changes in voltage. Therefore, mutation of R1 or R2 results in a misalignment of the S4 segments and the narrow hydrophobic constriction site (HCS), creating a secondary ion-conducting pathway through the affected voltage-sensing domain. The resulting small inward leak current, also called gating pore or omega current, flows at the resting membrane potential and stops when the membrane potential is depolarized, due to the outward movement of

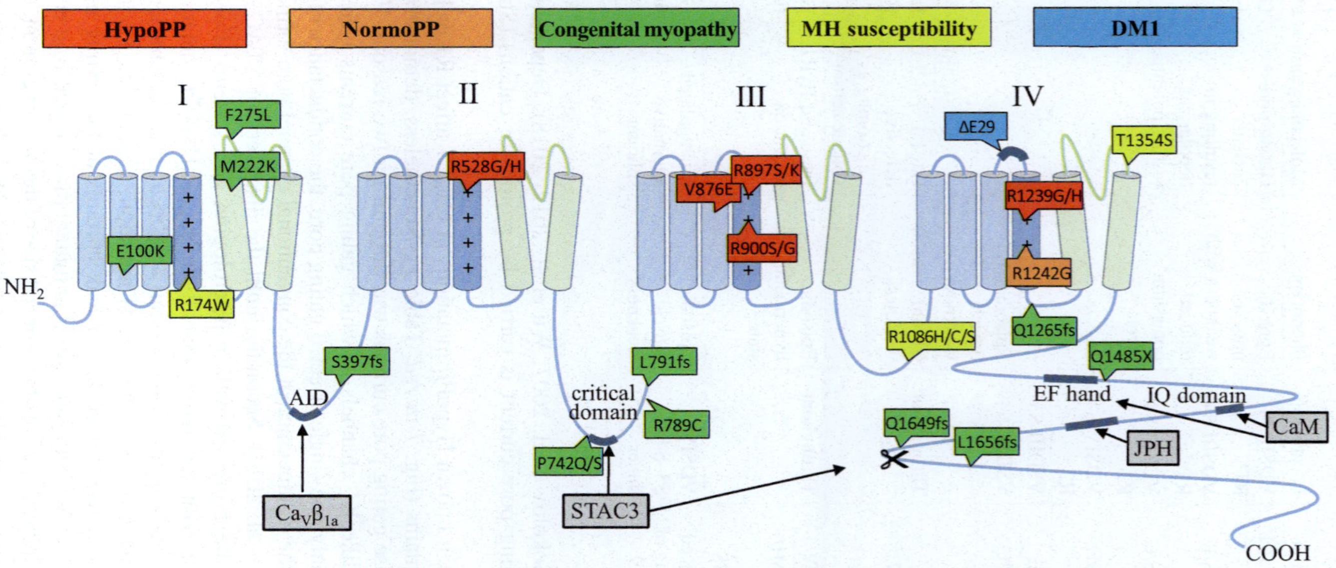

Fig. 7 $Ca_V1.1$ relative positions of the disease-causing mutations and splicing defects. $Ca_V1.1$ consists of four (I-IV) homologous repeats, each with a voltage-sensing part (S1-S4, shades of blue) and a pore forming region (S5-S6, gray), connected by cytoplasmic loops. The mutations causing hypokalemic periodic paralysis (HypoPP) are marked in red, normokalemic periodic paralysis (NormoPP) is in orange, congenital myopathy in green (fs for frameshift), malignant hyperthermia (MH) susceptibility in yellow, and myotonic dystrophy type 1 (DM1) in blue

Table 1 Summary of the $Ca_V1.1$ channelopathies discussed in this review

Disease	Mutation	Location	Phenotype	Molecular cause	Functional effects
HypoPP	R528G/H/C	VDSII S4 R1	Episodic muscle weakness with reduced potassium levels	Gating pore currents resulting in depolarized resting potential	Gain-of-function
	R897S/K	VDSIII S4 R1			
	R900S/G	VDSIII S4 R2			
	R1239H/G	VDSIV S4 R2			
	V876E	VSDIII S3			
Normo PP	R1242G	VSDIV S4	Episodic muscle weakness		
Congenital myopathy	Many (Fig. 7)	Throughout	Chronic muscle weakness	$Ca_V1.1$ defective functioning as voltage sensor	Loss-of-function
MH susceptibilty	R174W	Throughout	Episodic hypermetabolic state	Increased RyR1 activation	Gain-of-function
	R1086H/C/S				
	T1354S				
DM1	Enhanced splicing of exon29	VDSIV S3–S4 linker	Episodic muscle weakness	Misregulated alternative splicing	Gain-of-function

the S4 segments (Sokolov et al. 2007; Wu et al. 2012, 2018). Depending on the substitution the gating pore current is carried by protons or cations (Struyk and Cannon 2007).

Nine out of the ten known HypoPP mutations in $Ca_V1.1$ affect R1 or R2 of the voltage-sensing domains (Fig. 7 *in red*, Table 1). Nonetheless, demonstrating that these mutations cause gating pore currents was difficult for two reasons. First, each $Ca_V1.1$ mutant, unlike K^+ channels in which gating pore currents were initially characterized, has only one single affected gating pore, the conductance of which is much smaller than the conductance of the conventional channel pore formed by the pore domains (Mi et al. 2014). Secondly, until the discovery of STAC3, $Ca_V1.1$ channels could not be expressed well in heterologous cell systems (Polster et al. 2015). However, following the finding that all the $Na_V1.4$ HypoPP mutations result in anomalous gating pore currents, progress has also been made in characterizing the $Ca_V1.1$ HypoPP mutations. First, the R528H mutation, neutralizing R2 in VSDII, was demonstrated to cause inward cation leak currents at the resting membrane potential in patient muscle fibers and a knockin mutant mice (Jurkat-Rott et al. 2009; Wu et al. 2012). Nonetheless, due to the reduced size of the inward current, it was not possible to determine whether the current was carried by protons or sodium. More recently, co-expression of STAC3 of the R528H mutant in Xenopus oocytes revealed that the gating pore current is not carried by protons, but primarily by sodium. However, the mutation of the same residue to a glycine (R528G) results in a gating pore current carried by protons (F. Wu et al. 2018). Similarly, the mutant R1239H, neutralizing R2 in VSDIV, was shown to cause a proton-gating pore current in mouse muscle (Fuster et al. 2017a). Over the last 2 years, the four known HypoPP mutations of R1 and R2 in VSDIII were also expressed in Xenopus oocytes and characterized. Unlike R897K and R900S, the R897S and R900G give rise to substantial gating pore currents. In particular, the R897S mutant, which causes a severe clinical phenotype, generates the greatest gating pore current measured so far (Kubota et al. 2020; Wu et al. 2021).

Finally, the one atypical $Ca_V1.1$ HypoPP mutation that does not affect an S4 arginine, resides in the S3 domain of VSDIII and mutates a valine to a glutamate (V876E). The V876E mutant does not alter the activation of $Ca_V1.1$ currents, but still displays a sodium gating pore current (Fuster et al. 2017b). One possible explanation is that the substitution of the valine for a negatively charged residue alters the conformation of the neighboring residues R897 and R900, which are also implicated in the etiology of HypoPP (Pancaroglu and Van Petegem 2018). The fact that this non-canonical HypoPP mutation leaves the $Ca_V1.1$ calcium current unaltered, but yet generates a sodium leak current, supports the notion that gating pore currents consistently are the pathomechanism causing HypoPP. But how do these gating pore currents lead to HypoPP? And how does low extracellular K^+, which should cause a hyperpolarization of the plasma membrane rather than a depolarization paradoxically contribute to these episodes of weakness and paralysis? Skeletal muscle myotubes display a resting potential near the equilibrium potential of K^+ (–90 mV), arising from the balance of the inward rectifying K^+ channel current and the linear leak inward current (mainly Na^+). The currently proposed model for HypoPP suggests that the increased inward gating pore current due to the HypoPP mutation destabilizes this balance. At the same time, as extracellular K^+ is lowered, the voltage-dependence of inward rectifying K^+ channels shifts to more negative potentials, reducing K^+ conductance. As a result, the inward rectifying K^+ channels are not able to equate the inward current and the muscle membrane potential stabilizes at −50/60 mV, leading to $Na_v1.4$ channels inactivation and causing paralysis (Cannon 2015).

3.1.2 Normokalemic Periodic Paralysis (NormoPP)

As HypoPP mutations affect the R1 and R2 charges of the voltage sensor, similarly, mutations of the R3 charge in $Na_V1.4$ and $Ca_V1.1$ cause a disease characterized by similar symptoms. However, as R3 mutations result in attacks taking place at normal levels of serum K^+, the deriving muscle pathology is called normokalemic periodic paralysis (NormoPP) (Vicart et al. 2004). First, Vicart et al. (2004) reported 4 families carrying 3 distinct mutations of the R3 gating charge in VSDII of $Na_V1.4$ (R675G/Q/W) (Fig. 7, *in orange*). Further studies showed that, in contrast to HypoPP mutations, which cause gating pore currents at resting state potential, $Na_V1.4$ NormoPP mutations result in cation selective currents activated by depolarization and deactivated by hyperpolarization (Jiang et al. 2018; Sokolov et al. 2008).

For $Ca_V1.1$, Fan et al. (2013) reported a family with a mutation of the R3 gating charge in VSDIV (R1242G) causing NormoPP. Reconstitution of the mutant $Ca_V1.1$-R1242G in dysgenic myotubes revealed reduced current amplitude and left-shifted steady-state inactivation, which are a logical consequence of the outward gating pore current activated by depolarization. Surprisingly, in addition to the outward gating pore current, the R1242G mutation also allows an inward component at resting membrane potential after prolonged depolarization. In the case of this NormoPP mutation, the position of R1242G relative to the HCS determines the gating pore conductance. In the resting state, the residue is positioned under the HCS, and $Ca_V1.1$ conducts an outward gating pore current. In the depolarized state, the residue is located above the HCS, and $Ca_V1.1$ conducts an inward pore current. These bidirectional gating pore currents result in a complex muscle pathology, characterized by simultaneous paralysis due to $Na_V1.4$ inactivation at resting potential, caused by the outward gating pore currents, and weakness at depolarized potentials, caused by continuous triggering of calcium-induced calcium release by the inward gating pore currents.

3.1.3 $Ca_V1.1$ Related Myopathies

Congenital myopathies are a heterogeneous group of muscle diseases of variable severity. They are characterized by fetal, neonatal, or early onset symptoms, ranging from fetal akinesia to mild muscle weakness and hypotonia. These symptoms are accompanied by morphological abnormalities in skeletal muscle biopsies (Jungbluth et al. 2018). So far, mutations of about 30 genes, encoding proteins regulating EC coupling and calcium handling, have been implicated in the etiology of congenital myopathies, including $Ca_V1.1$ (Romero and Clarke 2013). First, a whole exome sequencing study associated two $Ca_V1.1$ variants, a frameshift mutation, and one affecting splicing, as the cause of the congenital myopathy phenotype in a 9 months old patient (Hunter et al. 2015). Subsequently, Schartner et al. (2017) identified several congenital myopathy mutations in $Ca_V1.1$ through exome sequencing of seven distinct affected families. More recently, two additional $Ca_V1.1$ missense mutations leading to fetal akinesia were reported (Ravenscroft et al. 2021). Many of the identified variants introduce a stop codon or a frameshift mutation leading to a premature truncation, with consequent loss of $Ca_V1.1$ function (Fig. 7, *in green*). Other mutations, on the other hand, lead to single amino acid changes. Even if not yet experimentally validated, the consequence of some of these missense mutations can be predicted based on their position. For example, two different residues in the II-III loop of $Ca_V1.1$ (P742Q/S and R789C) were reported to cause congenital myopathy when mutated (Ravenscroft et al. 2021; Schartner et al. 2017). The II-III loop of $Ca_V1.1$ contains the critical region for skeletal muscle EC coupling, i.e. for the mechanical coupling with the RyR1 (Kugler et al. 2004; Tanabe et al. 1990). Importantly, a crystal structure of the SH3 domains of STAC3 was solved in a complex with a $Ca_V1.1$ II-III loop peptide (Rufenach et al. 2020; Wong King Yuen et al. 2017). Therefore, these mutations in the II-III loop of $Ca_V1.1$ may specifically interfere with the interaction with STAC3 and obstruct EC coupling. Accordingly, mutation of P742 to a threonine was indeed shown to diminish voltage-induced calcium release from the RyR1 (Kugler et al. 2004), supporting the notion that mutations in the II-III loop can disrupt the communication with the RyR1.

Similarly, the E100 residue in the transmembrane S3 of repeat I is a critical counter charge for the translocation of the S4 helix through the membrane during depolarization (Chanda and Bezanilla 2008). The E100K mutation substitutes the countercharge with a positively charged residue (Schartner et al. 2017), and is therefore expected to interfere with the $Ca_V1.1$ voltage-sensing function (Flucher 2020).

Another missense mutation, M222K, affects a residue in the IS5 and it is five amino acids downstream of E216 (Ravenscroft et al. 2021). We recently demonstrated that E216 is an ion interaction partner of R1 and R2 of VSDIV, which is important for stabilizing the activated state of the channel (El Ghaleb et al. 2021). Therefore, the non-conservative M222K mutation may destabilize the IS5 structure and disrupt the interactions established by E216.

The F275L missense mutation affects a residue in the extracellular S5-S6 loop of the first repeat, which contributes to the ion selectivity filter (Wu et al. 2016). Although F275 is not directly part of the selectivity filter, it is a conserved residue and its mutation may alter the local secondary structure and impair the calcium permeation through the pore.

Several frameshift/nonsense mutations cause a premature stop codon, resulting in a non-functional channel and in congenital myopathy, e.g. S397Pfs*3, L791Cfs*37, Q1265Hfs*57 (Schartner et al. 2017). Some of these mutations affect the carboxyl-terminus of $Ca_V1.1$. For example, the nonsense mutation Q1485X results in a truncation of the $Ca_V1.1$ carboxyl-terminus (Schartner et al. 2017), that deletes the IQ domain, crucial for the interaction with STAC3 (Campiglio et al. 2018a), and the triad targeting region, which interacts with junctophilin (Nakada et al. 2018; Yang et al. 2022). Therefore, the latter truncation impairs $Ca_V1.1$ triad targeting and function. However, mutations resulting in frameshifts at a more distal position, such as the deletions Q1649Qfs*72 and L1656Rfs*67, are not predicted to interfere with $Ca_V1.1$ function, as they are located close to $Ca_V1.1$ intrinsic cleavage site. Nonetheless, myotubes from patients carrying the L1656R mutation showed reduced EC coupling (Schartner et al. 2017).

In conclusion, although many $Ca_V1.1$ mutations causing congenital myopathy have been reported, a functional characterization is necessary to fully understand the etiology of the disease. However, prediction of the effects of some mutations, laying in domains with known function, indicate that both altered EC coupling and calcium influx may contribute to the disease.

3.2 RyR1 Channelopathies

RyR1 is the calcium release channel and, due to the central role of calcium in skeletal muscle cells, mutations in the RyR1 gene can lead to severe disorders. RyR1 mutations have been mostly associated with malignant hyperthermia susceptibility (MHS), discussed in the paragraph of diseases caused by mutations in multiple EC coupling proteins, and congenital myopathies with cores, the most frequent of which is central core disease (CCD).

3.2.1 Central Core Disease (CCD) and Dusty Core Disease (DuCD)

Central core disease (CCD) is a congenital myopathy, clinically characterized by hypotonia, muscle weakness, delayed motor development, and sporadically hip dislocation and scoliosis (Jungbluth 2007a). CCD is named after the characteristic dense cores of metabolically inactive tissue, lacking mitochondria, localized at the center of the fibers. CCD fibers also show extreme slow-twitch fiber predominance (Wu et al. 2006). CCD is mostly autosomal dominantly inherited, although it can occasionally be recessive, and the disease-causing mutations typically affect the carboxyl-terminal region of the RyR1 protein (Jungbluth et al. 2002). Recently, Garibaldi et al. (2019) discerned between CCD, dominantly inherited, and the Dusty core disease (DuCD), caused by recessive mutation in the RyR1 gene. Clinically, patients with DuCD have a similar, but more severe phenotype compared to CCD-affected individuals and sometimes show fetal akinesia syndrome. However, unlike the fibers of CCD patients, which present central cores with clear ovoidal shapes lacking oxidative enzymatic staining, the metabolically inactive regions in Dusty core fibers are characterized by no clear borders, size, or shape.

CCD-associated mutations can affect RyR1 by causing either a gain-of-function in gating or a loss-of-function in conductance. The gain-of-function mutants display similar single-channel conductance, but a higher open probability, resulting in elevated intracellular calcium concentration and SR store depletion. For example, reconstitution of dyspedic myotubes with CCD-associated RyR1 mutants (R164C, I404M, Y523S, R2163H, R2435H, ΔR4215-F4217 in the rabbit sequence) caused elevated cytoplasmic calcium concentration at rest and SR calcium depletion, due to a leaky channel behavior (Avila and Dirksen 2001; Lyfenko et al. 2007). The elevated cytoplasmic calcium concentration at rest is also a hallmark of malignant hyperthermia susceptibility (MHS).

The loss-of-function mutants, conversely, show low or absent calcium conductance. For example, reconstitution of dyspedic myotubes with RyR1-ΔV4926-I4927 resulted in complete loss of calcium release from the RyR1 in response to sarcolemma depolarization or caffeine (Lyfenko et al. 2007). Similarly, calcium release was abolished in homozygous RyR1-I4895T knockin mice and severely reduced in heterozygous animals, suggesting a dominant-negative effect of the mutation (Loy et al. 2011). Analysis of the Q3970K mutation provides more mechanistic insights into how a mutation causes loss of RyR1 channel function. H3895 and Q3970 interact with the three residues responsible for binding calcium (Q3893, Q3967, and T5001). The Q3970K mutation results in functional calcium conducting channels, but in low open probability in the presence of calcium, ATP, and caffeine, suggesting a defective regulation of the channel activity by the three activating RyR1 ligands (Chirasani et al. 2019).

3.2.2 Other Congenital Myopathies Caused by RyR1 Mutations

Besides CCD, other congenital myopathies are known to arise from mutations in the RyR1 gene. Similarly to CCD, these myopathies are characterized by different muscle weakness levels and are associated with malignant hyperthermia.

Multi-minicore disease (MmD) is a recessively inherited neuromuscular disorder, characterized by muscle weakness, early spinal rigidity, scoliosis, external ophthalmoplegia, and respiratory impairment. Mutations in several genes can cause MmD, including selenoprotein N (SEPN-1), myosin 7 (MXH7), and titin (TTN), and, most frequently, RyR1 (Jungbluth et al. 2018). In MmD, muscle fibers are characterized by multiple areas with reduced oxidative activity. Unlike CCD, which is defined by well-defined centrally located cores running along a substantial length of the fiber, and unlike DuCD, which is characterized by dusty undefined cores, MmD muscle fibers present multiple minicores running along only a limited extent of the longitudinal fiber (Garibaldi et al. 2019; Jungbluth 2007b).

Core-rod myopathy is characterized by the presence of cores, typical of CCD, and rods, typical of nemaline rod myopathy, in distinct locations within the same or different fibers. Core-rod myopathy can be caused by mutations in several genes (NEB, ACTA1, KBTBC13, CFL2, TRIP4, TNNT1), including RyR1. RyR1-related core-rod myopathy is autosomal dominantly inherited (Monnier et al. 2009; Scacheri et al. 2000; von der Hagen et al. 2008) and clinically characterized by severe, sometimes fatal, hypotonia.

Centronuclear myopathy (CNM) is characterized by muscle fibers with centralized nuclei. It is mainly associated with mutations in the myotubularin gene MTM1 (on chromosome X), but also in dynamin 2 (DNM2), amphiphysin II (Bin1), titin (TTN), and RyR1 (Jungbluth et al. 2018). The RyR1-associated CNM is characterized by extraocular movement, muscle weakness, and moderate bulbar and respiratory impairment (Jungbluth et al. 2008).

King-Denborough syndrome (KDS) is characterized by dysmorphic facial features, short stature, spinal rigidity, scoliosis, and various histopathological features. At least three distinct mutations in RyR1 have been reported to cause KDS (Dowling et al. 2011).

3.3 STAC3 Channelopathies

Mutations in the STAC3 gene can result in a rare congenital myopathy, initially called Native American myopathy (NAM), as the first cases were identified in the Lumbee Tribe of North Carolina (Stamm et al. 2008). However, subsequently, multiple cases have been identified in patients worldwide (Grzybowski et al. 2017; Telegrafi et al. 2017; Zaharieva et al. 2018), and the name NAM became inaccurate. For this reason, NAM is also referred to as STAC3 disorder or Bailey–Bloch syndrome, after the first case report of a three-month-old patient in 1987 (Bailey and Bloch 1987).

3.3.1 Native American Myopathy (NAM)

Mutations in the human STAC3 gene have been linked to the congenital Native American Myopathy (NAM), a disorder characterized by multiple clinical features, including muscle weakness, short stature, delayed motor development, scoliosis, dysmorphic facial features, and susceptibility to malignant hyperthermia (MH) (Horstick et al. 2013; Webb et al. 1993). NAM is an extremely debilitating autosomal recessive disease, causing death in one-third of the affected individuals before the age of 18, due to respiratory failure or MH episodes (Stamm et al. 2008).

The first and more commonly reported mutation leading to NAM is a tryptophan to serine substitution at residue 284 of STAC3, affecting a conserved residue of the first SH3 domain, critical for the interaction with proline-rich regions (Horstick et al. 2013) (Fig. 8). Importantly, W284 is crucial for interacting with the II-III loop of $Ca_V1.1$ and isothermal calorimetry experiments demonstrated that the mutation to serine abolishes this interaction (Wong King Yuen et al. 2017). Reconstitution of STAC3 KO mouse myotubes with the STAC3-NAM mutant resulted in only a mild reduction in $Ca_V1.1$ charge movement compared to the control myotubes (20%), suggesting that STAC3-NAM can support the voltage-sensor function of $Ca_V1.1$ (Polster et al. 2016). In contrast to these findings, reconstitution of zebrafish KO myotubes with the W284S mutation highlighted the importance of the SH3-1/II-III loop interaction for the function of $Ca_V1.1$ as the voltage sensor of EC coupling. In fact, myotubes reconstituted with STAC3-W284S displayed negligible charge movement, indicating that $Ca_V1.1$ is not fully functional (Linsley et al. 2017).

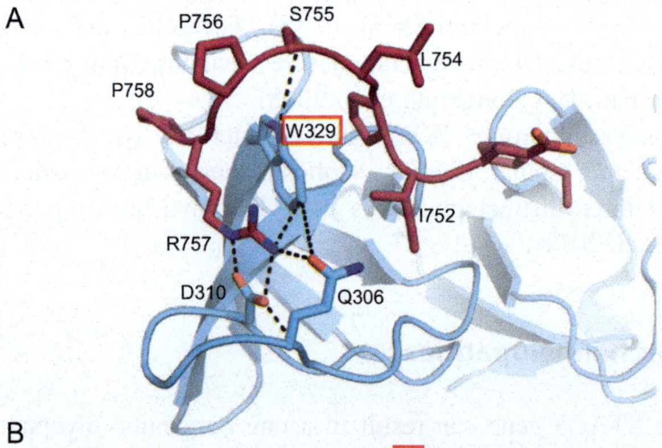

B

human	K I T V I D D S N E E W W R G K I G E K V G F F P
mouse	K I T V I D D S N E E W W R G K I G E K V G F F P
rat	K I T V I D D S N E E W W R G K I G E K V G F F P
rabbit	K I T V I D D S N E E W W R G K I G E K V G F F P
zebrafish	R I T V L D D S N E E W W R G K I G E K T G Y L P

Fig. 8 Position of the STAC3 W284 residue. (**a**) Crystal structure of the critical domain peptide (747–760) in the II–III loop of $Ca_V1.1$ and the STAC2 SH3 domains (PDB 6B27). The $Ca_V1.1$ peptide is colored in magenta, while STAC2 in light blue. The critical W329 residue, corresponding to W284 in STAC3, is circled in red. (**b**) Alignment of the sequences of the region of the SH3-1 domain of STAC3 shows the high conservation of W284 (circled in red) across species

Conversely, the voltage-induced calcium release from the RyR1 was severely impaired compared to WT STAC3 in both models (90% reduction) (Linsley et al. 2017; Polster et al. 2016). Together, these studies demonstrate the importance of the interaction of the SH3-1 domain of STAC3 with the II-III loop of $Ca_V1.1$, which is disrupted in the W284S NAM mutation. However, while the study in mice suggests that the mutation results in impaired conformational coupling between $Ca_V1.1$ and the RyR1, the study in zebrafish implies that it results in impaired $Ca_V1.1$ voltage sensing.

In addition to W284S, several STAC3 mutations resulting in premature stop codons (K288*, L255I*58) or deletions (Δ333IVVQ336) were reported to cause NAM (Grzybowski et al. 2017; Telegrafi et al. 2017; Zaharieva et al. 2018), likely due to loss-of-function or, in the case of the deletion, protein misfolding (Rufenach et al. 2020).

In clinical databases additional variants were reported, but the clinical significance and gravity are unclear. Some of these, such as F295L and K329N reduce EC coupling by weakening the interaction with the II-III loop of $Ca_V1.1$, similarly to W284S, but to a smaller extent. Other variants do not have any effect. Interestingly, the H311R variant affects the second SH3 domain of STAC3 and reduces EC coupling without interfering with the binding to the II-III loop, suggesting that this region may interact with another protein, e.g. the RyR1 (Rufenach et al. 2020).

Altogether, the molecular basis and the clinical features of NAM demonstrate that the SH3 domains of STAC3 have a crucial role in the conformational coupling of $Ca_V1.1$ and RyR1. While it has been established that the SH3-1 domain of STAC3 interacts with the II-III loop of $Ca_V1.1$, it is still unclear whether STAC3 also interacts with RyR1. Future research will need to clarify whether STAC3 functions as an adaptor protein between $Ca_V1.1$ and RyR1, mediating the coupling between the two, or if its binding to $Ca_V1.1$ results in an allosteric conformational change that allows the interaction with RyR1.

3.4 Channelopathies Caused by Several EC Coupling Proteins

Some channelopathies can be caused by mutations in several EC coupling proteins. The most glaring example is malignant hyperthermia susceptibility, due to a gain-of-function of the RyR1, which can be caused by mutations in RyR1, $Ca_V1.1$, or STAC3. This is because all three proteins contribute to the same signalosome; therefore, mutations in any of the three proteins result in a similar phenotype. Similarly, the muscle weakness displayed by DM1 patients seems to originate from aberrant splicing of both RyR1 and $Ca_V1.1$.

3.4.1 Malignant Hyperthermia (MH) Susceptibility

Malignant hyperthermia (MH) susceptibility is a pharmacogenetic disorder, because it requires both a genetic mutation and a pharmacological trigger. It is autosomal dominantly transmitted and the estimate prevalence of genetic abnormalities leading to MH susceptibility (MHS) ranges from 1:3,000 to 1:8,500 (Rosenberg et al. 2007).

MH episodes present in response to volatile anesthetics (halothane, isoflurane, desflurane) or a depolarizing muscle relaxant (succinylcholine) as rapid and uncontrollable intracellular calcium release in skeletal muscle cells, consequently leading to massive muscle contractions (Dowling et al. 2014). The symptoms of an MH episode include muscle rigidity, metabolic acidosis, and rapid raise of body temperature. This condition is potentially fatal if not recognized and timely treated with dantrolene, a skeletal muscle relaxant, which binds the RyR1 and inhibits calcium release (Noda et al. 2022; Paul-Pletzer et al. 2002).

MHS-causing mutations typically induce a gain-of-function in the calcium release channel RyR1, by increasing the sensitivity of RyR1 to triggering agents such as calcium, caffeine, and volatile anesthetics and/or by elevating resting intracellular calcium concentration through a leaky RyR1 (Yang et al. 2007a, b). Recently, the notion that MHS needs a pharmacological trigger to manifest symptoms was challenged by the report of an unexpected increased frequency of hyperglycemia in MHS patients. The MHS patients with hyperglycemia show a positive correlation between fasting blood glucose levels and contracture response to caffeine, which suggests that the severity of the increase in intracellular calcium may correlate with alterations in the glucose homeostasis. Furthermore, dantrolene treatment for 3 days normalizes the values of glucose tolerance test when MHS mice are compared to control mice, which indicates that the increased intracellular calcium concentration at rest is likely the cause of hyperglycemia (Altamirano et al. 2019). A follow-up study suggested that the hyperglycemia of MHS patients is caused by alterations in metabolism that promote glycogen breakdown and hamper the glucose uptake in skeletal muscle (Tammineni et al. 2020). Therefore, beyond acute MH attacks triggered by anesthetics and relaxants, the resting elevated calcium concentration in skeletal muscle promotes hyperglycemia and type II diabetes.

Most MHS patients (>70%) have functional missense mutations in the RyR1 gene. The first MHS RyR1 mutation, R164C, was identified after the discovery that the corresponding mutation in pigs caused porcine stress syndrome (Fujii et al. 1991; Gillard et al. 1991). To date, nearly 200 different RyR1 mutations leading to MHS have been identified (Bauerova-Hlinkova et al. 2020). These gain-of-function mutations induce RyR1 misfolding or affect domain–domain interfaces making RyR1 hypersensitive to opening or leaky. This leads to a depletion of the SR calcium stores and chronically increased intracellular calcium levels at rest (Van Petegem 2015).

In addition to the many MH mutations in the RyR1 gene, mutation of three different $Ca_V1.1$ residues has been linked to MHS (Fig. 7, *in yellow*). In fact, the voltage-sensor $Ca_V1.1$ does not only open the RyR1 in response to depolarization, but also stabilizes the closed state of RyR1 at resting conditions. As Eltit et al. (2011) demonstrated in dysgenic myotubes, the absence of $Ca_V1.1$ leads to a leaky RyR1, resulting in increased intracellular calcium concentration at rest. Accordingly, $Ca_V1.1$ MH-causing mutations appear to change the conformation of $Ca_V1.1$ in a way that it does not stabilize the closed state of RyR1. The R174W mutation affects the innermost gating charge in the S4 of VSDI and, while it does not alter EC coupling, it abolishes L-type calcium currents (Carpenter et al. 2009; Eltit et al.

2012). Nevertheless, dysgenic myotubes reconstituted with $Ca_V1.1$-R174W display increased sensitivity to caffeine and volatile anesthetics, as well as depleted SR calcium stores and increased cytosolic calcium concentration (Eltit et al. 2012). These diverse effects indicate that the R174W mutation stabilizes the affected VSDI in the closed state, without interfering with the orthograde coupling with the RyR1. However, at the same time, it disrupts the ability of $Ca_V1.1$ to prevent the RyR1 from leaking.

The R1086H/C/S mutations are located in the cytoplasmic III-IV loop of $Ca_V1.1$ and affect a highly conserved arginine (Jurkat-Rott et al. 2000; Monnier et al. 1997; Toppin et al. 2010). The R1086H mutation has been characterized in cultured myotubes and results in smaller $Ca_V1.1$ currents and in activation of EC coupling at more negative voltages (Weiss et al. 2004). Finally, the third MHS-linked $Ca_V1.1$ mutation T1354S is located in the extracellular S5-S6 pore loop of VSDIV. When expressed in dysgenic myotubes, $Ca_V1.1$-T1354S displayed faster activation of the L-type calcium current and shifted the sensitivity of calcium release to caffeine to more negative potentials (Pirone et al. 2010). In conclusion, the three MHS mutations in $Ca_V1.1$ affect very diverse domains but exert a common gain-of-function effect on the RyR1. In all these cases, the exact molecular mechanisms by which they cause this effect on the RyR1 remain elusive.

As previously mentioned, mutations in STAC3 can cause NAM, which is characterized, among other symptoms, by MHS (Stamm et al. 2008; Zaharieva et al. 2018). How mutations in STAC3 can cause MHS remains elusive. Whereas lack of $Ca_V1.1$ has been shown to cause a leaky RyR1, in STAC3 knockout mice the intracellular stores are not depleted (Nelson et al. 2013). However, STAC3 knockout zebrafish myotubes reconstituted with STAC3 carrying the most common NAM mutation (W284S) display increased caffeine sensitivity, which is a hallmark of MHS (Linsley et al. 2017). As the W284S mutation in STAC3 was shown to abolish or reduce the affinity of the SH3 domains of STAC3 for the $Ca_V1.1$ II-III loop (Rufenach et al. 2020; Wong King Yuen et al. 2017), one possibility is that the reduced interaction impairs the ability of $Ca_V1.1$ to prevent the RyR1 leak. Alternatively, as STAC3 has been suggested to directly interact with the RyR1 (Campiglio et al. 2018b), STAC3 may directly contribute to suppressing the RyR1 leak. However, in both scenarios, the lack of STAC3 would also be expected to increase caffeine sensitivity and intracellular calcium concentration. Additional experiments are needed to clarify the molecular mechanism by which NAM mutations cause MHS.

3.4.2 Myotonic Dystrophy Type 1 (DM1)

Myotonic dystrophy type 1 (DM1) is an autosomal dominant multisystemic disorder characterized by myotonia, progressive muscle weakness, cardiac arrhythmia, cataracts, insulin resistance, and other symptoms. Genetically, DM1 is caused by unstable expansion of a $(CTG)_n$ repeat tract in the 3'untranslated region of the myotonic dystrophy protein kinase (DMPK) gene (Mahadevan et al. 1992). Mechanistically, the toxic DMPK mRNAs with expanded CUG repeats bind to the Muscleblind-like 1 (MBNL1) protein with high affinity, resulting in MBNL1 sequestration in nuclear foci and loss of its activity as a splicing regulator (Buj-Bello et al. 2002; Jiang et al. 2004). Consequently, alternative splicing of many genes is misregulated in affected muscles.

Among the muscle symptoms of DM1, myotonia is associated with aberrant splicing of the CLCN1 chloride channel, which causes a frameshift and loss-of-function and triggers involuntary runs of muscle action potentials (Charlet et al. 2002; Lueck et al. 2007; Mankodi et al. 2002; Zielonka et al. 2012). However, muscle weakness and degeneration, which is the most aggravating symptom, seems to be a consequence of altered intracellular calcium homeostasis, as cultured DM1 muscle cells display an elevated intracellular calcium concentration (Jacobs et al. 1990). Therefore, the muscle weakness and degeneration may derive from the misregulated splicing of several genes involved in calcium regulation and EC coupling. Indeed, aberrant splicing of both RyR1 and $Ca_V1.1$ has been reported.

The splicing defect in RyR1 concerns the inclusion of exon 70 (residues 3481–2485). The embryonic variant-ASI, lacking exon 70, constitutes 100% of RyR1 in fetal muscle, but only 30% in adult muscle. Kimura et al. (2005) demonstrated that, in skeletal muscle from both DM1 patients and a DM1 mouse model, the expression of the embryonic (-ASI) RyR1 variant is significantly increased. Reconstitution of the (-ASI) variant in dyspedic myotubes resulted in decreased RyR1 single-channel activity and caffeine response compared to the adult isoform, suggesting that the upregulation of the embryonic RyR1 isoform in DM1 muscle may reduce RyR1 calcium transients (Kimura et al. 2005; Santoro et al. 2014).

The aberrant splicing of $Ca_V1.1$ concerns exon 29 (Fig. 7, *in blue*). During normal development, the $Ca_V1.1e$ embryonic splice variant, lacking exon 29, in adult muscle is completely replaced by the adult $Ca_V1.1a$ isoform (Sultana et al. 2016; Tuluc et al. 2009). However, Tang et al. (2012) reported that DM1 patients aberrantly express high levels of the embryonic $Ca_V1.1e$ variant. Deletion of exon 29 in $Ca_V1.1e$ variant causes a 30 mV leftward shift of the voltage-dependence of activation and an eightfold increase of current density in skeletal myotubes. Therefore, upon activation, DM1 patient skeletal muscle experiences an increased calcium influx through $Ca_V1.1$, which may contribute to the muscle degeneration. Indeed, a mouse model lacking exon 29, therefore expressing only $Ca_V1.1e$ at any developmental stage, progressively showed mitochondrial damage and reduced endurance (Sultana et al. 2016), suggesting that aberrant expression of $Ca_V1.1e$ indeed contributes to the muscle weakness and degeneration observed in DM1 patients. Consistent with this notion, the severity of the muscle weakness in DM1 patients correlates with the fraction of $Ca_V1.1e$ expressed (Tang et al. 2012).

In conclusion, in DM1 patients, muscle weakness and degeneration are at least partially caused by misregulated splicing of $Ca_V1.1$ and RyR1, among other proteins, leading to increased and slower calcium entry from $Ca_V1.1$ and reduced calcium release from the SR.

4 Summary and Perspective

Despite increasing progress in understanding the molecular mechanisms of skeletal muscle EC coupling, how the signal is transmitted from the voltage-sensor $Ca_V1.1$ to the calcium release channel RyR1 is still elusive. Particularly, the contributions of the essential STAC3 and $Ca_V\beta_{1a}$ proteins are still enigmatic. Do they play a direct role in the conformational coupling between $Ca_V1.1$ and RyR1, or are they merely important for $Ca_V1.1$ structure and function? Intriguingly, mutations in three different proteins, i.e. $Ca_V1.1$, RyR1, and STAC3, result in similar congenital myopathies and MHS, highlighting that they all contribute to the same signalosome.

Due to the recent advances in molecular genetics, skeletal muscle EC coupling channelopathies are increasingly being reported. The functional characterization of these disease-causing mutations can greatly contribute to dissecting the function of the interactions within the $Ca_V1.1$ signalosome, provide mechanistic insights and, in some cases, open the possibility to treatment strategies that focus on the precise effect of the mutations.

Acknowledgements We thank M. Fernandez-Quintero for supplying the structure models shown in Figs. 4 and 8 (generated with PyMol and Chimera) and Dr. Bernhard E. Flucher for critical reading and feedback on the manuscript. Research in the laboratory is funded by grants from the Austrian Science Fund (FWF); current projects: P33776 and the FWF-funded graduate program on Calcium Channels in Excitable Cells (Ca_VX) DOC30.

References

Adams BA, Beam KG (1990) Muscular dysgenesis in mice: a model system for studying excitation-contraction coupling. FASEB J 4(10):2809–2816. https://doi.org/10.1096/fasebj.4.10.2165014

Altamirano F, Riazi S, Ibarra Moreno CA, Kraeva N, Uryash A, Allen PD et al (2019) Is malignant hyperthermia associated with hyperglycaemia? Br J Anaesth 122(1):e3–e5. https://doi.org/10.1016/j.bja.2018.09.014

Armstrong CM, Bezanilla FM, Horowicz P (1972) Twitches in the presence of ethylene glycol bis (–aminoethyl ether)-NN'-tetraacetic acid. Biochim Biophys Acta 267(3):605–608. https://doi.org/10.1016/0005-2728(72)90194-6

Avila G, Dirksen RT (2000) Functional impact of the ryanodine receptor on the skeletal muscle L-type Ca(2+) channel. J Gen Physiol 115(4):467–480. https://doi.org/10.1085/jgp.115.4.467

Avila G, Dirksen RT (2001) Functional effects of central core disease mutations in the cytoplasmic region of the skeletal muscle ryanodine receptor. J Gen Physiol 118(3):277–290. https://doi.org/10.1085/jgp.118.3.277

Bailey AG, Bloch EC (1987) Malignant hyperthermia in a three-month-old American Indian infant. Anesth Analg 66(10):1043–1045. Retrieved from https://www.ncbi.nlm.nih.gov/pubmed/3631569

Bauerova-Hlinkova V, Hajduchova D, Bauer JA (2020) Structure and function of the human ryanodine receptors and their association with myopathies-present state, challenges, and perspectives. Molecules 25(18). https://doi.org/10.3390/molecules25184040

Beam KG, Knudson CM, Powell JA (1986) A lethal mutation in mice eliminates the slow calcium current in skeletal muscle cells. Nature 320(6058):168–170. https://doi.org/10.1038/320168a0

Benedetti B, Tuluc P, Mastrolia V, Dlaska C, Flucher BE (2015) Physiological and pharmacological modulation of the embryonic skeletal muscle calcium channel splice variant CaV1.1e. Biophys J 108(5):1072–1080. https://doi.org/10.1016/j.bpj.2015.01.026

Bleunven C, Treves S, Jinyu X, Leo E, Ronjat M, De Waard M et al (2008) SRP-27 is a novel component of the supramolecular signalling complex involved in skeletal muscle excitation-contraction coupling. Biochem J 411(2):343–349. https://doi.org/10.1042/BJ20070906

Bower NI, de la Serrana DG, Cole NJ, Hollway GE, Lee HT, Assinder S, Johnston IA (2012) Stac3 is required for myotube formation and myogenic differentiation in vertebrate skeletal muscle. J Biol Chem 287(52):43936–43949. https://doi.org/10.1074/jbc.M112.361311

Buj-Bello A, Furling D, Tronchere H, Laporte J, Lerouge T, Butler-Browne GS, Mandel JL (2002) Muscle-specific alternative splicing of myotubularin-related 1 gene is impaired in DM1 muscle cells. Hum Mol Genet 11(19):2297–2307. https://doi.org/10.1093/hmg/11.19.2297

Buratti R, Prestipino G, Menegazzi P, Treves S, Zorzato F (1995) Calcium dependent activation of skeletal muscle Ca2+ release channel (ryanodine receptor) by calmodulin. Biochem Biophys Res Commun 213(3):1082–1090. https://doi.org/10.1006/bbrc.1995.2238

Campiglio M, Flucher BE (2017) STAC3 stably interacts through its C1 domain with CaV1.1 in skeletal muscle triads. Sci Rep 7:41003. https://doi.org/10.1038/srep41003

Campiglio M, Coste de Bagneaux P, Ortner NJ, Tuluc P, Van Petegem F, Flucher BE (2018a) STAC proteins associate to the IQ domain of CaV1.2 and inhibit calcium-dependent inactivation. Proc Natl Acad Sci U S A 115(6):1376–1381. https://doi.org/10.1073/pnas.1715997115

Campiglio M, Kaplan MM, Flucher BE (2018b) STAC3 incorporation into skeletal muscle triads occurs independent of the dihydropyridine receptor. J Cell Physiol 233(12):9045–9051. https://doi.org/10.1002/jcp.26767

Cannon SC (2015) Channelopathies of skeletal muscle excitability. Compr Physiol 5(2):761–790. https://doi.org/10.1002/cphy.c140062

Cannon SC (2017) An atypical CaV1.1 mutation reveals a common mechanism for hypokalemic periodic paralysis. J Gen Physiol 149(12):1061–1064. https://doi.org/10.1085/jgp.201711923

Carpenter D, Ringrose C, Leo V, Morris A, Robinson RL, Halsall PJ et al (2009) The role of CACNA1S in predisposition to malignant hyperthermia. BMC Med Genet 10:104. https://doi.org/10.1186/1471-2350-10-104

Catterall WA (2011) Voltage-gated calcium channels. Cold Spring Harb Perspect Biol 3(8): a003947. https://doi.org/10.1101/cshperspect.a003947

Chanda B, Bezanilla F (2008) A common pathway for charge transport through voltage-sensing domains. Neuron 57(3):345–351. https://doi.org/10.1016/j.neuron.2008.01.015

Charlet BN, Savkur RS, Singh G, Philips AV, Grice EA, Cooper TA (2002) Loss of the muscle-specific chloride channel in type 1 myotonic dystrophy due to misregulated alternative splicing. Mol Cell 10(1):45–53. https://doi.org/10.1016/s1097-2765(02)00572-5

Chaudhari N, Beam KG (1989) The muscular dysgenesis mutation in mice leads to arrest of the genetic program for muscle differentiation. Dev Biol 133(2):456–467. https://doi.org/10.1016/0012-1606(89)90049-3

Chen YH, Li MH, Zhang Y, He LL, Yamada Y, Fitzmaurice A et al (2004) Structural basis of the alpha1-beta subunit interaction of voltage-gated Ca2+ channels. Nature 429(6992):675–680. https://doi.org/10.1038/nature02641

Chirasani VR, Xu L, Addis HG, Pasek DA, Dokholyan NV, Meissner G, Yamaguchi N (2019) A central core disease mutation in the Ca(2+)-binding site of skeletal muscle ryanodine receptor impairs single-channel regulation. Am J Physiol Cell Physiol 317(2):C358–C365. https://doi.org/10.1152/ajpcell.00052.2019

Curtis BM, Catterall WA (1984) Purification of the calcium antagonist receptor of the voltage-sensitive calcium channel from skeletal muscle transverse tubules. Biochemistry 23(10): 2113–2118. https://doi.org/10.1021/bi00305a001

Dayal A, Schredelseker J, Franzini-Armstrong C, Grabner M (2010) Skeletal muscle excitation-contraction coupling is independent of a conserved heptad repeat motif in the C-terminus of the

DHPRbeta(1a) subunit. Cell Calcium 47(6):500–506. https://doi.org/10.1016/j.ceca.2010.04.003

Dayal A, Bhat V, Franzini-Armstrong C, Grabner M (2013) Domain cooperativity in the beta1a subunit is essential for dihydropyridine receptor voltage sensing in skeletal muscle. Proc Natl Acad Sci U S A 110(18):7488–7493. https://doi.org/10.1073/pnas.1301087110

Dayal A, Schrotter K, Pan Y, Fohr K, Melzer W, Grabner M (2017) The Ca(2+) influx through the mammalian skeletal muscle dihydropyridine receptor is irrelevant for muscle performance. Nat Commun 8(1):475. https://doi.org/10.1038/s41467-017-00629-x

Dayal A, Fernandez-Quintero ML, Liedl KR, Grabner M (2021) Pore mutation N617D in the skeletal muscle DHPR blocks Ca(2+) influx due to atypical high-affinity Ca(2+) binding. eLife 10. https://doi.org/10.7554/eLife.63435

des Georges A, Clarke OB, Zalk R, Yuan Q, Condon KJ, Grassucci RA et al (2016) Structural basis for gating and activation of RyR1. Cell 167(1):145–157 e117. https://doi.org/10.1016/j.cell.2016.08.075

DiFranco M, Cannon SC (2022) Voltage-dependent Ca(2+) release is impaired in hypokalemic periodic paralysis caused by CaV1.1-R528H but not by NaV1.4-R669H. Am J Physiol Cell Physiol 323(2):C478–C485. https://doi.org/10.1152/ajpcell.00209.2022

Dowling JJ, Lillis S, Amburgey K, Zhou H, Al-Sarraj S, Buk SJ et al (2011) King-Denborough syndrome with and without mutations in the skeletal muscle ryanodine receptor (RYR1) gene. Neuromuscul Disord 21(6):420–427. https://doi.org/10.1016/j.nmd.2011.03.006

Dowling JJ, Lawlor MW, Dirksen RT (2014) Triadopathies: an emerging class of skeletal muscle diseases. Neurotherapeutics 11(4):773–785. https://doi.org/10.1007/s13311-014-0300-3

Dulhunty AF, Gage PW (1988) Effects of extracellular calcium concentration and dihydropyridines on contraction in mammalian skeletal muscle. J Physiol 399:63–80. https://doi.org/10.1113/jphysiol.1988.sp017068

Efremov RG, Leitner A, Aebersold R, Raunser S (2015) Architecture and conformational switch mechanism of the ryanodine receptor. Nature 517(7532):39–43. https://doi.org/10.1038/nature13916

El Ghaleb Y, Fernandez-Quintero ML, Monteleone S, Tuluc P, Campiglio M, Liedl KR, Flucher BE (2021) Ion-pair interactions between voltage-sensing domain IV and pore domain I regulate CaV1.1 gating. Biophys J. https://doi.org/10.1016/j.bpj.2021.09.004

El Ghaleb Y, Ortner NJ, Posch W, Fernández-Quintero ML, Tuinte WE, Monteleone S et al (2022) Calcium current modulation by the γ1 subunit depends on alternative splicing of CaV1.1. J Gen Physiol 154(9). https://doi.org/10.1085/jgp.202113028

Eltit JM, Li H, Ward CW, Molinski T, Pessah IN, Allen PD, Lopez JR (2011) Orthograde dihydropyridine receptor signal regulates ryanodine receptor passive leak. Proc Natl Acad Sci U S A 108(17):7046–7051. https://doi.org/10.1073/pnas.1018380108

Eltit JM, Bannister RA, Moua O, Altamirano F, Hopkins PM, Pessah IN et al (2012) Malignant hyperthermia susceptibility arising from altered resting coupling between the skeletal muscle L-type Ca2+ channel and the type 1 ryanodine receptor. Proc Natl Acad Sci U S A 109(20): 7923–7928. https://doi.org/10.1073/pnas.1119207109

Fan C, Lehmann-Horn F, Weber MA, Bednarz M, Groome JR, Jonsson MK, Jurkat-Rott K (2013) Transient compartment-like syndrome and normokalaemic periodic paralysis due to a Ca(v)1.1 mutation. Brain 136(Pt 12):3775–3786. https://doi.org/10.1093/brain/awt300

Flucher BE (2020) Skeletal muscle CaV1.1 channelopathies. Pflugers Arch 472(7):739–754. https://doi.org/10.1007/s00424-020-02368-3

Flucher BE, Kasielke N, Grabner M (2000) The triad targeting signal of the skeletal muscle calcium channel is localized in the COOH terminus of the alpha(1S) subunit. J Cell Biol 151(2): 467–478. https://doi.org/10.1083/jcb.151.2.467

Franzini-Armstrong C, Protasi F, Ramesh V (1998) Comparative ultrastructure of Ca2+ release units in skeletal and cardiac muscle. Ann N Y Acad Sci 853:20–30. Retrieved from http://www.ncbi.nlm.nih.gov/pubmed/10603933

Freise D, Held B, Wissenbach U, Pfeifer A, Trost C, Himmerkus N et al (2000) Absence of the gamma subunit of the skeletal muscle dihydropyridine receptor increases L-type Ca2+ currents and alters channel inactivation properties. J Biol Chem 275(19):14476–14481. https://doi.org/10.1074/jbc.275.19.14476

Fujii J, Otsu K, Zorzato F, de Leon S, Khanna VK, Weiler JE et al (1991) Identification of a mutation in porcine ryanodine receptor associated with malignant hyperthermia. Science 253(5018):448–451. https://doi.org/10.1126/science.1862346

Fuller-Bicer GA, Varadi G, Koch SE, Ishii M, Bodi I, Kadeer N et al (2009) Targeted disruption of the voltage-dependent calcium channel alpha2/delta-1-subunit. Am J Physiol Heart Circ Physiol 297(1):H117–H124. https://doi.org/10.1152/ajpheart.00122.2009

Fuster C, Perrot J, Berthier C, Jacquemond V, Allard B (2017a) Elevated resting H(+) current in the R1239H type 1 hypokalaemic periodic paralysis mutated Ca(2+) channel. J Physiol 595(20): 6417–6428. https://doi.org/10.1113/JP274638

Fuster C, Perrot J, Berthier C, Jacquemond V, Charnet P, Allard B (2017b) Na leak with gating pore properties in hypokalemic periodic paralysis V876E mutant muscle Ca channel. J Gen Physiol 149(12):1139–1148. https://doi.org/10.1085/jgp.201711834

Garibaldi M, Rendu J, Brocard J, Lacene E, Faure J, Brochier G et al (2019) 'Dusty core disease' (DuCD): expanding morphological spectrum of RYR1 recessive myopathies. Acta Neuropathol Commun 7(1):3. https://doi.org/10.1186/s40478-018-0655-5

Gillard EF, Otsu K, Fujii J, Khanna VK, de Leon S, Derdemezi J et al (1991) A substitution of cysteine for arginine 614 in the ryanodine receptor is potentially causative of human malignant hyperthermia. Genomics 11(3):751–755. https://doi.org/10.1016/0888-7543(91)90084-r

Grabner M, Dirksen RT, Suda N, Beam KG (1999) The II-III loop of the skeletal muscle dihydropyridine receptor is responsible for the bi-directional coupling with the ryanodine receptor. J Biol Chem 274(31):21913–21919. https://doi.org/10.1074/jbc.274.31.21913

Gregg RG, Messing A, Strube C, Beurg M, Moss R, Behan M et al (1996) Absence of the beta subunit (cchb1) of the skeletal muscle dihydropyridine receptor alters expression of the alpha 1 subunit and eliminates excitation-contraction coupling. Proc Natl Acad Sci U S A 93(24): 13961–13966. Retrieved from http://www.ncbi.nlm.nih.gov/pubmed/8943043

Grzybowski M, Schanzer A, Pepler A, Heller C, Neubauer BA, Hahn A (2017) Novel STAC3 mutations in the first non-Amerindian patient with native American myopathy. Neuropediatrics 48(6):451–455. https://doi.org/10.1055/s-0037-1601868

Hakamata Y, Nakai J, Takeshima H, Imoto K (1992) Primary structure and distribution of a novel ryanodine receptor/calcium release channel from rabbit brain. FEBS Lett 312(2–3):229–235. https://doi.org/10.1016/0014-5793(92)80941-9

Horstick EJ, Linsley JW, Dowling JJ, Hauser MA, McDonald KK, Ashley-Koch A et al (2013) Stac3 is a component of the excitation-contraction coupling machinery and mutated in native American myopathy. Nat Commun 4:1952. https://doi.org/10.1038/ncomms2952

Hunter JM, Ahearn ME, Balak CD, Liang WS, Kurdoglu A, Corneveaux JJ et al (2015) Novel pathogenic variants and genes for myopathies identified by whole exome sequencing. Mol Genet Genomic Med 3(4):283–301. https://doi.org/10.1002/mgg3.142

Jacobs AE, Benders AA, Oosterhof A, Veerkamp JH, van Mier P, Wevers RA, Joosten EM (1990) The calcium homeostasis and the membrane potential of cultured muscle cells from patients with myotonic dystrophy. Biochim Biophys Acta 1096(1):14–19. https://doi.org/10.1016/0925-4439(90)90006-b

Jiang H, Mankodi A, Swanson MS, Moxley RT, Thornton CA (2004) Myotonic dystrophy type 1 is associated with nuclear foci of mutant RNA, sequestration of muscleblind proteins and deregulated alternative splicing in neurons. Hum Mol Genet 13(24):3079–3088. https://doi.org/10.1093/hmg/ddh327

Jiang D, Gamal El-Din TM, Ing C, Lu P, Pomes R, Zheng N, Catterall WA (2018) Structural basis for gating pore current in periodic paralysis. Nature 557(7706):590–594. https://doi.org/10.1038/s41586-018-0120-4

Jones JL, Reynolds DF, Lai FA, Blayney LM (2005) Ryanodine receptor binding to FKBP12 is modulated by channel activation state. J Cell Sci 118(Pt 20):4613–4619. https://doi.org/10.1242/jcs.02582

Jung DH, Mo SH, Kim DH (2006) Calumenin, a multiple EF-hands Ca2+−binding protein, interacts with ryanodine receptor-1 in rabbit skeletal sarcoplasmic reticulum. Biochem Biophys Res Commun 343(1):34–42. https://doi.org/10.1016/j.bbrc.2006.02.115

Jungbluth H (2007a) Central core disease. Orphanet J Rare Dis 2:25. https://doi.org/10.1186/1750-1172-2-25

Jungbluth H (2007b) Multi-minicore disease. Orphanet J Rare Dis 2:31. https://doi.org/10.1186/1750-1172-2-31

Jungbluth H, Muller CR, Halliger-Keller B, Brockington M, Brown SC, Feng L et al (2002) Autosomal recessive inheritance of RYR1 mutations in a congenital myopathy with cores. Neurology 59(2):284–287. https://doi.org/10.1212/wnl.59.2.284

Jungbluth H, Wallgren-Pettersson C, Laporte J (2008) Centronuclear (myotubular) myopathy. Orphanet J Rare Dis 3:26. https://doi.org/10.1186/1750-1172-3-26

Jungbluth H, Treves S, Zorzato F, Sarkozy A, Ochala J, Sewry C et al (2018) Congenital myopathies: disorders of excitation-contraction coupling and muscle contraction. Nat Rev Neurol 14(3):151–167. https://doi.org/10.1038/nrneurol.2017.191

Jurkat-Rott K, McCarthy T, Lehmann-Horn F (2000) Genetics and pathogenesis of malignant hyperthermia. Muscle Nerve 23(1):4–17. https://doi.org/10.1002/(sici)1097-4598(200001)23:1<4::aid-mus3>3.0.co;2-d

Jurkat-Rott K, Weber MA, Fauler M, Guo XH, Holzherr BD, Paczulla A et al (2009) K+-dependent paradoxical membrane depolarization and Na+ overload, major and reversible contributors to weakness by ion channel leaks. Proc Natl Acad Sci U S A 106(10):4036–4041. https://doi.org/10.1073/pnas.0811277106

Jurkat-Rott K, Groome J, Lehmann-Horn F (2012) Pathophysiological role of omega pore current in channelopathies. Front Pharmacol 3:112. https://doi.org/10.3389/fphar.2012.00112

Kimura T, Nakamori M, Lueck JD, Pouliquin P, Aoike F, Fujimura H et al (2005) Altered mRNA splicing of the skeletal muscle ryanodine receptor and sarcoplasmic/endoplasmic reticulum Ca2+-ATPase in myotonic dystrophy type 1. Hum Mol Genet 14(15):2189–2200. https://doi.org/10.1093/hmg/ddi223

Kubota T, Wu F, Vicart S, Nakaza M, Sternberg D, Watanabe D et al (2020) Hypokalaemic periodic paralysis with a charge-retaining substitution in the voltage sensor. Brain Commun 2(2):fcaa103. https://doi.org/10.1093/braincomms/fcaa103

Kugler G, Weiss RG, Flucher BE, Grabner M (2004) Structural requirements of the dihydropyridine receptor alpha1S II-III loop for skeletal-type excitation-contraction coupling. J Biol Chem 279(6):4721–4728. https://doi.org/10.1074/jbc.M307538200

Lai FA, Anderson K, Rousseau E, Liu QY, Meissner G (1988) Evidence for a Ca2+ channel within the ryanodine receptor complex from cardiac sarcoplasmic reticulum. Biochem Biophys Res Commun 151(1):441–449. https://doi.org/10.1016/0006-291x(88)90613-4

Lau K, Van Petegem F (2014) Crystal structures of wild type and disease mutant forms of the ryanodine receptor SPRY2 domain. Nat Commun 5:5397. https://doi.org/10.1038/ncomms6397

Laver DR, Lamb GD (1998) Inactivation of Ca2+ release channels (ryanodine receptors RyR1 and RyR2) with rapid steps in [Ca2+] and voltage. Biophys J 74(5):2352–2364. https://doi.org/10.1016/S0006-3495(98)77944-5

Leong P, MacLennan DH (1998) The cytoplasmic loops between domains II and III and domains III and IV in the skeletal muscle dihydropyridine receptor bind to a contiguous site in the skeletal muscle ryanodine receptor. J Biol Chem 273(45):29958–29964. https://doi.org/10.1074/jbc.273.45.29958

Linsley JW, Hsu IU, Groom L, Yarotskyy V, Lavorato M, Horstick EJ et al (2017) Congenital myopathy results from misregulation of a muscle Ca2+ channel by mutant Stac3. Proc Natl Acad Sci U S A 114(2):E228–E236. https://doi.org/10.1073/pnas.1619238114

Loy RE, Orynbayev M, Xu L, Andronache Z, Apostol S, Zvaritch E et al (2011) Muscle weakness in Ryr1I4895T/WT knock-in mice as a result of reduced ryanodine receptor Ca2+ ion permeation and release from the sarcoplasmic reticulum. J Gen Physiol 137(1):43–57. https://doi.org/10.1085/jgp.201010523

Lueck JD, Mankodi A, Swanson MS, Thornton CA, Dirksen RT (2007) Muscle chloride channel dysfunction in two mouse models of myotonic dystrophy. J Gen Physiol 129(1):79–94. https://doi.org/10.1085/jgp.200609635

Lyfenko AD, Ducreux S, Wang Y, Xu L, Zorzato F, Ferreiro A et al (2007) Two central core disease (CCD) deletions in the C-terminal region of RYR1 alter muscle excitation-contraction (EC) coupling by distinct mechanisms. Hum Mutat 28(1):61–68. https://doi.org/10.1002/humu.20409

Mahadevan M, Tsilfidis C, Sabourin L, Shutler G, Amemiya C, Jansen G et al (1992) Myotonic dystrophy mutation: an unstable CTG repeat in the 3′ untranslated region of the gene. Science 255(5049):1253–1255. https://doi.org/10.1126/science.1546325

Mankodi A, Takahashi MP, Jiang H, Beck CL, Bowers WJ, Moxley RT et al (2002) Expanded CUG repeats trigger aberrant splicing of ClC-1 chloride channel pre-mRNA and hyperexcitability of skeletal muscle in myotonic dystrophy. Mol Cell 10(1):35–44. https://doi.org/10.1016/s1097-2765(02)00563-4

Matthews E, Labrum R, Sweeney MG, Sud R, Haworth A, Chinnery PF et al (2009) Voltage sensor charge loss accounts for most cases of hypokalemic periodic paralysis. Neurology 72(18):1544–1547. https://doi.org/10.1212/01.wnl.0000342387.65477.46

McPherson PS, Campbell KP (1993) Characterization of the major brain form of the ryanodine receptor/Ca2+ release channel. J Biol Chem 268(26):19785–19790. Retrieved from https://www.ncbi.nlm.nih.gov/pubmed/7690041

Meissner G (1984) Adenine nucleotide stimulation of Ca2+−induced Ca2+ release in sarcoplasmic reticulum. J Biol Chem 259(4):2365–2374. Retrieved from https://www.ncbi.nlm.nih.gov/pubmed/6698971

Mi W, Rybalchenko V, Cannon SC (2014) Disrupted coupling of gating charge displacement to Na + current activation for DIIS4 mutations in hypokalemic periodic paralysis. J Gen Physiol 144(2):137–145. https://doi.org/10.1085/jgp.201411199

Monnier N, Procaccio V, Stieglitz P, Lunardi J (1997) Malignant-hyperthermia susceptibility is associated with a mutation of the alpha 1-subunit of the human dihydropyridine-sensitive L-type voltage-dependent calcium-channel receptor in skeletal muscle. Am J Hum Genet 60(6):1316–1325. https://doi.org/10.1086/515454

Monnier N, Laquerriere A, Marret S, Goldenberg A, Marty I, Nivoche Y, Lunardi J (2009) First genomic rearrangement of the RYR1 gene associated with an atypical presentation of lethal neonatal hypotonia. Neuromuscul Disord 19(10):680–684. https://doi.org/10.1016/j.nmd.2009.07.007

Nakada T, Kashihara T, Komatsu M, Kojima K, Takeshita T, Yamada M (2018) Physical interaction of junctophilin and the CaV1.1 C terminus is crucial for skeletal muscle contraction. Proc Natl Acad Sci U S A 115(17):4507–4512. https://doi.org/10.1073/pnas.1716649115

Nakai J, Imagawa T, Hakamat Y, Shigekawa M, Takeshima H, Numa S (1990) Primary structure and functional expression from cDNA of the cardiac ryanodine receptor/calcium release channel. FEBS Lett 271(1–2):169–177. https://doi.org/10.1016/0014-5793(90)80399-4

Nakai J, Adams BA, Imoto K, Beam KG (1994) Critical roles of the S3 segment and S3-S4 linker of repeat I in activation of L-type calcium channels. Proc Natl Acad Sci U S A 91(3):1014–1018. https://doi.org/10.1073/pnas.91.3.1014

Nakai J, Dirksen RT, Nguyen HT, Pessah IN, Beam KG, Allen PD (1996) Enhanced dihydropyridine receptor channel activity in the presence of ryanodine receptor. Nature 380(6569):72–75. https://doi.org/10.1038/380072a0

Nelson BR, Wu F, Liu Y, Anderson DM, McAnally J, Lin W et al (2013) Skeletal muscle-specific T-tubule protein STAC3 mediates voltage-induced Ca2+ release and contractility. Proc Natl Acad Sci U S A 110(29):11881–11886. https://doi.org/10.1073/pnas.1310571110

Niu J, Dick IE, Yang W, Bamgboye MA, Yue DT, Tomaselli G et al (2018a) Allosteric regulators selectively prevent Ca(2+)-feedback of CaV and NaV channels. eLife 7. https://doi.org/10.7554/eLife.35222

Niu J, Yang W, Yue DT, Inoue T, Ben-Johny M (2018b) Duplex signaling by CaM and Stac3 enhances CaV1.1 function and provides insights into congenital myopathy. J Gen Physiol 150(8):1145–1161. https://doi.org/10.1085/jgp.201812005

Noda M, Ikeda T, Suzuki H, Takeshima H, Takahashi T, Kuno M, Numa S (1986) Expression of functional sodium channels from cloned cDNA. Nature 322(6082):826–828. https://doi.org/10.1038/322826a0

Noda Y, Mukaida K, Miyoshi H, Nakamura R, Yasuda T, Saeki N et al (2022) The effects of dantrolene in the presence or absence of ryanodine receptor type 1 variants in individuals predisposed to malignant hyperthermia. Anaesth Intensive Care 50(4):312–319. https://doi.org/10.1177/0310057X211053644

Obermair GJ, Kugler G, Baumgartner S, Tuluc P, Grabner M, Flucher BE (2005) The Ca2+ channel alpha2delta-1 subunit determines Ca2+ current kinetics in skeletal muscle but not targeting of alpha1S or excitation-contraction coupling. J Biol Chem 280(3):2229–2237. https://doi.org/10.1074/jbc.M411501200

Opatowsky Y, Chen CC, Campbell KP, Hirsch JA (2004) Structural analysis of the voltage-dependent calcium channel beta subunit functional core and its complex with the alpha 1 interaction domain. Neuron 42(3):387–399. https://doi.org/10.1016/s0896-6273(04)00250-8

Otsu K, Willard HF, Khanna VK, Zorzato F, Green NM, MacLennan DH (1990) Molecular cloning of cDNA encoding the Ca2+ release channel (ryanodine receptor) of rabbit cardiac muscle sarcoplasmic reticulum. J Biol Chem 265(23):13472–13483. Retrieved from https://www.ncbi.nlm.nih.gov/pubmed/2380170

Pai AC (1965) Developmental genetics of a lethal mutation, muscular dysgenesis (Mdg), in the mouse. I. Genetic analysis and gross morphology. Dev Biol 11:82–92. https://doi.org/10.1016/0012-1606(65)90038-2

Pancaroglu R, Van Petegem F (2018) Calcium channelopathies: structural insights into disorders of the muscle excitation-contraction complex. Annu Rev Genet 52:373–396. https://doi.org/10.1146/annurev-genet-120417-031311

Paul-Pletzer K, Yamamoto T, Bhat MB, Ma J, Ikemoto N, Jimenez LS et al (2002) Identification of a dantrolene-binding sequence on the skeletal muscle ryanodine receptor. J Biol Chem 277(38):34918–34923. https://doi.org/10.1074/jbc.M205487200

Perni S, Lavorato M, Beam KG (2017) De novo reconstitution reveals the proteins required for skeletal muscle voltage-induced Ca(2+) release. Proc Natl Acad Sci U S A 114(52):13822–13827. https://doi.org/10.1073/pnas.1716461115

Pirone A, Schredelseker J, Tuluc P, Gravino E, Fortunato G, Flucher BE et al (2010) Identification and functional characterization of malignant hyperthermia mutation T1354S in the outer pore of the Cavalpha1S-subunit. Am J Physiol Cell Physiol 299(6):C1345–C1354. https://doi.org/10.1152/ajpcell.00008.2010

Polster A, Perni S, Bichraoui H, Beam KG (2015) Stac adaptor proteins regulate trafficking and function of muscle and neuronal L-type Ca2+ channels. Proc Natl Acad Sci U S A 112(2):602–606. https://doi.org/10.1073/pnas.1423113112

Polster A, Nelson BR, Olson EN, Beam KG (2016) Stac3 has a direct role in skeletal muscle-type excitation-contraction coupling that is disrupted by a myopathy-causing mutation. Proc Natl Acad Sci U S A 113(39):10986–10991. https://doi.org/10.1073/pnas.1612441113

Polster A, Dittmer PJ, Perni S, Bichraoui H, Sather WA, Beam KG (2018a) Stac proteins suppress Ca(2+)-dependent inactivation of neuronal l-type Ca(2+) channels. J Neurosci 38(43):9215–9227. https://doi.org/10.1523/JNEUROSCI.0695-18.2018

Polster A, Nelson BR, Papadopoulos S, Olson EN, Beam KG (2018b) Stac proteins associate with the critical domain for excitation-contraction coupling in the II-III loop of CaV1.1. J Gen Physiol. https://doi.org/10.1085/jgp.201711917

Powell JA, Fambrough DM (1973) Electrical properties of normal and dysgenic mouse skeletal muscle in culture. J Cell Physiol 82(1):21–38. https://doi.org/10.1002/jcp.1040820104

Pragnell M, De Waard M, Mori Y, Tanabe T, Snutch TP, Campbell KP (1994) Calcium channel beta-subunit binds to a conserved motif in the I-II cytoplasmic linker of the alpha 1-subunit. Nature 368(6466):67–70. https://doi.org/10.1038/368067a0

Prosser BL, Wright NT, Hernandez-Ochoa EO, Varney KM, Liu Y, Olojo RO et al (2008) S100A1 binds to the calmodulin-binding site of ryanodine receptor and modulates skeletal muscle excitation-contraction coupling. J Biol Chem 283(8):5046–5057. https://doi.org/10.1074/jbc.M709231200

Ravenscroft G, Clayton JS, Faiz F, Sivadorai P, Milnes D, Cincotta R et al (2021) Neurogenetic fetal akinesia and arthrogryposis: genetics, expanding genotype-phenotypes and functional genomics. J Med Genet 58(9):609–618. https://doi.org/10.1136/jmedgenet-2020-106901

Rebbeck RT, Karunasekara Y, Gallant EM, Board PG, Beard NA, Casarotto MG, Dulhunty AF (2011) The beta(1a) subunit of the skeletal DHPR binds to skeletal RyR1 and activates the channel via its 35-residue C-terminal tail. Biophys J 100(4):922–930. https://doi.org/10.1016/j.bpj.2011.01.022

Rios E, Brum G (1987) Involvement of dihydropyridine receptors in excitation-contraction coupling in skeletal muscle. Nature 325(6106):717–720. https://doi.org/10.1038/325717a0

Rogers EF, Koniuszy FR et al (1948) Plant insecticides; ryanodine, a new alkaloid from Ryania speciosa Vahl. J Am Chem Soc 70(9):3086–3088. https://doi.org/10.1021/ja01189a074

Romero NB, Clarke NF (2013) Congenital myopathies. Handb Clin Neurol 113:1321–1336. https://doi.org/10.1016/B978-0-444-59565-2.00004-6

Rosenberg H, Davis M, James D, Pollock N, Stowell K (2007) Malignant hyperthermia. Orphanet J Rare Dis 2:21. https://doi.org/10.1186/1750-1172-2-21

Rufenach B, Van Petegem F (2021) Structure and function of STAC proteins: calcium channel modulators and critical components of muscle excitation-contraction coupling. J Biol Chem 297(1):100874. https://doi.org/10.1016/j.jbc.2021.100874

Rufenach B, Christy D, Flucher BE, Bui JM, Gsponer J, Campiglio M, Van Petegem F (2020) Multiple sequence variants in STAC3 affect interactions with CaV1.1 and excitation-contraction coupling. Structure 28(8):922–932 e925. https://doi.org/10.1016/j.str.2020.05.005

Santoro M, Piacentini R, Masciullo M, Bianchi ML, Modoni A, Podda MV et al (2014) Alternative splicing alterations of Ca2+ handling genes are associated with Ca2+ signal dysregulation in myotonic dystrophy type 1 (DM1) and type 2 (DM2) myotubes. Neuropathol Appl Neurobiol 40(4):464–476. https://doi.org/10.1111/nan.12076

Scacheri PC, Hoffman EP, Fratkin JD, Semino-Mora C, Senchak A, Davis MR et al (2000) A novel ryanodine receptor gene mutation causing both cores and rods in congenital myopathy. Neurology 55(11):1689–1696. https://doi.org/10.1212/wnl.55.11.1689

Schartner V, Romero NB, Donkervoort S, Treves S, Munot P, Pierson TM et al (2017) Dihydropyridine receptor (DHPR, CACNA1S) congenital myopathy. Acta Neuropathol 133(4):517–533. https://doi.org/10.1007/s00401-016-1656-8

Schneider MF, Chandler WK (1973) Voltage dependent charge movement of skeletal muscle: a possible step in excitation-contraction coupling. Nature 242(5395):244–246. https://doi.org/10.1038/242244a0

Schredelseker J, Di Biase V, Obermair GJ, Felder ET, Flucher BE, Franzini-Armstrong C, Grabner M (2005) The beta 1a subunit is essential for the assembly of dihydropyridine-receptor arrays in skeletal muscle. Proc Natl Acad Sci U S A 102(47):17219–17224. https://doi.org/10.1073/pnas.0508710102

Schredelseker J, Dayal A, Schwerte T, Franzini-Armstrong C, Grabner M (2009) Proper restoration of excitation-contraction coupling in the dihydropyridine receptor beta1-null zebrafish relaxed is an exclusive function of the beta1a subunit. J Biol Chem 284(2):1242–1251. https://doi.org/10.1074/jbc.M807767200

Schredelseker J, Shrivastav M, Dayal A, Grabner M (2010) Non-Ca2+–conducting Ca2+ channels in fish skeletal muscle excitation-contraction coupling. Proc Natl Acad Sci U S A 107(12): 5658–5663. https://doi.org/10.1073/pnas.0912153107

Sheridan DC, Cheng W, Carbonneau L, Ahern CA, Coronado R (2004) Involvement of a heptad repeat in the carboxyl terminus of the dihydropyridine receptor beta1a subunit in the mechanism of excitation-contraction coupling in skeletal muscle. Biophys J 87(2):929–942. https://doi.org/10.1529/biophysj.104.043810

Sokolov S, Scheuer T, Catterall WA (2007) Gating pore current in an inherited ion channelopathy. Nature 446(7131):76–78. https://doi.org/10.1038/nature05598

Sokolov S, Scheuer T, Catterall WA (2008) Depolarization-activated gating pore current conducted by mutant sodium channels in potassium-sensitive normokalemic periodic paralysis. Proc Natl Acad Sci U S A 105(50):19980–19985. https://doi.org/10.1073/pnas.0810562105

Stamm DS, Aylsworth AS, Stajich JM, Kahler SG, Thorne LB, Speer MC, Powell CM (2008) Native American myopathy: congenital myopathy with cleft palate, skeletal anomalies, and susceptibility to malignant hyperthermia. Am J Med Genet A 146A(14):1832–1841. https://doi.org/10.1002/ajmg.a.32370

Sternberg D, Maisonobe T, Jurkat-Rott K, Nicole S, Launay E, Chauveau D et al (2001) Hypokalaemic periodic paralysis type 2 caused by mutations at codon 672 in the muscle sodium channel gene SCN4A. Brain 124(Pt 6):1091–1099. https://doi.org/10.1093/brain/124.6.1091

Struyk AF, Cannon SC (2007) A Na+ channel mutation linked to hypokalemic periodic paralysis exposes a proton-selective gating pore. J Gen Physiol 130(1):11–20. https://doi.org/10.1085/jgp.200709755

Sultana N, Dienes B, Benedetti A, Tuluc P, Szentesi P, Sztretye M et al (2016) Restricting calcium currents is required for correct fiber type specification in skeletal muscle. Development 143(9): 1547–1559. https://doi.org/10.1242/dev.129676

Suzuki H, Kawai J, Taga C, Yaoi T, Hara A, Hirose K et al (1996) Stac, a novel neuron-specific protein with cysteine-rich and SH3 domains. Biochem Biophys Res Commun 229(3):902–909. https://doi.org/10.1006/bbrc.1996.1900

Tae HS, Norris NC, Cui Y, Karunasekara Y, Board PG, Dulhunty AF, Casarotto MG (2009) Molecular recognition of the disordered dihydropyridine receptor II-III loop by a conserved spry domain of the type 1 ryanodine receptor. Clin Exp Pharmacol Physiol 36(3):346–349. https://doi.org/10.1111/j.1440-1681.2008.05130.x

Takeshima H, Iino M, Takekura H, Nishi M, Kuno J, Minowa O et al (1994) Excitation-contraction uncoupling and muscular degeneration in mice lacking functional skeletal muscle ryanodine-receptor gene. Nature 369(6481):556–559. https://doi.org/10.1038/369556a0

Tammineni ER, Kraeva N, Figueroa L, Manno C, Ibarra CA, Klip A et al (2020) Intracellular calcium leak lowers glucose storage in human muscle, promoting hyperglycemia and diabetes. eLife 9. https://doi.org/10.7554/eLife.53999

Tanabe T, Takeshima H, Mikami A, Flockerzi V, Takahashi H, Kangawa K et al (1987) Primary structure of the receptor for calcium channel blockers from skeletal muscle. Nature 328(6128): 313–318. https://doi.org/10.1038/328313a0

Tanabe T, Beam KG, Powell JA, Numa S (1988) Restoration of excitation-contraction coupling and slow calcium current in dysgenic muscle by dihydropyridine receptor complementary DNA. Nature 336(6195):134–139. https://doi.org/10.1038/336134a0

Tanabe T, Beam KG, Adams BA, Niidome T, Numa S (1990) Regions of the skeletal muscle dihydropyridine receptor critical for excitation-contraction coupling. Nature 346(6284): 567–569. https://doi.org/10.1038/346567a0

Tang ZZ, Yarotskyy V, Wei L, Sobczak K, Nakamori M, Eichinger K et al (2012) Muscle weakness in myotonic dystrophy associated with misregulated splicing and altered gating of Ca(V)1.1 calcium channel. Hum Mol Genet 21(6):1312–1324. https://doi.org/10.1093/hmg/ddr568

Telegrafi A, Webb BD, Robbins SM, Speck-Martins CE, FitzPatrick D, Fleming L et al (2017) Identification of STAC3 variants in non-native American families with overlapping features of

Carey-Fineman-Ziter syndrome and moebius syndrome. Am J Med Genet A 173(10): 2763–2771. https://doi.org/10.1002/ajmg.a.38375

Toppin PJ, Chandy TT, Ghanekar A, Kraeva N, Beattie WS, Riazi S (2010) A report of fulminant malignant hyperthermia in a patient with a novel mutation of the CACNA1S gene. Can J Anaesth 57(7):689–693. https://doi.org/10.1007/s12630-010-9314-4

Tripathy A, Xu L, Mann G, Meissner G (1995) Calmodulin activation and inhibition of skeletal muscle Ca2+ release channel (ryanodine receptor). Biophys J 69(1):106–119. https://doi.org/10.1016/S0006-3495(95)79880-0

Tuinte WE, Torok E, Mahlknecht I, Tuluc P, Flucher BE, Campiglio M (2022) STAC3 determines the slow activation kinetics of CaV 1.1 currents and inhibits its voltage-dependent inactivation. J Cell Physiol. https://doi.org/10.1002/jcp.30870

Tuluc P, Molenda N, Schlick B, Obermair GJ, Flucher BE, Jurkat-Rott K (2009) A CaV1.1 Ca2+ channel splice variant with high conductance and voltage-sensitivity alters EC coupling in developing skeletal muscle. Biophys J 96(1):35–44. https://doi.org/10.1016/j.bpj.2008.09.027

Tuluc P, Benedetti B, Coste de Bagneaux P, Grabner M, Flucher BE (2016) Two distinct voltage-sensing domains control voltage sensitivity and kinetics of current activation in CaV1.1 calcium channels. J Gen Physiol 147(6):437–449. https://doi.org/10.1085/jgp.201611568

Van Petegem F (2015) Ryanodine receptors: allosteric ion channel giants. J Mol Biol 427(1):31–53. https://doi.org/10.1016/j.jmb.2014.08.004

Van Petegem F, Clark KA, Chatelain FC, Minor DL Jr (2004) Structure of a complex between a voltage-gated calcium channel beta-subunit and an alpha-subunit domain. Nature 429(6992): 671–675. https://doi.org/10.1038/nature02588

Vicart S, Sternberg D, Fournier E, Ochsner F, Laforet P, Kuntzer T et al (2004) New mutations of SCN4A cause a potassium-sensitive normokalemic periodic paralysis. Neurology 63(11): 2120–2127. https://doi.org/10.1212/01.wnl.0000145768.09934.ec

von der Hagen M, Kress W, Hahn G, Brocke KS, Mitzscherling P, Huebner A et al (2008) Novel RYR1 missense mutation causes core rod myopathy. Eur J Neurol 15(4):e31–e32. https://doi.org/10.1111/j.1468-1331.2008.02094.x

Webb BD, Manoli I, Jabs EW (1993) STAC3 disorder. In: Adam MP, Ardinger HH, Pagon RA, Wallace SE, Bean LJH, Mirzaa G, Amemiya A (eds) GeneReviews((R)). University of Washington, Seattle

Weber F, Lehmann-Horn F (1993) Hypokalemic periodic paralysis. In: Adam MP, Everman DB, Mirzaa GM, Pagon RA, Wallace SE, Bean LJH, Gripp KW, Amemiya A (eds) GeneReviews ((R)). University of Washington, Seattle

Wei L, Varsanyi M, Dulhunty AF, Beard NA (2006) The conformation of calsequestrin determines its ability to regulate skeletal ryanodine receptors. Biophys J 91(4):1288–1301. https://doi.org/10.1529/biophysj.106.082610

Weiss RG, O'Connell KM, Flucher BE, Allen PD, Grabner M, Dirksen RT (2004) Functional analysis of the R1086H malignant hyperthermia mutation in the DHPR reveals an unexpected influence of the III-IV loop on skeletal muscle EC coupling. Am J Physiol Cell Physiol 287(4): C1094–C1102. https://doi.org/10.1152/ajpcell.00173.2004

Wium E, Dulhunty AF, Beard NA (2012) A skeletal muscle ryanodine receptor interaction domain in triadin. PLoS One 7(8):e43817. https://doi.org/10.1371/journal.pone.0043817

Wong King Yuen SM, Campiglio M, Tung CC, Flucher BE, Van Petegem F (2017) Structural insights into binding of STAC proteins to voltage-gated calcium channels. Proc Natl Acad Sci U S A 114(45):E9520–E9528. https://doi.org/10.1073/pnas.1708852114

Wu S, Ibarra MC, Malicdan MC, Murayama K, Ichihara Y, Kikuchi H et al (2006) Central core disease is due to RYR1 mutations in more than 90% of patients. Brain 129(Pt 6):1470–1480. https://doi.org/10.1093/brain/awl077

Wu F, Mi W, Hernandez-Ochoa EO, Burns DK, Fu Y, Gray HF et al (2012) A calcium channel mutant mouse model of hypokalemic periodic paralysis. J Clin Invest 122(12):4580–4591. https://doi.org/10.1172/JCI66091

Wu J, Yan Z, Li Z, Qian X, Lu S, Dong M et al (2016) Structure of the voltage-gated calcium channel Ca(v)1.1 at 3.6 A resolution. Nature 537(7619):191–196. https://doi.org/10.1038/nature19321

Wu F, Quinonez M, DiFranco M, Cannon SC (2018) Stac3 enhances expression of human CaV1.1 in xenopus oocytes and reveals gating pore currents in HypoPP mutant channels. J Gen Physiol 150(3):475–489. https://doi.org/10.1085/jgp.201711962

Wu F, Quinonez M, Cannon SC (2021) Gating pore currents occur in CaV1.1 domain III mutants associated with HypoPP. J Gen Physiol 153(11). https://doi.org/10.1085/jgp.202112946

Yan Z, Bai X, Yan C, Wu J, Li Z, Xie T et al (2015) Structure of the rabbit ryanodine receptor RyR1 at near-atomic resolution. Nature 517(7532):50–55. https://doi.org/10.1038/nature14063

Yang T, Allen PD, Pessah IN, Lopez JR (2007a) Enhanced excitation-coupled calcium entry in myotubes is associated with expression of RyR1 malignant hyperthermia mutations. J Biol Chem 282(52):37471–37478. https://doi.org/10.1074/jbc.M701379200

Yang T, Esteve E, Pessah IN, Molinski TF, Allen PD, Lopez JR (2007b) Elevated resting [Ca(2+)](i) in myotubes expressing malignant hyperthermia RyR1 cDNAs is partially restored by modulation of passive calcium leak from the SR. Am J Physiol Cell Physiol 292(5):C1591–C1598. https://doi.org/10.1152/ajpcell.00133.2006

Yang ZF, Panwar P, McFarlane CR, Tuinte WE, Campiglio M, Van Petegem F (2022) Structures of the junctophilin/voltage-gated calcium channel interface reveal hot spot for cardiomyopathy mutations. Proc Natl Acad Sci U S A 119(10):e2120416119. https://doi.org/10.1073/pnas.2120416119

Zaharieva IT, Sarkozy A, Munot P, Manzur A, O'Grady G, Rendu J et al (2018) STAC3 variants cause a congenital myopathy with distinctive dysmorphic features and malignant hyperthermia susceptibility. Hum Mutat 39(12):1980–1994. https://doi.org/10.1002/humu.23635

Zalk R, Clarke OB, des Georges A, Grassucci RA, Reiken S, Mancia F et al (2015) Structure of a mammalian ryanodine receptor. Nature 517(7532):44–49. https://doi.org/10.1038/nature13950

Zielonka D, Jurkat-Rott K, Stachowiak P, Bryl A, Marcinkowski JT, Lehmann-Horn F (2012) A Becker myotonia patient with compound heterozygosity for CLCN1 mutations and Prinzmetal angina pectoris. Neuromuscul Disord 22(4):355–360. https://doi.org/10.1016/j.nmd.2011.10.024

Mechanisms of Vascular $Ca_V1.2$ Channel Regulation During Diabetic Hyperglycemia

Miguel Martín-Aragón Baudel, Junyoung Hong, Johannes W. Hell, Madeline Nieves-Cintrón, and Manuel F. Navedo

Contents

Abstract

Diabetes is a leading cause of disability and mortality worldwide. A major underlying factor in diabetes is the excessive glucose levels in the bloodstream (e.g., hyperglycemia). Vascular complications directly result from this metabolic abnormality, leading to disabling and life-threatening conditions. Dysfunction of vascular smooth muscle cells is a well-recognized factor mediating vascular complications during diabetic hyperglycemia. The function of vascular smooth muscle cells is exquisitely controlled by different ion channels. Among the ion channels, the L-type $Ca_V1.2$ channel plays a key role as it is the main Ca^{2+} entry pathway regulating vascular smooth muscle contractile state. The activity of $Ca_V1.2$ channels in vascular smooth muscle is altered by diabetic hyperglycemia, which may contribute to vascular complications. In this chapter, we summarize the current understanding of the regulation of $Ca_V1.2$ channels in vascular smooth muscle by different signaling pathways. We place special attention on the regulation of $Ca_V1.2$ channel activity in vascular smooth muscle by a newly

M. Martín-Aragón Baudel (✉) · J. Hong · J. W. Hell · M. Nieves-Cintrón · M. F. Navedo (✉)
Department of Pharmacology, University of California Davis, Davis, CA, USA
e-mail: martinaragon@ucdavis.edu; mfnavedo@ucdavis.edu

J. Striessnig (ed.), *Voltage-gated Ca^{2+} Channels: Pharmacology, Modulation and their Role in Human Disease*, Handbook of Experimental Pharmacology 279,
https://doi.org/10.1007/164_2022_628

uncovered AKAP5/P2Y_{11}/AC5/PKA/$Ca_V1.2$ axis that is engaged during diabetic hyperglycemia. We further describe the pathophysiological implications of activation of this axis as it relates to myogenic tone and vascular reactivity and propose that this complex may be targeted for developing therapies to treat diabetic vascular complications.

Keywords

Calcium · Diabetes · Ion channel · Smooth muscle · Vasculature

1 Introduction

The cases of people with diabetes are on the rise. According to the World Health Organization, diabetes is reaching epidemic proportions, with approximately 1 in 11 adults worldwide suffering from this chronic disease (Organization 2016). Diabetes is a multifactorial metabolic disorder that occurs due to the inability of the body to produce or effectively use insulin. Hyperglycemia (e.g., elevated blood sugar levels) is the immediate metabolic abnormality that results from insulin dysfunction, and it is associated with cardiovascular complications such as hypertension, stroke, heart attack, retinopathy, nephropathy, or decreased cognitive function (Raghavan et al. 2019; Stehouwer 2018; Brown et al. 2010; Scherer and Hill 2016). Vascular complications are responsible for most hospitalizations and morbidity and mortality in patients with diabetes (Beckman and Creager 2016). Small resistance arteries (<150 μm in diameter), in their ability to regulate efficient nutrient delivery, blood flow, tissue perfusion, and total peripheral resistance, are key contributors to the pathophysiological framework of this disease (Serne et al. 2007; Stehouwer 2018). Therefore, it is important to understand the key components of vascular dysfunction in diabetes to develop novel therapeutic approaches.

The myogenic tone is an intrinsic property of vascular smooth muscle cells lining small resistance arteries and arterioles (<150 μm in diameter). This mechanism translates surges in intravascular pressure into increases in the contractility (Bayliss 1902; Davis and Hill 1999; Tykocki et al. 2017). Different inputs from endothelial cells and nerve terminals, together with changes in intravascular pressure, dictate the function and contractile state of vascular smooth muscle cells, therefore influencing overall vascular reactivity. The combined activity of several ion channels and Rho-associated kinase (ROK)-dependent mechanisms modulates the myosin/actin cross-bridge formation resulting in smooth muscle contraction or dilation (Somlyo and Somlyo 2003; El-Yazbi and Abd-Elrahman 2017; Knot and Nelson 1998). The activity of ion channels facilitates the control of membrane potential (V_M) and the magnitude of intracellular Ca^{2+} concentration ($[Ca^{2+}]_i$) that controls the contractile state of smooth muscle cells (Knot and Nelson 1998; Tykocki et al. 2017). The different ionic conductance elicited by voltage-gated (K_V), Ca^{2+}-activated (BK_{Ca}) and ATP-sensitive (K_{ATP}) K^+ channels, ryanodine receptors (RyR), IP_3 receptors (IP_3R), and voltage-gated Ca^{2+} channels is critical for regulation of V_M and $[Ca^{2+}]_i$

(Jackson 2021a; Tykocki et al. 2017). Therefore, changes in the expression and/or function of these channels have been associated with vascular dysfunction in different pathological states such as hyperglycemia and diabetes (Nieves-Cintrón et al. 2021). In this book chapter, we will focus our narrative on the role of the L-type Ca^{2+} channel $Ca_V1.2$ in controlling vascular smooth muscle function in health and the pathophysiology of diabetic hyperglycemia.

2 Role of $Ca_V1.2$ in Smooth Muscle Cells

The L-type Ca^{2+} channel $Ca_V1.2$ has a specific subunit and splice variant composition in vascular smooth muscle: a pore-forming $Ca_V1.2$ α_{1C} subunit with 9* exon and auxiliary β3 (predominant isoform), $\alpha_2\delta$ and γ subunits that regulate trafficking and channel function (Ghosh et al. 2017; Kharade et al. 2013; Liao et al. 2004; Bannister et al. 2011). In smooth muscle cells, L-type Ca^{2+} channel $Ca_V1.2$ is the major Ca^{2+} entry pathway regulating cell excitability, muscle contraction, proliferation, and gene expression (Tykocki et al. 2017; Brozovich et al. 2016; Keef et al. 2001). Intravascular pressure- and agonist-mediated membrane depolarization activates $Ca_V1.2$ channels leading to cell-wide increases in global $[Ca^{2+}]_i$ that activate the contractile machinery (Tykocki et al. 2017). Accordingly, pressure- or agonist-induced cation entry through TRPM4 and TRPP2 channels along with Cl^- efflux through Ca^{2+}-activated Cl^- channels (CaCCs) has been suggested as the key ion channels involved in membrane depolarization and subsequent $Ca_V1.2$ activation in smooth muscle cells (Jackson 2021b). The resulting increase in $[Ca^{2+}]_i$ is converted into smooth muscle contraction mainly through the binding of Ca^{2+} to calmodulin (CaM), and Ca^{2+}/CaM-dependent activation of myosin light-chain kinase (MLCK). This kinase phosphorylates the 20 kD myosin light chains, stimulating myosin–actin interaction which triggers force production and contraction in vascular smooth muscle cells (Cole and Welsh 2011; Walsh 2011).

Different vasoconstrictor agonists such as phenylephrine, norepinephrine, serotonin, histamine, angiotensin II, and endothelin have shown a $Ca_V1.2$-direct functional dependence (Tykocki et al. 2017). Agonist-induced regulation of $Ca_V1.2$ in smooth muscle has traditionally been linked to engagement of $G_{q/11}$-coupled receptors leading to PKC activation (Keef et al. 2001). In particular, hyperactive $Ca_V1.2$ channels that increase global $[Ca^{2+}]_i$ and myogenic tone in response to angiotensin II are regulated by PKCα targeting to $Ca_V1.2$ via AKAP5 in smooth muscle cells from murine cerebral and mesenteric arteries (Navedo et al. 2008; Navedo and Santana 2013). Activation of α_1- (coupled to $G_{q/11}$) and α_2- (coupled to G_i) adrenoreceptors in smooth muscle cells from rat portal vein has also been shown to activate $Ca_V1.2$ through a PKC-dependent mechanism (Lepretre et al. 1994), although the mechanisms for the α_2-G_i adrenoreceptor-dependent regulation of $Ca_V1.2$ channels was not established. The tyrosine-protein kinase Src (c-SRC) has been reported to modulate the activity of $Ca_V1.2$ through G_q-PKC-dependent (Callaghan et al. 2004) and independent (Gulia et al. 2013; Gui et al. 2006) mechanisms. More recently, a PKA-dependent activation of $Ca_V1.2$ in discrete nanodomains in the context of diabetic hyperglycemia was described (Nystoriak et al. 2017; Prada et al. 2019, 2020b; Syed et al. 2019). Paradoxically, PKA-dependent regulation of $Ca_V1.2$ channels in vascular smooth muscle in response to hyperglycemia and diabetes was associated with vasoconstriction. The mechanisms and further implications of these observations are described below.

In addition to their role in regulating arterial contractility, fully differentiated vascular smooth muscle cells can switch their phenotype, thus retaining remarkable plasticity. De-differentiated vascular smooth muscle cells lose their contractile properties to promote neo-vascularization by expressing extracellular matrix components and remodeling enzymes as well as increased levels of secretory organelles and pro-inflammatory cytokines (Clarke et al. 2008). This plasticity is necessary for promoting vascularization during growth or following vascular injury (Kocher et al. 1991). However, the pathogenic migration and proliferation of vascular smooth muscle cells is a hallmark of vascular disorders such as atherosclerosis and restenosis. Calcium channel blockers (CCBs) have long been known to exert anti-atherogenic (Mason 2002) and anti-migratory effects (Ruiz-Torres et al. 2003) and also induce a reduction in plaque formation (Koshiyama et al. 1999; Yamashita et al. 2001). However, the precise mechanisms behind these vasoprotective effects remain elusive. Changes in $[Ca^{2+}]_i$ are coupled to promigratory effectors via calcium-dependent protein kinases, Ca^{2+}/calmodulin-dependent kinases (CaMK), and/or activation of the β isoform of platelet-derived growth factor receptors (PDGFR-β), which result in actin polymerization and myosin II force generation (Gerthoffer 2007). The involvement of $[Ca^{2+}]_i$ in the proliferation of vascular smooth muscle cells is linked to the Ca^{2+}-dependent activation of different transcription factors (e.g., NFAT, CREB), a process known as excitation-transcription coupling (Kudryavtseva et al. 2013). Despite some evidence pointing to the role that $Ca_V1.2$ channels play in these processes, precise mechanisms still need to be elucidated. Recent evidence suggests that intranuclear CREB phosphorylation, leading to macrophage accumulation and media remodeling, is dependent on a $Ca_V1.2$/CaMKK2/CaMKK1a complex localized to caveolae (Suzuki et al. 2022). This study showed that direct activation of CaMKK2 by Ca^{2+} influx through $Ca_V1.2$ localized to caveolae leads to CaMK1a phosphorylation and translocation to the nucleus. Sustained depolarization of a mesenteric artery induced the transcription of genes involved in chemotaxis, leukocyte adhesion, and inflammation, and these effects were reversed by inhibitors of $Ca_V1.2$, CaMKK2, and CaMK or caveolae disruption. Therefore $Ca_V1.2$ in vascular smooth muscle could represent a target to treat vascular remodeling during pathological conditions.

Serum deprivation of cultured human vascular smooth muscle cells increases functional $Ca_V1.2$-mediated raises in $[Ca^{2+}]_i$ leading to activation of PDGF-induced migration and chemotaxis (Patel et al. 2005). This PDGF-induced migration of synthetic vascular smooth muscle cells has been recently shown to be reliant on c-Src tyrosine kinase-dependent phosphorylation of $Ca_V1.2$ (Guo et al. 2018). In these cells, PDGF and c-Src exposure significantly increased $Ca_V1.2$ current-voltage relationship and migration, these effects being blocked with a selective inhibitor of SFKs. In addition, alternative splice variants of $Ca_V1.2$ α_{1C} that change the electrophysiological properties of the channel have been found in atherosclerotic regions of human arteries, possibly affecting smooth muscle cell proliferation and providing a $Ca_V1.2$-specific molecular signature of atherosclerosis (Tiwari et al. 2006). Another study found that PDGF inhibited $Ca_V1.2$ expression by activating RhoA, ERK1/2, and p38 MAPK pathways (Ouyang et al. 2014). These effects were also observed in proliferating smooth muscle cells and were abolished with the statin Fluvastatin. These observations may provide a possible explanation as to why CCBs lose efficacy as vascular disease progresses and smooth muscle cells become more differentiated. In addition, a recent study has shown how three major classes of CCBs activate the endoplasmic reticulum (ER) stromal-interacting molecule (STIM) Ca^{2+} sensor protein and the plasma membrane ORAI Ca^{2+} channels, which could promote Ca^{2+} signaling-dependent vascular smooth muscle cell remodeling and worsen advanced vascular disease prognosis (Johnson et al. 2020). Therefore, further research is needed to understand $Ca_V1.2$ involvement in the pathological proliferation and migration of vascular smooth muscle cells.

3 Effects of Diabetic Hyperglycemia on Vascular $Ca_V1.2$

The influence that hyperglycemia has on modulating the activity of $Ca_V1.2$ channels in vascular smooth muscle cells is well recognized. In the pathological context of diabetic hyperglycemia, some studies have reported either decreased or unmodified $Ca_V1.2$ channel activity in vascular smooth muscle cells (Fulton et al. 1991; Wang et al. 2000; Carmines et al. 1996; Abebe and MacLeod 1990; Mulhern and Docherty 1989; Dong et al. 2008). However, the majority of studies indicate an increase in $Ca_V1.2$ channel activity in vascular smooth muscle in a wide variety of hyperglycemic stages and models of diabetes (Nystoriak et al. 2017; Ungvari 1999; Prada et al. 2019; White and Carrier 1990; Navedo et al. 2010; Pinho et al. 2010; Wilde et al. 2000; Ma et al. 2017; Syed et al. 2019). This functional dichotomy could be attributed to the use of different vascular beds, freshly isolated or cultured cells, and the type and duration of the hyperglycemic insult. Moreover, $Ca_V1.2$ channel function could be different with the use of different animal models of disease that present concomitant specific metabolic characteristics such as hyperinsulinemia (McCarty 2006) or hypercholesterolemia (Clunn et al. 1999; Bowles et al. 2004). However, a recent study using vascular smooth muscle cells from diabetic patients also reported an increase in the activity of $Ca_V1.2$ channels (Nystoriak et al. 2017), indicating similarities between animal models and human disease.

In mammalian cells, glucose is the main source of energy, and it is facilitated across the plasma membrane by glucose transporters (GLUTs). Although different isoforms have been found in vascular smooth muscle cells, the predominant isoforms described are GLUT1 and GLUT4 (Atkins et al. 2001; Park et al. 2005). GLUT1 is constitutively expressed on the plasma membrane for basal glucose uptake (Pyla et al. 2013), whereas GLUT4 translocation to the plasma membrane requires insulin stimulation, and its expression seems to be decreased in vascular diseases caused by diabetes and hypertension (Park et al. 2005). However, the relative contribution of the different GLUTs in vascular smooth muscle during basal versus pathological conditions, as well as the existence of compensatory mechanisms amongst different isoforms, remains unclear and an opportunity for future studies.

In intriguing studies consistent with a potential role for GLUT in glucose regulation of $Ca_V1.2$ channels, potentiation of $Ca_V1.2$ activity in vascular smooth muscle by elevated extracellular glucose was found to depend on a mechanism requiring glucose uptake and metabolization (Nystoriak et al. 2017). Accordingly, pretreatment of vascular smooth muscle cells with indinavir (a GLUT4 inhibitor (Park et al. 2005)) prevented the elevation of $Ca_V1.2$ channel activity elicited by elevated extracellular glucose (Nystoriak et al. 2017). This result correlates with independent observations that facilitative glucose transport may be responsible for the modulation of vascular reactivity (Park et al. 2005). Indeed, alterations in L-type Ca^{2+} channel activity in vascular smooth muscle during diabetic hyperglycemia may contribute to altering myogenic tone and vascular reactivity, with direct consequences on tissue perfusion, blood pressure, and the modulation of the contractile and proliferative gene program (Hien et al. 2016; Wamhoff et al. 2004; Tykocki et al. 2017; Navedo and Amberg 2013), as discussed below.

4 Mechanisms

A key question is what is the mechanism by which diabetic hyperglycemia stimulates $Ca_V1.2$ channel activity in vascular smooth muscle? The abundance of the $Ca_V1.2$ channel protein in cerebral, mesenteric, and adipose arteries from high-fat diet (HFD, type 2 diabetic model) mice and diabetic patients remains unchanged, indicating that changes in $Ca_V1.2$ channel activity are likely driven by post-translational modifications (Nystoriak et al. 2014, 2017). In aortic vascular smooth muscle from a mouse model of type 1 diabetes (Streptozotocin – STZ), an increase in the expression of the phosphatidylinositol 3-kinase δ isoform (PI3Kδ) was found to be responsible for the potentiation of L-type Ca^{2+} currents (Pinho et al. 2010). However, the precise mechanism of this modulation was not examined. In COS-7 cells and neurons, PI3Kδ was found to phosphorylate the accessory Ca_V β subunit via Akt/protein kinase B (PKB) to promote channel trafficking to the membrane and the subsequent stimulation of channel activity (Viard et al. 2004). In a different study, the elevation of plasma fatty acids in a rat model of HFD was correlated with increases in $Ca_V1.2$ activity in cerebral vascular smooth muscle cells (Wilde et al. 2000). Yet, the precise mechanisms by which PI3Kδ and plasma fatty acids regulated $Ca_V1.2$ channels during diabetic hyperglycemia were not comprehensively explored, which is an opportunity for future research.

While increased PKC activity has been reported in vascular smooth muscle cells during diabetic hyperglycemia (Hashim et al. 2006; Patel et al. 1999; Williams et al. 1997; Williams and Schrier 1992), and this kinase has been shown to increase $Ca_V1.2$ channel activity (Navedo et al. 2005, 2006; Nieves-Cintron et al. 2008; Payne and Smeda 2002; Ruan and Arendshorst 1996; Salamanca and Khalil 2005),

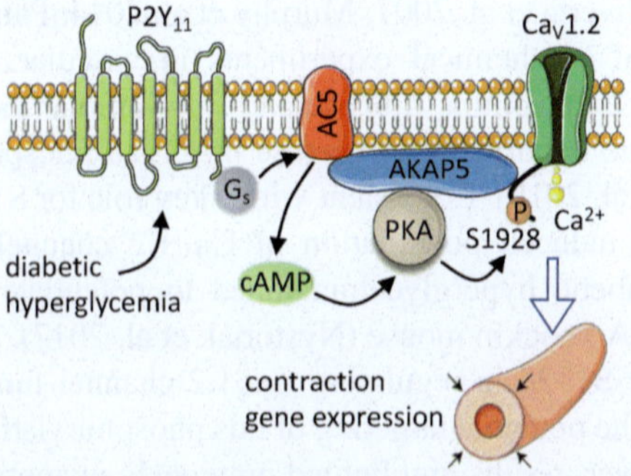

Fig. 1 Schematic of the model for regulation of vascular $Ca_V1.2$ channels during diabetic hyperglycemia. Diabetic hyperglycemia induces the activation of $P2Y_{11}$ receptors, which promotes AC5 activity and localized cAMP production. This cAMP microdomain can enable a pool of PKA that is intimately associated with L-type $Ca_V1.2$ channels to increase their phosphorylation at S1928, which will potentiate channel activity. AKAP5 serves as the scaffold to intimately bind these proteins in discrete nanodomains. Hyperactive $Ca_V1.2$ channels result in increased global $[Ca^{2+}]_i$ leading to changes in gene expression and contraction of vascular smooth muscle. These changes may influence arterial diameter, myogenic tone, blood flow, and blood pressure

the hyperglycemic effects on vascular $Ca_V1.2$ channel activity do not seem to depend on this kinase. This notion initially arose because elevated extracellular glucose could still potentiate $Ca_V1.2$ channel activity in vascular smooth muscle from a PKC knockout mouse (Navedo et al. 2010). Intriguingly, pretreatment of vascular smooth muscle cells with a PKA inhibitor prevented the elevated glucose stimulation of the $Ca_V1.2$ channel (Fig. 1) (Navedo et al. 2010; Nystoriak et al. 2017). These results implicated PKA in elevated glucose-mediated potentiation of vascular $Ca_V1.2$ channels, which could result in vasoconstriction. These observations were unexpected, as PKA activity has traditionally been linked to vasorelaxation (Porter et al. 1998; Wellman et al. 1996, 2001; Aiello et al. 1995; Moore et al. 2014). The above results raised the key question of how PKA underlies the elevated glucose-induced potentiation of $Ca_V1.2$ channels in vascular smooth muscle.

In comprehensive studies, glucose-induced PKA regulation of $Ca_V1.2$ channels in vascular smooth muscle was found to be mediated by the phosphorylation of the $Ca_V1.2$ pore-forming α_{1C} subunit at serine 1928 (S1928) (Fig. 1) (Nystoriak et al. 2017). S1928 is a highly conserved PKA phosphorylation site (Dai et al. 2009; Davare and Hell 2003; Murphy et al. 2014; Hulme et al. 2006; De Jongh et al. 1996; Gao et al. 1997), which was thought to potentially modulate $Ca_V1.2$ channel activity in cardiac, vascular smooth muscle, and neuronal cells. However, cardiac cells from mice with disrupted phosphorylation of S1928 (e.g., S1928A) have normal PKA-dependent regulation of $Ca_V1.2$ channels (Lemke et al. 2008; Katchman et al. 2017; Ganesan et al. 2006). These results indicated that increased phosphorylation of S1928 was not functionally relevant for $Ca_V1.2$ channel regulation in the heart. In stark contrast, subsequent studies found that S1928 phosphorylation was needed for β2 adrenergic receptor-mediated $Ca_V1.2$ channel stimulation in neurons

(Qian et al. 2017; Oliveria et al. 2007; Murphy et al. 2014; Patriarchi et al. 2016). In addition, a series of biochemical experiments in vascular smooth muscle cells exposed to elevated extracellular glucose and from animal models of diabetes and diabetic human tissue confirmed an increase in S1928 phosphorylation (Nystoriak et al. 2017; Prada et al. 2019). Consistent with a key role for S1928 phosphorylation in elevated glucose-induced potentiation of $Ca_V1.2$ channel activity in vascular smooth muscle, diabetic hyperglycemia failed to potentiate channel function in cells from an S1928A knockin mouse (Nystoriak et al. 2017). Thus, results strongly supported a role for S1928 in regulating $Ca_V1.2$ channel function during diabetic hyperglycemia and the potential targeting of this phosphorylation site for therapeutic intervention. Moreover, results highlighted previously unappreciated tissue-specific differences in $Ca_V1.2$ channel regulation, which deserves further exploration.

The role of PKA in mediating vascular smooth muscle relaxation or contractile effects could be explained if PKA molecules were compartmentalized into different subpopulations along with their upstream regulators to activate distinct mechanisms. For example, a subpopulation of PKA activated by elevated glucose can be in close contact with $Ca_V1.2$ channels, whereas another subpopulation may be more closely associated with K^+ channels and relaxation pathways in the contractile machinery. The distinct distribution of PKA to any of these different compartments is likely orchestrated by scaffold proteins (Rhee and Rusch 2018; Syed et al. 2020). A-kinase anchoring proteins (AKAPs) are scaffolding proteins that can anchor the regulatory subunits of PKA and guide cellular localization via various targeting domains (Langeberg and Scott 2015; Johnstone et al. 2018; Man et al. 2019). In particular, AKAP5 (murine AKAP150 and human AKAP79) interacts with adenylyl cyclase (AC), PKA, PKC, and $Ca_V1.2$ (Gao et al. 1997; Bauman et al. 2006; Navedo et al. 2008; Efendiev et al. 2010; Murphy et al. 2014; Zhang et al. 2013; Davare et al. 1999), and is, therefore, an archetype candidate to arbitrate the PKA-driven signaling nanodomain in response to diabetic hyperglycemia. Indeed, super-resolution localization maps revealed a close association between PKA and $Ca_V1.2$ in native human adipose and mouse cerebral vascular smooth muscle cells (Nystoriak et al. 2017). Genetic depletion, inhibition, or modification of AKAP5 reduced the magnitude of these associations and prevented the stimulation of $Ca_V1.2$ channels by elevated glucose and diabetes in mouse cerebral vascular smooth muscle (Nystoriak et al. 2017). These exciting results provided a mechanism by which hyperglycemia/diabetes activates signaling within nanostructural domains via AKAP5 to recruit PKA and phosphorylate $Ca_V1.2$ at S1928 to potentiate channel activity (Fig. 1).

The activation of PKA and its recruitment into the AKAP5/$Ca_V1.2$ complex requires the activity of upstream second messengers, regulatory enzymes, and signal generators. Previous work by independent groups had linked the elevation in glucose levels with the production of the second messenger cAMP in yeast and murine pancreatic β cells (Munder and Kuntzel 1989; Dyachok et al. 2008). Consistent with this, it was found that elevating extracellular glucose also led to increased cAMP levels in vascular smooth muscle cells (Prada et al. 2019; Syed et al. 2019). Different adenylate cyclase (ACs) isoforms can mediate cAMP production in vascular smooth muscle. Indeed, during diabetes, alterations in AC signaling have been reported in vascular smooth muscle cells, which may contribute to cell proliferation, increased oxidative stress, impaired relaxation, and enhanced expression of contractile genes (Hashim et al. 2004; Li et al. 2008; Matsumoto et al. 2005; Hien et al. 2016). In the context of mechanisms underlying elevated glucose/diabetes-induced cAMP synthesis, activation of PKA, and potentiation of $Ca_V1.2$ channels, AC5 was identified as the key AC isoform (Fig. 1) (Syed et al. 2019). Super-resolution imaging revealed close proximity between subpopulations of AC5 and the $Ca_V1.2$ pore-forming α_{1C} subunit, and this association was increased in vascular smooth muscle cells from diabetic mice (Syed et al. 2019). Such an association was not observed between AC6 and α_{1C} subunits. Further supporting the role of AC5 in regulating vascular smooth muscle during diabetic hyperglycemia, an independent microarray analysis found that AC5 expression was upregulated upon chronic elevations of extracellular glucose (Hien et al. 2016). Although other AC isoforms have been reported to participate in the regulation of arterial contractility in diabetes (Matsumoto et al. 2005), the specific activation of AC5 leading to localized AC5-dependent cAMP synthesis (Fig. 1) seems to be critical for the specific engagement of the PKA/$Ca_V1.2$ nanodomain in vascular smooth muscle during diabetic hyperglycemia.

The generation of second messengers by ACs (as in glucose-triggered, AC5-mediated increases in cAMP levels) is dependent upon the activation of G protein-coupled receptors (GPCRs) (Johnstone et al. 2018). Purinergic receptors are GPCRs that can be activated by nucleotides/nucleosides (Huang et al. 2021). In vascular smooth muscle, elevations in extracellular glucose are known to stimulate the autocrine release of nucleotides, such as ATP (Nilsson et al. 2006; Parodi et al. 2002). These observations led to the hypothesis that glucose-mediated activation of the AKAP5/AC5/PKA axis leading to potentiation of vascular $Ca_V1.2$ channels was mediated by nucleotide activation of a purinergic receptor. Amongst the different purinergic receptor subtypes, $P2Y_{11}$(human)/$P2Y_{11}$-like (murine) receptors were sought as a potential GPCR candidate as they are activated by ATP and are the only purinergic receptors that can couple to G_s proteins to trigger AC5/PKA signaling and vascular $Ca_V1.2$ regulation (Kennedy 2017; Martin-Aragon Baudel et al. 2020). Consistent with the hypothesis, super-resolution imaging found that $P2Y_{11}$/$P2Y_{11}$-like receptors were in close proximity to $Ca_V1.2$ and PKA (Prada et al. 2019). In addition, inhibition of $P2Y_{11}$/$P2Y_{11}$-like receptors, but not $P2Y_1$ or $P2Y_6$ receptors, prevented the elevations in cAMP levels and the potentiation of $Ca_V1.2$ channels in response to elevated glucose in human and mouse vascular smooth muscle cells (Prada et al. 2019). The selective $P2Y_{11}$ agonist NF546 increased cAMP levels and $Ca_V1.2$ channel activity in human and mouse vascular smooth muscle cells, similar to the elevated glucose effects (Prada et al. 2019). Important, NF546 and elevated glucose effects on vascular smooth muscle cAMP levels and $Ca_V1.2$ channel activity were not additive, suggesting that both stimuli are likely acting via the same pathway. These results strongly implicated $P2Y_{11}$/$P2Y_{11}$-like receptors as the signal generators triggering the AKAP5/AC5/PKA axis to potentiate vascular $Ca_V1.2$ channel activity in vascular smooth muscle in response to diabetic hyperglycemia (Fig. 1).

The functional association of $P2Y_{11}/P2Y_{11}$-like receptors with other members of the signaling axis was mediated by AKAP5 in human and murine vascular smooth muscle (Prada et al. 2020b). Accordingly, genetic ablation of AKAP5 (e.g., $AKAP5^{-/-}$) prevented the increase in cAMP levels and $Ca_V1.2$ channel activity in response to elevated glucose and the $P2Y_{11}$ agonist NF546 in vascular smooth muscle. Moreover, the association of $P2Y_{11}$ receptors with PKA and the $Ca_V1.2$ α_{1C} subunit, as well as the $Ca_V1.2$ α_{1C} subunit with AC5 and PKA with AC5 were completely abrogated in $AKAP5^{-/-}$ vascular smooth muscle cells (Prada et al. 2020b). Unexpectedly, the $P2Y_{11}/P2Y_{11}$-like receptors remained in close association with AC5 in $AKAP5^{-/-}$ vascular smooth muscle cells. These observations are consistent with a recent independent study indicating that GPCRs can be in a pre-coupled state with AC isoforms to sustain cell signaling (Navarro et al. 2018). However, the $P2Y_{11}$ receptors/AC5 complex in $AKAP5^{-/-}$ vascular smooth muscle cells was not functional as elevated glucose did not stimulate the activity of vascular $Ca_V1.2$ channels (Prada et al. 2020b). The reasons for this observation were correlated with (1) an increase in phosphodiesterase (PDEs) activity that prevented elevations in cAMP levels, and (2) the broader disruption of the signaling components that prevented optimal activation of pools of PKA in close proximity with $Ca_V1.2$ channels in $AKAP5^{-/-}$ cells upon elevated extracellular glucose. Altogether, the results suggested the formation of an $AKAP5/P2Y_{11}/AC5/PKA/Ca_V1.2$ signaling complex that is essential for the regulation of $Ca_V1.2$ channels in vascular smooth muscle during diabetic hyperglycemia. Thus, this complex may have profound implications for the regulation of vascular function during diabetic hyperglycemia. Moreover, this complex may be targeted for the development of therapies to treat vascular complications during diabetes.

5 Pathophysiological Impact

Alterations in $Ca_V1.2$ channel activity in vascular smooth muscle during diabetic hyperglycemia may have profound implications for vascular function. For example, a reduction in smooth muscle $Ca_V1.2$ channel activity during diabetes was found to reduce vascular responsiveness to vasoconstrictors (Fulton et al. 1991; Carmines et al. 1996), and one study suggested this may contribute to impaired glomerular filtration (Carmines et al. 1996). Conversely, potentiation of smooth muscle $Ca_V1.2$ channel activity during diabetic hyperglycemia has been linked with elevations in $[Ca^{2+}]_i$ and contraction (Nystoriak et al. 2014, 2017; Syed et al. 2019; Ungvari et al. 1999; Prada et al. 2019, 2020b; White and Carrier 1990; Wilde et al. 2000; Pinho et al. 2010). The increase in vascular smooth muscle $[Ca^{2+}]_i$ and contraction in response to diabetic hyperglycemia can contribute to elevating the myogenic response, altering tissue perfusion, and increasing blood pressure. In addition, alterations in $Ca_V1.2$ channel activity during diabetic hyperglycemia may also influence gene programs involved in regulating the contractile and proliferative state of vascular smooth muscle (Richard et al. 1992; Hien et al. 2016; Wamhoff et al. 2004). These changes in gene programs in vascular smooth muscle can be driven by several transcription factors, including NFAT (Nieves-Cintron et al. 2015; Nystoriak et al. 2014; Nilsson et al. 2006), Nrf2 (Lu et al. 2017; Velmurugan et al. 2013), and MRTF (Hien et al. 2016). Intriguingly, disruption of the AKAP5/P2Y$_{11}$/AC5/PKA/$Ca_V1.2$ signaling complex by genetic ablation of one or several of its components or pharmacological manipulation was shown to ameliorate the diabetic hyperglycemia-induced alterations in $Ca_V1.2$ channel activity, functional expression of several K^+ channels, smooth muscle $[Ca^{2+}]_i$ and contractile state, myogenic tone, blood flow, and blood pressure (Navedo et al. 2010; Nystoriak et al. 2014, 2017; Prada et al. 2019, 2020a; Syed et al. 2019; Nieves-Cintron et al. 2015). These results highlight the pathophysiological relevance of the AKAP5/P2Y$_{11}$/AC5/PKA/$Ca_V1.2$ signaling complex in impacting vascular function in response to diabetic hyperglycemia.

6 Conclusions/Future Directions

Diabetic hyperglycemia has a major impact on the vasculature. Uncovering mechanisms underlying vascular complications in response to diabetic hyperglycemia is important to understanding pathophysiology as well as for the development of rational therapies. In this chapter, we deliberately focused on the effects of diabetic hyperglycemia on vascular smooth muscle L-type $Ca_V1.2$ channels. The rationale is that these ion channels play a major role in regulating vascular smooth muscle function, and their activity is well known to be altered by diabetic hyperglycemia. Moreover, by exploring how diabetic hyperglycemia alters $Ca_V1.2$ channel function in vascular smooth muscle, a new nanomolecular signaling complex formed by AKAP5, P2Y$_{11}$, AC5, PKA, and $Ca_V1.2$ was uncovered. The activation of this signaling complex in response to diabetic hyperglycemia alters vascular smooth muscle contractile state. Thus, diabetic hyperglycemia-induced changes in vascular $Ca_V1.2$ channel activity have a key impact on the control of arterial diameter, development of myogenic tone, and regulation of blood flow and blood pressure. The targeting of the AKAP5/P2Y$_{11}$/AC5/PKA/$Ca_V1.2$ signaling complex represents an area of opportunity for the development of therapies that could help ameliorate vascular complications during diabetic hyperglycemia.

References

Abebe W, Macleod KM (1990) Protein kinase C-mediated contractile responses of arteries from diabetic rats. Br J Pharmacol 101:465–471

Aiello EA, Walsh MP, Cole WC (1995) Phosphorylation by protein kinase A enhances delayed rectifier K+ current in rabbit vascular smooth muscle cells. Am J Phys 268:H926–H934

Atkins KB, Johns D, Watts S, Clinton Webb R, Brosius FC III (2001) Decreased vascular glucose transporter expression and glucose uptake in DOCA-salt hypertension. J Hypertens 19:1581–1587

Bannister JP, Thomas-Gatewood CM, Neeb ZP, Adebiyi A, Cheng X, Jaggar JH (2011) Ca(V)1.2 channel N-terminal splice variants modulate functional surface expression in resistance size artery smooth muscle cells. J Biol Chem 286:15058–15066
Bauman AL, Soughayer J, Nguyen BT, Willoughby D, Carnegie GK, Wong W, Hoshi N, Langeberg LK, Cooper DM, Dessauer CW, Scott JD (2006) Dynamic regulation of camp synthesis through anchored Pka-adenylyl cyclase V/Vi complexes. Mol Cell 23:925–931
Bayliss WM (1902) On the local reaction of the arterial wall to changes in internal pressure. J Physiol 28:220–231
Beckman JA, Creager MA (2016) Vascular complications of diabetes. Circ Res 118:1771–1785
Bowles DK, Heaps CL, Turk JR, Maddali KK, Price EM (2004) Hypercholesterolemia inhibits L-type calcium current in coronary macro-, not microcirculation. J Appl Physiol 96:2240–2248
Brown A, Reynolds LR, Bruemmer D (2010) Intensive glycemic control and cardiovascular disease: an update. Nat Rev Cardiol 7:369–375
Brozovich FV, Nicholson CJ, Degen CV, Gao YZ, Aggarwal M, Morgan KG (2016) Mechanisms of vascular smooth muscle contraction and the basis for pharmacologic treatment of smooth muscle disorders. Pharmacol Rev 68:476–532
Callaghan B, Koh SD, Keef KD (2004) Muscarinic M2 receptor stimulation of Cav1.2b requires phosphatidylinositol 3-kinase, protein kinase C, and C-Src. Circ Res 94:626–633
Carmines PK, Ohishi K, Ikenaga H (1996) Functional impairment of renal afferent arteriolar voltage-gated calcium channels in rats with diabetes mellitus. J Clin Invest 98:2564–2571
Clarke MC, Littlewood TD, Figg N, Maguire JJ, Davenport AP, Goddard M, Bennett MR (2008) Chronic apoptosis of vascular smooth muscle cells accelerates atherosclerosis and promotes calcification and medial degeneration. Circ Res 102:1529–1538
Clunn GF, Wijetunge S, Hughes AD (1999) Effect of hypercholesterolaemia on voltage-operated calcium channel currents in rabbit arterial smooth muscle cells. J Hum Hypertens 13:849–853
Cole WC, Welsh DG (2011) Role of myosin light chain kinase and myosin light chain phosphatase in the resistance arterial myogenic response to intravascular pressure. Arch Biochem Biophys 510:160–173
Dai S, Hall DD, Hell JW (2009) Supramolecular assemblies and localized regulation of voltage-gated ion channels. Physiol Rev 89:411–452
Davare MA, Hell JW (2003) Increased phosphorylation of the neuronal L-type Ca(2+) channel Ca (V)1.2 during aging. Proc Natl Acad Sci U S A 100:16018–16023
Davare MA, Dong F, Rubin CS, Hell JW (1999) The A-kinase anchor protein Map2b and camp-dependent protein kinase are associated with class C L-type calcium channels in neurons. J Biol Chem 274:30280–30287
Davis MJ, Hill MA (1999) Signaling mechanisms underlying the vascular myogenic response. Physiol Rev 79:387–423
De Jongh KS, Murphy BJ, Colvin AA, Hell JW, Takahashi M, Catterall WA (1996) Specific phosphorylation of a site in the full-length form of the alpha 1 subunit of the cardiac L-type calcium channel by adenosine 3′,5′- cyclic monophosphate-dependent protein kinase. Biochemistry 35:10392–10402
Dong L, Zheng YM, Van Riper D, Rathore R, Liu QH, Singer HA, Wang YX (2008) Functional and molecular evidence for impairment of calcium-activated potassium channels in type-1 diabetic cerebral artery smooth muscle cells. J Cereb Blood Flow Metab 28:377–386
Dyachok O, Idevall-Hagren O, Sagetorp J, Tian G, Wuttke A, Arrieumerlou C, Akusjarvi G, Gylfe E, Tengholm A (2008) Glucose-induced cyclic amp oscillations regulate pulsatile insulin secretion. Cell Metab 8:26–37
Efendiev R, Samelson BK, Nguyen BT, Phatarpekar PV, Baameur F, Scott JD, Dessauer CW (2010) Akap79 interacts with multiple adenylyl cyclase (Ac) isoforms and scaffolds Ac5 and −6 to alpha-amino-3-hydroxyl-5-methyl-4-isoxazole-propionate (Ampa) receptors. J Biol Chem 285:14450–14458
El-Yazbi AF, Abd-Elrahman KS (2017) Rok and arteriolar myogenic tone generation: molecular evidence in health and disease. Front Pharmacol 8:87

Fulton DJ, Hodgson WC, Sikorski BW, King RG (1991) Attenuated responses to endothelin-1, Kcl and Cacl2, but not noradrenaline, of aortae from rats with streptozotocin-induced diabetes mellitus. Br J Pharmacol 104:928–932

Ganesan AN, Maack C, Johns DC, Sidor A, O'rourke B (2006) Beta-adrenergic stimulation of L-type Ca2+ channels in cardiac myocytes requires the distal carboxyl terminus of alpha1c but not serine 1928. Circ Res 98:E11–E18

Gao T, Yatani A, Dell'acqua ML, Sako H, Green SA, Dascal N, Scott JD, Hosey MM (1997) Camp-dependent regulation of cardiac L-type Ca2+ channels requires membrane targeting of Pka and phosphorylation of channel subunits. Neuron 19:185–196

Gerthoffer WT (2007) Mechanisms of vascular smooth muscle cell migration. Circ Res 100:607–621

Ghosh D, Syed AU, Prada MP, Nystoriak MA, Santana LF, Nieves-Cintron M, Navedo MF (2017) Calcium channels in vascular smooth muscle. Adv Pharmacol 78:49–87

Gui P, Wu X, Ling S, Stotz SC, Winkfein RJ, Wilson E, Davis GE, Braun AP, Zamponi GW, Davis MJ (2006) Integrin receptor activation triggers converging regulation of Cav1.2 calcium channels by C-Src and protein kinase A pathways. J Biol Chem 281:14015–14025

Gulia J, Navedo MF, Gui P, Chao JT, Mercado JL, Santana LF, Davis MJ (2013) Regulation of L-type calcium channel sparklet activity by C-Src and Pkc-A. Am J Physiol Cell Physiol 305: C568–C577

Guo X, Kashihara T, Nakada T, Aoyama T, Yamada M (2018) Pdgf-induced migration of synthetic vascular smooth muscle cells through C-Src-activated L-type Ca(2+) channels with full-length Ca(V)1.2 C-terminus. Pflugers Arch 470:909–921

Hashim S, Li Y, Nagakura A, Takeo S, Anand-Srivastava MB (2004) Modulation of G-protein expression and adenylyl cyclase signaling by high glucose in vascular smooth muscle. Cardiovasc Res 63:709–718

Hashim S, Li Y, Anand-Srivastava MB (2006) G protein-linked cell signaling and cardiovascular functions in diabetes/hyperglycemia. Cell Biochem Biophys 44:51–64

Hien TT, Turczynska KM, Dahan D, Ekman M, Grossi M, Sjogren J, Nilsson J, Braun T, Boettger T, Garcia-Vaz E, Stenkula K, Sward K, Gomez MF, Albinsson S (2016) Elevated glucose levels promote contractile and cytoskeletal gene expression in vascular smooth muscle via rho/protein kinase C and actin polymerization. J Biol Chem 291:3552–3568

Huang Z, Xie N, Illes P, Di Virgilio F, Ulrich H, Semyanov A, Verkhratsky A, Sperlagh B, Yu S-G, Huang C, Tang Y (2021) From purines to purinergic signalling: molecular functions and human diseases. Signal Transduct Target Ther 6:162

Hulme JT, Westenbroek RE, Scheuer T, Catterall WA (2006) Phosphorylation of serine 1928 in the distal C-terminal domain of cardiac Cav1.2 channels during beta1-adrenergic regulation. Proc Natl Acad Sci U S A 103:16574–16579

Jackson WF (2021a) Calcium-dependent ion channels and the regulation of arteriolar myogenic tone. Front Physiol 12

Jackson WF (2021b) Myogenic tone in peripheral resistance arteries and arterioles: the pressure is on! Front Physiol 12:699517–699517

Johnson MT, Gudlur A, Zhang X, Xin P, Emrich SM, Yoast RE, Courjaret R, Nwokonko RM, Li W, Hempel N, Machaca K, Gill DL, Hogan PG, Trebak M (2020) L-type Ca(2+) channel blockers promote vascular remodeling through activation of stim proteins. Proc Natl Acad Sci U S A 117:17369–17380

Johnstone TB, Agarwal SR, Harvey RD, Ostrom RS (2018) Camp signaling compartmentation: adenylyl cyclases as anchors of dynamic signaling complexes. Mol Pharmacol 93:270–276

Katchman A, Yang L, Zakharov SI, Kushner J, Abrams J, Chen BX, Liu G, Pitt GS, Colecraft HM, Marx SO (2017) Proteolytic cleavage and Pka phosphorylation of alpha1c subunit are not required for adrenergic regulation of Cav1.2 in the heart. Proc Natl Acad Sci U S A 114: 9194–9199

Keef KD, Hume JR, Zhong J (2001) Regulation of cardiac and smooth muscle Ca(2+) channels (Ca (V)1.2a,B) by protein kinases. Am J Physiol Cell Physiol 281:C1743–C1756

Kennedy C (2017) P2y11 receptors: properties, distribution and functions. Adv Exp Med Biol 1051:107–122

Kharade SV, Sonkusare SK, Srivastava AK, Thakali KM, Fletcher TW, Rhee SW, Rusch NJ (2013) The beta3 subunit contributes to vascular calcium channel upregulation and hypertension in angiotensin Ii-infused C57bl/6 mice. Hypertension 61:137–142

Knot HJ, Nelson MT (1998) Regulation of arterial diameter and wall [Ca^{2+}] in cerebral arteries of rat by membrane potential and intravascular pressure. J Physiol 508(Pt 1):199–209

Kocher O, Gabbiani F, Gabbiani G, Reidy MA, Cokay MS, Peters H, Hüttner I (1991) Phenotypic features of smooth muscle cells during the evolution of experimental carotid artery intimal thickening. Biochemical and morphologic studies. Lab Investig 65:459–470

Koshiyama H, Tanaka S, Minamikawa J (1999) Effect of calcium channel blocker amlodipine on the intimal-medial thickness of carotid arterial wall in type 2 diabetes. J Cardiovasc Pharmacol 33:894–896

Kudryavtseva O, Aalkjaer C, Matchkov VV (2013) Vascular smooth muscle cell phenotype is defined by Ca2+−dependent transcription factors. FEBS J 280:5488–5499

Langeberg LK, Scott JD (2015) Signalling scaffolds and local organization of cellular behaviour. Nat Rev Mol Cell Biol 16:232–244

Lemke T, Welling A, Christel CJ, Blaich A, Bernhard D, Lenhardt P, Hofmann F, Moosmang S (2008) Unchanged beta-adrenergic stimulation of cardiac L-type calcium channels in Ca V 1.2 phosphorylation site S1928a mutant mice. J Biol Chem 283:34738–34744

Lepretre N, Mironneau J, Morel JL (1994) Both alpha 1a- and alpha 2a-adrenoreceptor subtypes stimulate voltage-operated L-type calcium channels in rat portal vein myocytes. Evidence for two distinct transduction pathways. J Biol Chem 269:29546–29552

Li Y, Descorbeth M, Anand-Srivastava MB (2008) Role of oxidative stress in high glucose-induced decreased expression of gialpha proteins and adenylyl cyclase signaling in vascular smooth muscle cells. Am J Physiol Heart Circ Physiol 294:H2845–H2854

Liao P, Yu D, Lu S, Tang Z, Liang MC, Zeng S, Lin W, Soong TW (2004) Smooth muscle-selective alternatively spliced exon generates functional variation in Cav1.2 calcium channels. J Biol Chem 279:50329–50335

Lu T, Sun X, Li Y, Chai Q, Wang XL, Lee HC (2017) Role of Nrf2 signaling in the regulation of vascular Bk channel beta1 subunit expression and Bk channel function in high-fat diet-induced diabetic mice. Diabetes 66:2681–2690

Ma YG, Wang JW, Bai YG, Liu M, Xie MJ, Dai ZJ (2017) Salidroside contributes to reducing blood pressure and alleviating cerebrovascular contractile activity in diabetic Goto-Kakizaki rats by inhibition of L-type calcium channel in smooth muscle cells. BMC Pharmacol Toxicol 18:30

Man KNM, Navedo MF, Horne MC, Hell JW (2019) Beta2 adrenergic receptor complexes with the L-type Ca(2+) channel Cav1.2 and Ampa-type glutamate receptors: paradigms for pharmacological targeting of protein interactions. Annu Rev Pharmacol Toxicol 60:155

Martin-Aragon Baudel M, Espinosa-Tanguma R, Nieves-Cintron M, Navedo MF (2020) Purinergic signaling during hyperglycemia in vascular smooth muscle cells. Front Endocrinol (Lausanne) 11:329

Mason RP (2002) Mechanisms of plaque stabilization for the dihydropyridine calcium channel blocker amlodipine: review of the evidence. Atherosclerosis 165:191–199

Matsumoto T, Wakabayashi K, Kobayashi T, Kamata K (2005) Functional changes in adenylyl cyclases and associated decreases in relaxation responses in mesenteric arteries from diabetic rats. Am J Physiol Heart Circ Physiol 289:H2234–H2243

Mccarty MF (2006) Pkc-mediated modulation of L-type calcium channels may contribute to fat-induced insulin resistance. Med Hypotheses 66:824–831

Moore CL, Nelson PL, Parelkar NK, Rusch NJ, Rhee SW (2014) Protein kinase A-phosphorylated Kv1 channels in Psd95 signaling complex contribute to the resting membrane potential and diameter of cerebral arteries. Circ Res 114:1258–1267

Mulhern M, Docherty JR (1989) Effects of experimental diabetes on the responsiveness of rat aorta. Br J Pharmacol 97:1007–1012

Munder T, Kuntzel H (1989) Glucose-induced camp signaling in saccharomyces cerevisiae is mediated by the Cdc25 protein. FEBS Lett 242:341–345

Murphy JG, Sanderson JL, Gorski JA, Scott JD, Catterall WA, Sather WA, Dell'acqua ML (2014) Akap-anchored Pka maintains neuronal L-type calcium channel activity and Nfat transcriptional signaling. Cell Rep 7:1577–1588

Navarro G, Cordomi A, Casado-Anguera V, Moreno E, Cai NS, Cortes A, Canela EI, Dessauer CW, Casado V, Pardo L, Lluis C, Ferre S (2018) Evidence for functional pre-coupled complexes of receptor heteromers and adenylyl cyclase. Nat Commun 9:1242

Navedo MF, Amberg GC (2013) Local regulation of L-type Ca(2)(+) channel sparklets in arterial smooth muscle. Microcirculation 20:290–298

Navedo MF, Santana LF (2013) Cav1.2 sparklets in heart and vascular smooth muscle. J Mol Cell Cardiol 58:67–76

Navedo MF, Amberg GC, Votaw VS, Santana LF (2005) Constitutively active L-type Ca^{2+} channels. Proc Natl Acad Sci U S A 102:11112–11117

Navedo MF, Amberg GC, Nieves M, Molkentin JD, Santana LF (2006) Mechanisms underlying heterogeneous Ca2+ sparklet activity in arterial smooth muscle. J Gen Physiol 127:611–622

Navedo MF, Nieves-Cintron M, Amberg GC, Yuan C, Votaw VS, Lederer WJ, Mcknight GS, Santana LF (2008) Akap150 is required for stuttering persistent Ca2+ sparklets and angiotensin Ii-induced hypertension. Circ Res 102:E1–E11

Navedo MF, Takeda Y, Nieves-Cintron M, Molkentin JD, Santana LF (2010) Elevated Ca2+ sparklet activity during acute hyperglycemia and diabetes in cerebral arterial smooth muscle cells. Am J Physiol Cell Physiol 298:C211–C220

Nieves-Cintron M, Amberg GC, Navedo MF, Molkentin JD, Santana LF (2008) The control of Ca2+ influx and Nfatc3 signaling in arterial smooth muscle during hypertension. Proc Natl Acad Sci U S A 105:15623–15628

Nieves-Cintron M, Nystoriak MA, Prada MP, Johnson K, Fayer W, Dell'acqua ML, Scott JD, Navedo MF (2015) Selective downregulation of Kv2.1 function contributes to enhanced arterial tone during diabetes. J Biol Chem 290:7918–7929

Nieves-Cintrón M, Flores-Tamez VA, Le T, Baudel MM-A, Navedo MF (2021) Cellular and molecular effects of hyperglycemia on ion channels in vascular smooth muscle. Cell Mol Life Sci 78:31–61

Nilsson J, Nilsson LM, Chen YW, Molkentin JD, Erlinge D, Gomez MF (2006) High glucose activates nuclear factor of activated T cells in native vascular smooth muscle. Arterioscler Thromb Vasc Biol 26:794–800

Nystoriak MA, Nieves-Cintron M, Nygren PJ, Hinke SA, Nichols CB, Chen CY, Puglisi JL, Izu LT, Bers DM, Dell'acqua ML, Scott JD, Santana LF, Navedo MF (2014) Akap150 contributes to enhanced vascular tone by facilitating large-conductance Ca2+−activated K+ channel remodeling in hyperglycemia and diabetes mellitus. Circ Res 114:607–615

Nystoriak MA, Nieves-Cintron M, Patriarchi T, Buonarati OR, Prada MP, Morotti S, Grandi E, Fernandes JD, Forbush K, Hofmann F, Sasse KC, Scott JD, Ward SM, Hell JW, Navedo MF (2017) Ser1928 phosphorylation by Pka stimulates the L-type Ca2+ channel Cav1.2 and vasoconstriction during acute hyperglycemia and diabetes. Sci Signal 10

Oliveria SF, Dell'acqua ML, Sather WA (2007) Akap79/150 anchoring of calcineurin controls neuronal L-type Ca2+ channel activity and nuclear signaling. Neuron 55:261–275

Organization, W. H (2016) Global report on diabetes. World Health Organization, Geneva

Ouyang Q-F, Han Y, Lin Z-H, Xie H, Xu C-S, Xie L-D (2014) Fluvastatin upregulates the A_{1c} subunit of Cav1.2 channel expression in vascular smooth muscle cells via RhoA and ERK/P38 MAPK pathways. Dis Markers:237067

Park JL, Loberg RD, Duquaine D, Zhang H, Deo BK, Ardanaz N, Coyle J, Atkins KB, Schin M, Charron MJ, Kumagai AK, Pagano PJ, Brosius FC III (2005) Glut4 facilitative glucose transporter specifically and differentially contributes to agonist-induced vascular reactivity in mouse aorta. Arterioscler Thromb Vasc Biol 25:1596–1602

Parodi J, Flores C, Aguayo C, Rudolph MI, Casanello P, Sobrevia L (2002) Inhibition of nitrobenzylthioinosine-sensitive adenosine transport by elevated D-glucose involves activation of P2y2 purinoceptors in human umbilical vein endothelial cells. Circ Res 90:570–577

Patel NA, Chalfant CE, Yamamoto M, Watson JE, Eichler DC, Cooper DR (1999) Acute hyperglycemia regulates transcription and posttranscriptional stability of PKCbetaii mRNA in vascular smooth muscle cells. FASEB J 13:103–113

Patel MK, Clunn GF, Lymn JS, Austin O, Hughes AD (2005) Effect of serum withdrawal on the contribution of L-type calcium channels (Cav1.2) to intracellular Ca2+ responses and chemotaxis in cultured human vascular smooth muscle cells. Br J Pharmacol 145:811–817

Patriarchi T, Qian H, Di Biase V, Malik ZA, Chowdhury D, Price JL, Hammes EA, Buonarati OR, Westenbroek RE, Catterall WA, Hofmann F, Xiang YK, Murphy GG, Chen CY, Navedo MF, Hell JW (2016) Phosphorylation of Cav1.2 on S1928 uncouples the L-type Ca2+ channel from the beta2 adrenergic receptor. EMBO J 35:1330–1345

Payne GW, Smeda JS (2002) Cerebrovascular alterations in pressure and protein kinase C-mediated constriction in Dahl salt-sensitive rats. J Hypertens 20:1355–1363

Pinho JF, Medeiros MA, Capettini LS, Rezende BA, Campos PP, Andrade SP, Cortes SF, Cruz JS, Lemos VS (2010) Phosphatidylinositol 3-kinase-delta up-regulates L-type Ca2+ currents and increases vascular contractility in a mouse model of type 1 diabetes. Br J Pharmacol 161:1458–1471

Porter VA, Bonev AD, Knot HJ, Heppner TJ, Stevenson AS, Kleppisch T, Lederer WJ, Nelson MT (1998) Frequency modulation of Ca^{2+} sparks is involved in regulation of arterial diameter by cyclic nucleotides. Am J Phys 274:C1346–C1355

Prada MP, Syed AU, Buonarati OR, Reddy GR, Nystoriak MA, Ghosh D, Simo S, Sato D, Sasse KC, Ward SM, Santana LF, Xiang YK, Hell JW, Nieves-Cintron M, Navedo MF (2019) A Gs-coupled purinergic receptor boosts Ca(2+) influx and vascular contractility during diabetic hyperglycemia. elife 8

Prada MP, Syed AU, Reddy GR, Martin-Aragon Baudel M, Flores-Tamez VA, Sasse KC, Ward SM, Sirish P, Chiamvimonvat N, Bartels P, Dickson EJ, Hell JW, Scott JD, Santana LF, Xiang YK, Navedo MF, Nieves-Cintron M (2020a) Akap5 complex facilitates purinergic modulation of vascular L-type CRa(2+) channel Cav1.2. Nat Commun 11:5303

Prada MP, Syed AU, Reddy GR, Martín-Aragón Baudel M, Flores-Tamez VA, Sasse KC, Ward SM, Sirish P, Chiamvimonvat N, Bartels P, Dickson EJ, Hell JW, Scott JD, Santana LF, Xiang YK, Navedo MF, Nieves-Cintrón M (2020b) Akap5 complex facilitates purinergic modulation of vascular L-type Ca2+ channel Cav1.2. Nat Commun 11

Pyla R, Poulose N, Jun JY, Segar L (2013) Expression of conventional and novel glucose transporters, Glut1, −9, −10, and −12, in vascular smooth muscle cells. Am J Physiol Cell Physiol 304:C574–C589

Qian H, Patriarchi T, Price JL, Matt L, Lee B, Nieves-Cintron M, Buonarati OR, Chowdhury D, Nanou E, Nystoriak MA, Catterall WA, Poomvanicha M, Hofmann F, Navedo MF, Hell JW (2017) Phosphorylation of Ser1928 mediates the enhanced activity of the L-type Ca2+ channel Cav1.2 by the beta2-adrenergic receptor in neurons. Sci Signal 10:Eaaf9659

Raghavan S, Vassy JL, Ho YL, Song RJ, Gagnon DR, Cho K, Wilson PWF, Phillips LS (2019) Diabetes mellitus-related all-cause and cardiovascular mortality in a national cohort of adults. J Am Heart Assoc 8:E011295

Rhee SW, Rusch NJ (2018) Molecular determinants of beta-adrenergic signaling to voltage-gated K (+) channels in the cerebral circulation. Microcirculation 25. https://doi.org/10.1111/Micc.12425

Richard S, Neveu D, Carnac G, Bodin P, Travo P, Nargeot J (1992) Differential expression of voltage-gated Ca(2+)-currents in cultivated aortic myocytes. Biochim Biophys Acta 1160:95–104

Ruan X, Arendshorst WJ (1996) Role of protein kinase C in angiotensin Ii-induced renal vasoconstriction in genetically hypertensive rats. Am J Phys 270

Ruiz-Torres A, Lozano R, Melon J, Carraro R (2003) L-calcium channel blockade induced by diltiazem inhibits proliferation, migration and F-actin membrane rearrangements in human vascular smooth muscle cells stimulated with insulin and Igf-1. Int J Clin Pharmacol Ther 41: 386–391

Salamanca DA, Khalil RA (2005) Protein kinase C isoforms as specific targets for modulation of vascular smooth muscle function in hypertension. Biochem Pharmacol 70:1537–1547

Scherer PE, Hill JA (2016) Obesity, diabetes, and cardiovascular diseases. Circ Res 118:1703–1705

Serne EH, De Jongh RT, Eringa EC, Rg IJ, Stehouwer CD (2007) Microvascular dysfunction: a potential pathophysiological role in the metabolic syndrome. Hypertension 50:204–211

Somlyo AP, Somlyo AV (2003) Ca^{2+} sensitivity of smooth muscle and nonmuscle myosin Ii: modulated by G proteins, kinases, and myosin phosphatase. Physiol Rev 83:1325–1358

Stehouwer CDA (2018) Microvascular dysfunction and hyperglycemia: a vicious cycle with widespread consequences. Diabetes 67:1729–1741

Suzuki Y, Ozawa T, Kurata T, Nakajima N, Zamponi GW, Giles WR, Imaizumi Y, Yamamura H (2022) A molecular complex of Ca(V)1.2/Camkk2/Camk1a in caveolae is responsible for vascular remodeling via excitation-transcription coupling. Proc Natl Acad Sci U S A 119: E2117435119

Syed AU, Reddy GR, Ghosh D, Prada MP, Nystoriak MA, Morotti S, Grandi E, Sirish P, Chiamvimonvat N, Hell JW, Santana LF, Xiang YK, Nieves-Cintron M, Navedo MF (2019) Adenylyl cyclase 5-generated camp controls cerebral vascular reactivity during diabetic hyperglycemia. J Clin Invest 129:3140–3152

Syed AU, Le T, Navedo MF, Nieves-Cintrón M (2020) Ion channels and their regulation in vascular smooth muscle. In: Basic and clinical understanding of microcirculation. IntechOpen

Tiwari S, Zhang Y, Heller J, Abernethy DR, Soldatov NM (2006) Atherosclerosis-related molecular alteration of the human Cav1.2 calcium channel alpha1c subunit. Proc Natl Acad Sci U S A 103: 17024–17029

Tykocki NR, Boerman EM, Jackson WF (2017) Smooth muscle ion channels and regulation of vascular tone in resistance arteries and arterioles. Compr Physiol 7:485–581

Ungvari Z (1999) Increased myogenic tone in skeletal muscle arterioles of diabetic rats. Possible role of increased activity of smooth muscle Ca2+ channels and protein kinase C. Cardiovasc Res 43:1018–1028

Ungvari Z, Pacher P, Kecskemeti V, Papp G, Szollar L, Koller A (1999) Increased myogenic tone in skeletal muscle arterioles of diabetic rats. Possible role of increased activity of smooth muscle Ca2+ channels and protein kinase C. Cardiovasc Res 43:1018–1028

Velmurugan GV, Sundaresan NR, Gupta MP, White C (2013) Defective Nrf2-dependent redox signalling contributes to microvascular dysfunction in type 2 diabetes. Cardiovasc Res 100:143–150

Viard P, Butcher AJ, Halet G, Davies A, Nurnberg B, Heblich F, Dolphin AC (2004) Pi3k promotes voltage-dependent calcium channel trafficking to the plasma membrane. Nat Neurosci 7:939–946

Walsh MP (2011) Vascular smooth muscle myosin light chain diphosphorylation: mechanism, function, and pathological implications. IUBMB Life 63:987–1000

Wamhoff BR, Bowles DK, Mcdonald OG, Sinha S, Somlyo AP, Somlyo AV, Owens GK (2004) L-type voltage-gated Ca2+ channels modulate expression of smooth muscle differentiation marker genes via a rho kinase/myocardin/Srf-dependent mechanism. Circ Res 95:406–414

Wang R, Wu Y, Tang G, Wu L, Hanna ST (2000) Altered L-type Ca(2+) channel currents in vascular smooth muscle cells from experimental diabetic rats. Am J Physiol Heart Circ Physiol 278:H714–H722

Wellman GC, Bonev AD, Nelson MT, Brayden JE (1996) Gender differences in coronary artery diameter involve estrogen, nitric oxide, and Ca^{2+}-dependent K^+ channels. Circ Res 79:1024–1030

Wellman GC, Santana LF, Bonev AD, Nelson MT (2001) Role of phospholamban in the modulation of arterial Ca^{2+} sparks and Ca^{2+}-activated K^+ channels by camp. Am J Physiol Cell Physiol 281:C1029–C1037

White RE, Carrier GO (1990) Vascular contraction induced by activation of membrane calcium ion channels is enhanced in streptozotocin-diabetes. J Pharmacol Exp Ther 253:1057–1062

Wilde DW, Massey KD, Walker GK, Vollmer A, Grekin RJ (2000) High-fat diet elevates blood pressure and cerebrovascular muscle Ca(2+) current. Hypertension 35:832–837

Williams B, Schrier RW (1992) Characterization of glucose-induced in situ protein kinase C activity in cultured vascular smooth muscle cells. Diabetes 41:1464–1472

Williams B, Gallacher B, Patel H, Orme C (1997) Glucose-induced protein kinase C activation regulates vascular permeability factor mRNA expression and peptide production by human vascular smooth muscle cells in vitro. Diabetes 46:1497–1503

Yamashita T, Kawashima S, Ozaki M, Rikitake Y, Hirase T, Inoue N, Hirata K, Yokoyama M (2001) A calcium channel blocker, benidipine, inhibits intimal thickening in the carotid artery of mice by increasing nitric oxide production. J Hypertens 19:451–458

Zhang M, Patriarchi T, Stein IS, Qian H, Matt L, Nguyen M, Xiang YK, Hell JW (2013) Adenylyl cyclase anchoring by a kinase anchor protein Akap5 (Akap79/150) is important for postsynaptic beta-adrenergic signaling. J Biol Chem 288:17918–17931

Sympathetic Nervous System Regulation of Cardiac Calcium Channels

Pedro J. del Rivero Morfin, Steven O. Marx, and Manu Ben-Johny

Contents

Abstract

Calcium influx through voltage-gated calcium channels, $Ca_v1.2$, in cardiomyocytes initiates excitation-contraction coupling in the heart. The force and rate of cardiac contraction are modulated by the sympathetic nervous system,

P. J. del Rivero Morfin · M. Ben-Johny (✉)
Department of Physiology and Cellular Biophysics, Columbia University Vagelos College of Physicians and Surgeons, New York, NY, USA
e-mail: mbj2124@cumc.columbia.edu

S. O. Marx (✉)
Division of Cardiology, Department of Medicine, Columbia University Vagelos College of Physicians and Surgeons, New York, NY, USA

Department of Pharmacology and Molecular Signaling, Columbia University Vagelos College of Physicians and Surgeons, New York, NY, USA
e-mail: sm460@cumc.columbia.edu

J. Striessnig (ed.), *Voltage-gated Ca^{2+} Channels: Pharmacology, Modulation and their Role in Human Disease*, Handbook of Experimental Pharmacology 279,
https://doi.org/10.1007/164_2022_632

mediated substantially by changes in intracellular calcium. Norepinephrine released from sympathetic neurons innervating the heart and epinephrine secreted by the adrenal chromaffin cells bind to β-adrenergic receptors on the sarcolemma of cardiomyocytes initiating a signaling cascade that generates cAMP and activates protein kinase A, the targets of which control calcium influx. For decades, the mechanisms by which PKA regulated calcium channels in the heart were not known. Recently, these mechanisms have been elucidated. In this chapter, we will review the history of the field and the studies that led to the identification of the evolutionarily conserved process.

Keywords

Adrenergic · Calcium channels · Heart · Phosphorylation

1 Introduction

In nature, few biological mechanisms are more crucial to survival than the fight-or-flight response. In response to acute stress, such as when being pursued by a predator, or during exercise, activation of the sympathetic nervous system augments cardiac output by: (1) elevating the heart rate (HR), a process known as chronotropy, and (2) enhancing the force of myocardial contraction, known as inotropy (Bers 2002). This ensures rapid delivery of oxygen to the periphery. Not surprisingly, this physiological response is conserved in vertebrates and other animals (Taylor et al. 1999).

The subcellular mechanism underlying adrenergic response is highly regulated and organ specific. Norepinephrine and epinephrine released from the adrenal medulla and sympathetic nerve terminals bind and activate the G-protein-coupled adrenergic receptors that initiate a signaling cascade to phosphorylate various cellular proteins including ion channels, transporters, and contractile proteins, ultimately tuning excitation-contraction coupling. In cardiomyocytes, the β1 adrenergic receptor (β1-AR) is the predominant isoform, followed by the β2 adrenergic receptor (β2-AR) (Lefkowitz et al. 2000). Both receptors are coupled to $G_{\alpha s}$ proteins that activate adenylate cyclase, which converts ATP into the second messenger cyclic adenosine monophosphate (cAMP), which in turn activates protein kinase A (PKA), initiating phosphorylation of a bevy of downstream targets (Lefkowitz et al. 2000).

In pacemaking nodal cells, cellular automaticity is governed by a "voltage clock" and a coupled "Ca^{2+}-clock," as reviewed previously (Lakatta et al. 2010). The former is comprised of rhythmic oscillation of membrane potential driven by various ion channels, including the cyclic-nucleotide gated HCN channels that mediate the "funny" current, L-type Ca^{2+} channels, sodium-calcium exchangers (NCX) and K^+ currents (Tsutsui et al. 2018). The latter relies on oscillations in cytosolic Ca^{2+} driven by rhythmic local Ca^{2+} release from the sarcoplasmic reticulum through ryanodine receptors (RyR2), and Ca^{2+} extrusion via both the sarcoplasmic reticulum (SR) Ca^{2+}-ATPase (SERCA) and NCX. The "Ca^{2+} clock" is coupled to the "voltage clock" in part through electrogenic exchange of Ca^{2+} through NCX. During the fight-or-flight response, increased cytosolic cAMP enhances the "funny current" contributing to diastolic depolarization that alters the voltage clock, while PKA phosphorylation of targets including potentially L-type Ca^{2+} channel complexes, RyR2, and phospholamban alter Ca^{2+} oscillations, thereby increasing the spontaneous beating (Greene et al. 2012; Torre et al. 2021; Tsutsui et al. 2018). Together, these processes contribute to the chronotropic response.

In ventricular cardiomyocytes, membrane depolarization during the action potential evokes a transient elevation in cytosolic Ca^{2+} concentration that initiates muscle contraction (Fig. 1a), a process known as excitation-contraction coupling (Bers 2002). Specifically, voltage-dependent activation of sarcolemmal L-type Ca^{2+} channels conveys local Ca^{2+} influx into the dyadic space. This small increase in Ca^{2+} activates the closely-juxtaposed RyR2 in the SR to evoke a much larger Ca^{2+} release from the intracellular stores. Ca^{2+} binding to Troponin C and further conformational changes in Troponin I then initiates actin-myosin-dependent contraction. Cytosolic Ca^{2+} is extruded through SERCA and NCX, allowing relaxation of the muscle. During β-adrenergic stimulation, several of these proteins are phosphorylated by PKA (Fig. 1a), resulting in: (1) a marked upregulation of the Ca^{2+} transient resulting in enhanced strength of myocardial contraction (Fig. 1b), and (2) a hastening of Ca^{2+} efflux from the sarcoplasm and enhanced myocardial relaxation, a process known as lusitropy. The targets of PKA phosphorylation, as well as specific residues that are phosphorylated, have garnered considerable interest as summarized briefly: (1) RyR2 is directly phosphorylated by PKA and is thought to contribute to enhanced Ca^{2+} release (Reiken et al. 2003; Marx et al. 2000). (2) PKA phosphorylation of phospholamban, an inhibitor of SERCA, stimulates Ca^{2+}-uptake into the SR (Lindemann et al. 1983; Li et al. 2000), (3) Phosphorylation of KCNQ1/KCNE1/Yotiao complex upregulates I_{Ks} current (Marx et al. 2002). (4) Phosphorylation of troponin I enhances relaxation of the contractile apparatus (Kentish et al. 2001; Li et al. 2000). Beyond these, adrenergic stimulation yields a striking increase in the L-type Ca^{2+} current, a change that is likely to exert a dominant effect on increasing the Ca^{2+} transient. In response to β-adrenergic agonists, Ca^{2+} current is augmented by two- to threefold (Reuter 1983), marked by an increased open probability of the channel and a shift from inactive and low open probability (P_o) modes to a high P_o mode with long and frequent openings (Yue et al. 1990; Tsien et al. 1986; Bean et al. 1984). However, the molecular basis for this modulation has remained highly controversial for nearly 50 years. The obvious targets of PKA phosphorylation are the Ca^{2+} channel α- or β-subunits, which were the focus of many studies (Hell 2010; Fuller et al. 2010; Kamp and Hell 2000; Catterall 2015; Gerhardstein et al. 1999). Most notably, nearly all studies using heterologous expression systems proved to be unreliable in predicting the mechanism, because recapitulating the PKA-mediated activation of Ca^{2+} channels was not achievable. Recent advances, however, reveal a surprising mechanism involving a small G-protein, Rad (Ras associated with diabetes) (Papa et al. 2021; Liu et al. 2020). Basal association of Rad with the L-type channel complex inhibits channel function, while PKA phosphorylation relieves this inhibition and upregulates Ca^{2+}

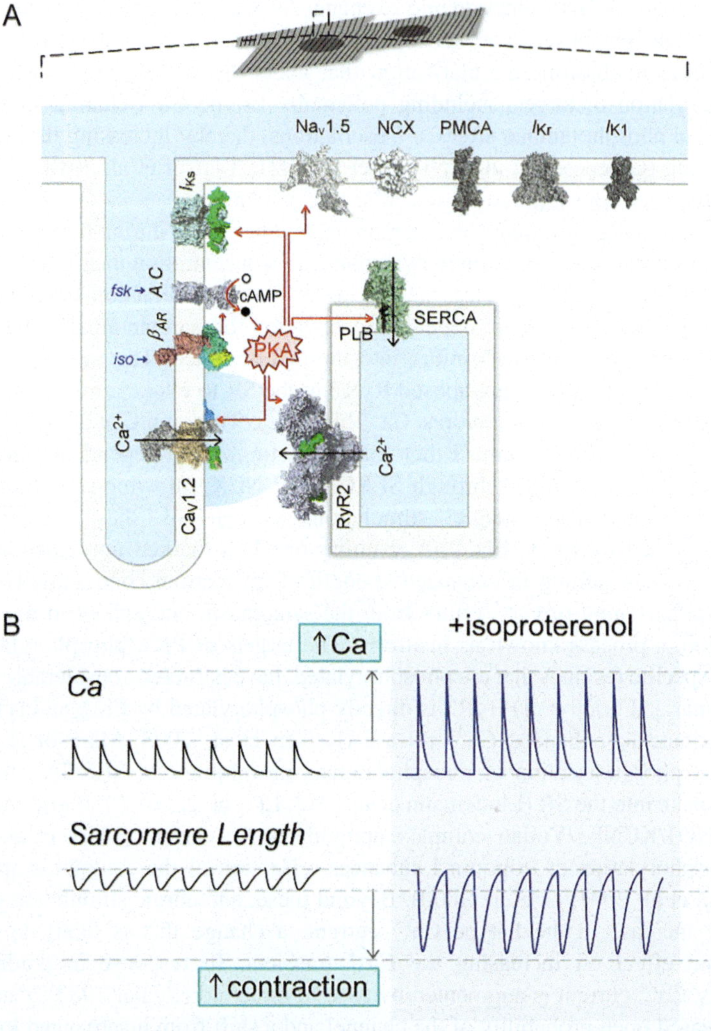

Fig. 1 Adrenergic regulation of cardiac ion channels. (**a**) Schematic shows that β-adrenergic stimulation results in activation of PKA which phosphorylates various cardiac ion channels. For L-type $Ca_V1.2$ channels, β-adrenergic stimulation results in increased Ca^{2+} entry, yet the precise mechanism that enhances channel function has remained elusive. (**b**) β-adrenergic stimulation results in an increase in the magnitude of the Ca^{2+} transient and increased contractility

entry into cardiomyocytes. This elegant mechanism sets the stage for resolving long-standing physiological and pathophysiological unknowns and opens new frontiers to develop next-generation therapeutics.

Here, we summarize fundamental features of the adrenergic regulation of L-type channels, review molecular mechanisms underlying channel regulation, and highlight key ambiguities that remain to be determined.

2 Biophysical Signature of Adrenergic Regulation of L-Type Channels

Macroscopic behavior – Early studies found that adrenaline increased the plateau height of the cardiac action potentials and enhanced the uptake of radiolabeled Ca^{2+} into myocytes, suggesting increased Ca^{2+} permeability of cardiac cells (Jork et al. 1967; Reuter 1987). Whole-cell voltage clamp recordings showed that acute application of adrenaline markedly increased the peak inward L-type Ca^{2+} current. Subsequent studies also confirmed this increase in Ca^{2+} current upon application of either (1) isoproterenol that activates β-adrenergic receptors, (2) forskolin that stimulates adenylyl cyclase, or (3) membrane-permeable 8-bromo-cAMP that directly activates PKA (Reuter 1983, 1974; Tsien et al. 1986). Figure 2a shows increased Ca^{2+} current in mouse ventricular cardiomyocytes in response to

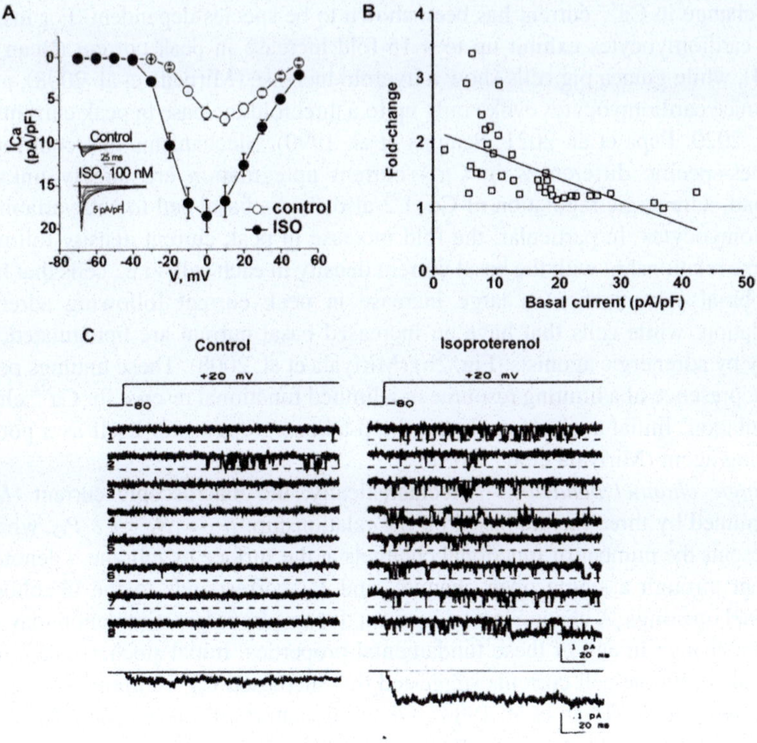

Fig. 2 Key biophysical features of adrenergic regulation of $Ca_V1.2$. (**a**) Exemplar traces show increased Ca^{2+} current in mouse cardiomyocytes following adrenergic stimulation using isoproterenol. (**b**) The fold increase in Ca^{2+} current amplitude follows an inverse relationship with the basal current density, suggesting a limited resource. Data in Panel (**b**) from (Yang et al. 2019). (**c**) Single Ca^{2+} channel exhibit increased activity following adrenergic stimulation. At baseline, channels largely exhibit no activity or sparse channel openings. Isoproterenol switches channels to high-activity gating modes. Panel (**c**) reproduced with permission (Yue et al. 1990)

isoproterenol. Furthermore, injection of the catalytic subunit of cAMP-dependent protein kinase A (PKA) into cardiomyocytes also yielded a similar effect (Osterrieder et al. 1982; Brum et al. 1983). These results pointed to a phosphorylation-dependent mechanism for adrenergic upregulation of Ca^{2+} current. Analysis of current-voltage relationship also showed a hyperpolarizing shift in the voltage-dependence of channel activation such that the fold increase in current is higher in the intermediate voltage range (Bean et al. 1984). This feature may be physiologically important to optimally enhance Ca^{2+}-induced Ca^{2+}-release through the RyR2 channels (Marks 2003; Wang et al. 2001; López-López et al. 1995). Furthermore, β-adrenergic regulation has also been reported to enhance Ca^{2+}-dependent inactivation and reduce voltage-dependent inactivation (Morales et al. 2019; van der Heyden et al. 2005). The former may be due to increased local Ca^{2+} near the channel stemming from enhanced Ca^{2+}-induced Ca^{2+} release.

Surprisingly, studies have revealed considerable variability in the magnitude of increase in whole-cell Ca^{2+} current following adrenergic stimulation. First, the net fold change in Ca^{2+} current has been shown to be species-dependent. For instance, frog cardiomyocytes exhibit up to a 15-fold increase in peak current (Bean et al. 1984), while guinea pig cells show a fivefold increase (Miriyala et al. 2008), and rat and mice cardiomyocytes evoke only up to a threefold increase in peak currents (Liu et al. 2020; Papa et al. 2021; Scamps et al. 1990). Mechanisms that contribute to species-specific differences in Ca^{2+} current upregulation are largely unknown. Second, adrenergic regulation of $Ca_V1.2$ also shows basal cell to cell variability in cardiomyocytes. In particular, the fold increase in peak current density follows an inverse relationship with the basal current density in each cell – i.e., cells that have a low basal current show a large increase in peak current following adrenergic regulation, while cells that have an increased basal current are upregulated minimally by adrenergic agonists (Fig. 2b) (Miriyala et al. 2008). These findings pointed to the presence of a limiting resource or a limited functional reserve for Ca^{2+} channel modulation. Initial analysis excluded the β-adrenergic receptor itself as a potential limiting agent (Miriyala et al. 2008).

Single channel behavior – Mechanistically, the macroscopic current (I_{Ca}) is determined by three factors based on the relationship: $I_{Ca} = N_f \times I \times P_O$, where N_f represents the number of functional channels at the surface membrane, I denotes the current through a single open channel, and P_O corresponds to the likelihood of channel openings. A change in peak current due to adrenergic stimulation may result from a change in any of these fundamental properties. Initial stochastic fluctuation analysis of whole-cell currents suggested that increased Ca^{2+} current resulted from an increase in N_f (Bean et al. 1984). Given the nature of this analysis, this change could reflect either an increase in the number of channels at the surface membrane, or an increase in the activity of a distinct pool of "sleepy" channels which have minimal activity at baseline (i.e., contributes minimally to the gating noise). Single channel studies probed these possibilities more directly (Yue et al. 1990; Hirano et al. 1994; Herzig et al. 1993). These studies confirmed no change in the single channel conductance (i), but instead reported discrete changes in the pattern of channel gating ("mode"). More specifically, analysis of individual sweeps revealed four distinct patterns of channel activity: (1) *mode* 0, denoting sweeps with no openings whatsoever, (2) *mode* 0_a, corresponding to sweeps with rare and brief openings, (3) *mode* 1, where channels open frequently and channel openings are longer lived, and (4) *mode* 2, corresponding to sweeps containing extraordinarily long openings similar to that observed with the dihydropyridine agonist Bay K8644 (Hess et al. 1984). At baseline, channel openings are rare (i.e., mode 0 dominates) but when channels do open, channels exhibit mode 0_a and mode 1 patterns of gating (Fig. 2c). However, in the presence of adrenergic agonists, the proportion of mode 0 and mode 0_a openings is reduced, and both mode 1 and mode 2 pattern of activity is enhanced (Fig. 2c) (Yue et al. 1990). Importantly, despite redistribution of the gating patterns, the kinetic properties of individual modes are largely unaffected. Further analysis employing discrete-time Markov models also showed that β-adrenergic stimulation increased both the forward as well as the reverse rate of transition from the putative dephosphorylated conformation to the phosphorylated conformation, a paradoxical finding as PKA activation would be expected to impact only the forward phosphorylation reaction (Herzig et al. 1993). In all, these findings are consistent with β-adrenergic regulation of $Ca_V1.2$ increases the P_O by largely enhancing channel availability (Herzig et al. 1993; Yue et al. 1990; Hirano et al. 1994).

3 Molecular Mechanisms Underlying L-Type Upregulation

Given its central role in cardiac physiology, the molecular mechanisms responsible for adrenergic upregulation of L-type channels in the heart have been of longstanding interest. Here we summarize potential mechanisms (Fig. 3) and highlight key evidence.

4 Direct Phosphorylation of α_{1C} and β_{2B} Subunits?

The traditional view hypothesized direct phosphorylation of the α_{1C} and β_{2B} subunits (Curtis and Catterall 1985). Indeed, consistent with this possibility, Ser1928 on the C-terminus of the pore-forming α_{1C} subunit was initially identified as the major PKA phosphorylated residue (De Jongh et al. 1996; Mitterdorfer et al. 1996). The role for phosphorylation of Ser1928 was explored using heterologous expression: in some studies, phosphorylation of Ser1928 was shown to be required for PKA-mediated upregulation of heterologously expressed Ca^{2+} current (Gao et al. 1997; Naguro et al. 2001), whereas in another study, activation of Ca^{2+} channels persisted in an α_{1C} subunit truncated at residue 1905 (Bunemann et al. 1999). Several residues on the C-terminal variable region of the β_2 subunit were also proposed to be PKA phosphorylation sites (Haase et al. 1996; Bunemann et al. 1999).

The O'Rourke laboratory used adenovirus to express in cardiomyocytes the α_{1C} subunit harboring alanine-substitution of the Ser1928 residue (Ganesan et al. 2006). To distinguish the native channels from the mutant transgenic channels, they inserted mutations in α_{1C} to make the channels dihydropyridine-resistant (Ganesan

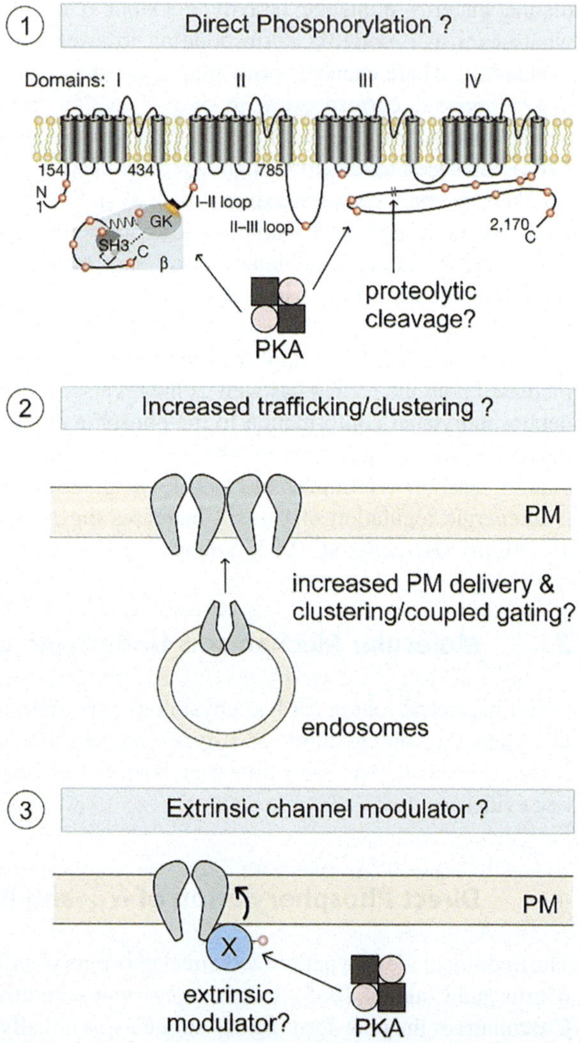

Fig. 3 Putative Mechanisms underlying adrenergic regulation of cardiac $Ca_V1.2$: (**1**) PKA activation may result in direct phosphorylation of either the pore-forming α subunit or the auxiliary β subunit, which could upregulate channel activity. Proteolytic cleavage of the channel is also proposed to be important in this process. (**2**) β-adrenergic stimulation may enhance channel trafficking to surface membrane and promote channel clustering and cooperative gating. (**3**) PKA may phosphorylate a modulator that binds to one of the $Ca_V1.2$ subunits

et al. 2005). They also used adenoviral overexpression of mutant β_2 subunits in cardiomyocytes to test the role of these phosphorylation sites. Remarkably, phosphorylation of Ser1928 in the α_{1C} subunit and phosphorylation of Ser478 and Ser479 in the β_2 subunit were not required for β-adrenergic stimulation of $Ca_V1.2$ (Ganesan et al. 2006). An additional phosphorylation site, Ser459, on β_2 subunit was shown by the Colecraft group to be non-essential (Miriyala et al. 2008). Similar findings were shown using genetically-altered mice: (1) Ser1928 was mutated to alanine, which had no effect on basal or β-adrenergic agonist stimulation of Ca^{2+} current in ventricular cardiomyocytes (Lemke et al. 2008); (2) deletion of the variable C-terminus of the β_2 subunit by insertion of a stop codon that effectively removed the known PKA phosphorylation. Additional phosphorylation sites were proposed including Ser^{1700} and Thr^{1704} in the $Ca_V1.2$ α_{1C} subunit (Fuller et al. 2010).

It is conceivable that a combination of phosphorylation sites on the α_{1C} and β_2 subunits could be the basis for adrenergic regulation of $Ca_V1.2$ channels. Based upon this premise, we established an approach by which dihydropyridine-resistant Ca^{2+} channels could be expressed in cardiomyocytes of mice (Yang and Colecraft 2013). The cardiac-specific expression of these channels could be induced by doxycycline. Two lines of transgenic mice were initially generated with N-terminal 3X FLAG-epitope–tagged, dihydropyridine-resistant α_{1C}, designated pseudo-wild-type (pWT) α_{1C} and mice in the same α_{1C} background with alanine-substitutions of Ser1700 and Thr1704 as well as deletion of the presumed proteolytic cleavage site. Phosphorylation of these two residues is not essential for β-adrenergic stimulation of Ca^{2+} currents (Yang and Colecraft 2013), which was later confirmed by others using a knock-in approach (Poomvanicha et al. 2017) and in additional transgenic mice (Liu et al. 2020; Yang et al. 2019; Katchman et al. 2017). Thereafter, we created additional transgenic mice in which all potential PKA phosphorylation sites were mutated based upon the known PKA consensus phosphorylation sequences. In transgenic mice in which 22 serine or threonine residues were mutated to alanine (17 conserved putative consensus PKA phosphorylation sites that were not previously studied, and 5 conserved PKA/CaMKII phosphorylation sites known to be non-essential, including Ser1700 and Thr1704), adrenergic regulation of $Ca_V1.2$ was normal (Katchman et al. 2017). To test the possibility that the functionally relevant PKA targets in α_{1C} could be different amongst species, we generated transgenic mice with alanine-substitutions at additional potential PKA sites in the mouse channel that were not conserved. A total of 51 alanine-substitutions at 35 sites were introduced. Adrenergic regulation of $Ca_V1.2$ channels was normal, indicating that phosphorylation of α_{1C} is not essential for adrenergic stimulation of $Ca_V1.2$ (Liu et al. 2020). We also generated transgenic mice expressing mutant human β_{2B} subunits in which 37 alanine-substitutions were made within 28 consensus PKA phosphorylation sites (Liu et al. 2020). The overexpression of mutant β subunits competitively replaced wild-type β_2 subunits in the $Ca_V1.2$ complex. Calcium channels with the mutant β subunit demonstrated normal adrenergic regulation. Thereafter, the mutant α_{1C} and β_2 mice were crossed, and adrenergic regulation was normal. These studies reveal that adrenergic regulation of $Ca_V1.2$ does not require phosphorylation of either α_{1C} or β_2.

5 Role of C-Terminal Proteolytic Cleavage of α_{1C}?

In the heart, post-translational proteolytic cleavage of α_{1C} yields a distal C-terminal truncated α_{1C} of ~210 kDa (Gao et al. 1997; Hell et al. 1993; De Jongh et al. 1991, 1996; Hulme et al. 2005). It has been speculated that PKA regulation of $Ca_V1.2$ requires the proteolytic cleavage of the distal C-terminus and subsequent phosphorylation-dependent release of the inhibition imparted by the distal C-terminus (Gerhardstein et al. 2000; Catterall 2015; Fuller et al. 2010). Consistent with this hypothesis, knock-in mice expressing a truncated α_{1C} at either Gly1796 or Asp1904 die at birth due to low membrane expression of $Ca_V1.2$ (Fu et al. 2011; Domes et al. 2011). It is important to note, however, that a truncated α_{1C} subunit is never expressed without its distal C-terminus in vivo. The functional relevance of proteolytic cleavage of α_{1C} was determined in cardiomyocytes by creating mice with deletion of the cleavage site. Sequence alignment of α_{1S} and α_{1C} was the basis for the hypothesis that Ala1800 within in the ^{1798}NNAN motif is the α_{1C} proteolytic site (Hulme et al. 2005). Initially, we created a transgenic mouse line with expression of α_{1C} with deletion of the ^{1798}NNAN motif. Cleavage of α_{1C} persisted in cardiomyocytes after deletion of the ^{1798}NNAN motif, likely due to the presence of a nearby similar motif, ^{1794}NANI^{1797}. Thereafter, we created a transgenic mouse line with inducible cardiomyocyte-specific expression of α_{1C} with deletion of ^{1794}NANINNANN^{1802}, along with deletion of a nearby PEST sequence (^{1769}DTESP) (Katchman et al. 2017). The α_{1C} subunit was resistant to proteolytic cleavage in cardiomyocytes with the deletion of both sites, but not the PEST sequence alone. β-adrenergic stimulation of $Ca_V1.2$ in cardiomyocytes was preserved in these mice, demonstrating that C-terminal proteolytic cleavage is not required for adrenergic agonist stimulation of Ca^{2+} channels.

6 Increased Trafficking and Clustering of $Ca_V1.2$?

Beyond a frank increase in open probability, adrenergic regulation has also been proposed to stimulate rapid sarcolemmal insertion of $Ca_V1.2$ channels (Jhun et al. 2012; Villar et al. 2021). This overall process enables the formation of large $Ca_V1.2$ "super-clusters" that exhibit cooperative gating and further amplifies Ca^{2+} influx (Ito et al. 2019). In this scheme, a subsarcolemmal pre-synthesized pool of channels are localized to early endosomes, recycling endosomes, and late endosomes. Isoproterenol and subsequent PKA activation is thought to mobilize and deliver the endosomal channels to the T-tubular sarcolemma as large channel clusters, with a time course of minutes. The clustering of channels is proposed to further enhance their activity through concerted gating observed with Ca^{2+} as charge carrier. Consistent with this scheme, disruption of cytoskeletal elements including both microtubules and actin filaments prevent mobilization of $Ca_V1.2$ from endosomal pools and also diminished adrenergic upregulation of $Ca_V1.2$ (Villar et al. 2021). Few caveats merit further attention. First, these experiments largely inferred $Ca_V1.2$ localization in myocytes by monitoring a fluorescently labeled β_{2A} subunit that is overexpressed. Unlike β_{2B} that is most prevalent in cardiomyocytes, the β_{2A} subunit is palmitoylated, which alters its subcellular localization and trafficking (Chien et al. 1996). Second, single channel recordings demonstrate that individual $Ca_V1.2$ channels can dynamically undergo a PKA-dependent upregulation in their P_O that is of similar magnitude as macroscopic measurements (Yue et al. 1990; Hirano et al. 1994). This suggests that PKA modulation does not require clustering of channels or cooperative gating but is an elementary feature of the channel complex. Third, introduction of triple glycine mutations in domain I S6 gate of $Ca_V1.2$

(GGG-mutant) that disrupts channel gating (Findeisen and Minor 2009) also prevented PKA-dependent increase in Ca^{2+} current (Papa et al. 2021). The increased flexibility of the S6 gate stabilized a low P_O gating configuration. If increased trafficking to the sarcolemma sufficed for adrenergic upregulation, then GGG-mutant channels should still support some residual current increase by increasing the number of channels, even if its gating was disrupted. Thus, although it is possible that PKA activation could broadly impact channel trafficking or enhance clustering, the relevance of such a mechanism for upregulating Ca^{2+} current during sympathetic response is not fully clear.

7 Involvement of an Extrinsic Modulator?

Having excluded direct PKA phosphorylation of the channel α_{1C} and β_{2B} subunits, one possibility is that the channels may recruit an extrinsic modulator that alters channel function in a phosphorylation-dependent manner. One tantalizing clue in support of this possibility emerged from transgenic mice expressing mutant $Ca_V1.2$ incapable of binding to the β subunit (Yang et al. 2019). In particular, the $Ca_V1.2$ interacts with the guanylate kinase domain of the β subunit through a well-established AID motif localized to the domain I–II intracellular linker (Van Petegem et al. 2004). Disruption of this binding interface in the AID-mutant channels prevents association of the β subunit. In heterologous expression systems, this prevents functional expression of the $Ca_V1.2$ channels. By contrast, in cardiomyocytes, these mutant channels demonstrate robust current with minor changes in the baseline Ca^{2+} current properties. Surprisingly, these mutant channels were non-responsive to forskolin and isoproterenol, suggesting that the binding of the β subunit is essential for adrenergic upregulation (Yang et al. 2019). That is, the interaction of the β subunit to the α_{1C} subunit is critical for PKA regulation, although the phosphorylation of this subunit per se is not necessary. Taken together, a potential channel modulator that interacts with the β subunit may be the ultimate target of PKA phosphorylation.

What could be this missing link? The $Ca_V1.2$ channels are localized to the cardiac dyad, a specialized Ca^{2+} signaling hub composed of a multitude of proteins – e.g., juntophilin 2 (Jph2), RyR2, A-kinase anchoring proteins (AKAP), etc. – that all come together to coordinate excitation-contraction coupling (Harvey and Hell 2013; Lu and Pu 2020; Nakada et al. 2018; Yu et al. 2018). Given this rich diversity of proteins in the vicinity of $Ca_V1.2$, identifying a potential PKA-responsive modulator posed a major challenge. As we expect a transient interaction, standard approaches such as co-immunoprecipitation are often insufficient to identify candidate modulators particularly within the intracellular milieu.

To overcome this challenge, we utilized a proximity proteomics approach to obtain an in-depth map of the $Ca_V1.2$ neighborhood in cardiomyocytes. To do so, the engineered peroxidase APEX2 (Hung et al. 2016) was covalently tethered to either the α_{1C} or β_{2B} subunit and expressed in transgenic mice using a doxycycline-inducible cardiac-specific promoter (Liu et al. 2020). This approach is ideally suited for monitoring relatively rapid changes in subcellular neighborhood as the labeling time is relatively short. Upon treatment of either isolated cardiomyocytes or perfusing whole heart with hydrogen peroxide after incubation with biotin-phenol, biotin-phenoxyl radicals are generated and they attach covalently to proteins within an approximate 20 nm range of $Ca_V1.2$. Biotinylated proteins are then affinity purified and triple-stage mass spectrometry (TMT SPS MS^3) is used to identify biotinylated proteins and their relative abundance. This overall process provided the first comprehensive view of proteins in the vicinity of $Ca_V1.2$ throughout its lifecycle. These included hundreds of proteins that are localized to the membrane, cytoskeletal, or sarcomeric compartments, with some also associated with intracellular compartments, likely reflecting proteins encountered during channel synthesis, maturation, and trafficking. To further identify a relevant subset of proteins responsible for PKA-dependent regulation of $Ca_V1.2$, changes in the neighborhood were assessed following treatment with isoproterenol. Not surprisingly, this process revealed that recruitment of the PKA catalytic subunit to the channel complex, lending further confidence to this strategy (Fig. 4A). Remarkably, analysis of proteins displaying isoproterenol-induced changes identified the monomeric Ras-like G-protein, Rad, whose abundance near the $Ca_V1.2$ neighborhood was reduced 30–50%. Despite this change in the vicinity of the channel, the total level of Rad in the whole cell still remained the same. This localized change in the neighborhood suggests that Rad could be a long-sought mediator of adrenergic regulation of $Ca_V1.2$ channels. Importantly, these findings also highlight the power of proximity proteins as a generalizable approach to determine highly specific changes in ion channel signaling complexes.

8 Rad, a Central Element for PKA Regulation of $Ca_V1.2$

First discovered as a protein overexpressed in the skeletal muscle of type II diabetic humans, Rad belongs to a four-member subfamily of Ras-like monomeric G-proteins known as RGK (Rad, Rem, Rem2, Gem/Kir) proteins (Yang and Colecraft 2013). RGK proteins are differentially expressed in various tissues including the heart where they subserve a wide range of physiological functions including regulation of cytoskeletal dynamics, synapse development, and Ca^{2+} channel modulation (Finlin and Andres 1997; Finlin et al. 2003; Béguin et al. 2001; Manning et al. 2013). Structurally, RGKs share a common architecture containing a guanine nucleotide-binding domain (G-domain), a variable-length N-terminal region, and a C-terminal region responsible for membrane association (Correll et al. 2008). Three key aspects of RGK modulation merit further attention. First, RGK proteins serve as endogenous inhibitors of Ca_V1 and Ca_V2 channels. Second, although the G-domain of RGKs bind guanine nucleotides, their affinity for interaction varies amongst various members and they undergo minimal structural changes when bound to GTP versus GDP (Yang and Colecraft 2013). Instead, this domain interacts with the β subunit through electrostatic interactions between a conserved Arg^{208} in Rad protein and three Asp residues (Asp^{244}, Asp^{320}, and Asp^{322}) in the guanylate kinase

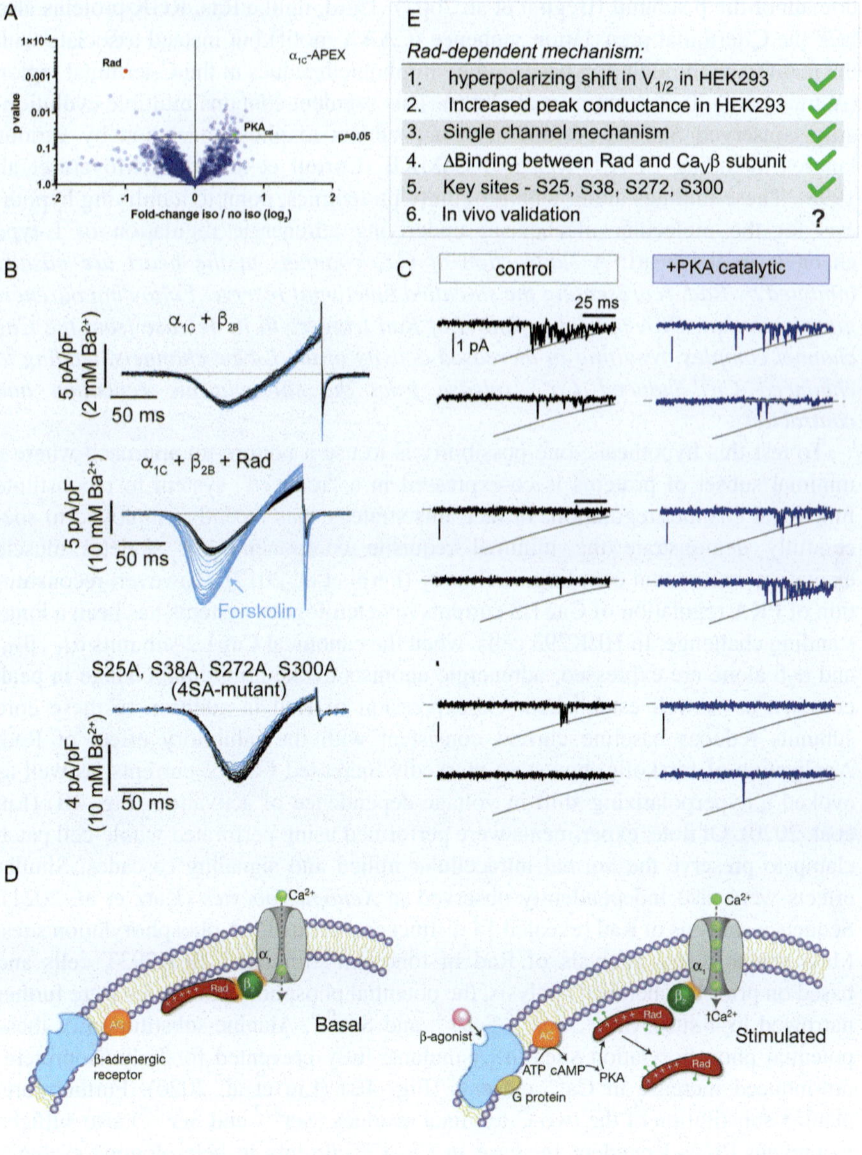

Fig. 4 Rad-dependent mechanism recapitulates key features of cardiac adrenergic regulation. (**a**) Proximity proteomics analysis demonstrated reduced labeling of Rad, a Ca_V channel inhibitor, from the L-type channel neighborhood. (**b**) Whole-cell perforated patch recordings show that co-expression of Rad reconstitutes adrenergic regulation of $Ca_V1.2$ in heterologous expression systems. Alanine-substitution of key phosphorylated residues disables adrenergic modulation. (**c**) Single channel recordings show functional reconstitution of adrenergic modulation of $Ca_V1.2$ in heterologous cells. Changes in gating pattern closely resemble early recordings of endogenous Ca^{2+} channels in cardiac myocytes (see Fig. 2c). (**d**) Schematic summarizes mechanism of adrenergic regulation of $Ca_V1.2$ channels. Rad basally associates with a subpopulation of $Ca_V1.2$ channels through the $Ca_V\beta$ subunit to inhibit channel activity. Adrenergic stimulation and subsequent phosphorylation of Rad results in its dissociation from the $Ca_V\beta$ subunit, there relieving channel inhibition. Panels (**a**–**d**) from Liu et al. (2020)

domain of the β subunit (Béguin et al. 2007). Third, unlike Ras, RGK proteins also lack the C-terminal prenylation sequence (CAAX motif) but instead associate with the membrane through polybasic and hydrophobic residues in the C-terminal region (Astrain et al. 2022; Yanuar et al. 2006). This segment contains multiple evolutionarily conserved Ser/Thr residues that are subject to phosphorylation by various kinases including PKA, PKC, and CaMKII (Correll et al. 2008; Moyers et al. 1998). These findings along with proximity proteomics, point to a unifying hypothesis for the molecular mechanism underlying adrenergic regulation of L-type channels in the heart: *A subfraction of Ca_V channels in the heart are basally inhibited by Rad, representing the so-called functional reserve. Following adrenergic stimulation, PKA phosphorylation of Rad triggers in its release from the Ca_V channel complex, resulting in increased activity of the L-type channels, leading to enhanced Ca^{2+}-induced Ca^{2+} release from the sarcoplasmic reticulum and contractility.*

To test this hypothesis, one possibility is to use a bottom-up approach where a minimal subset of proteins is co-expressed in a "reduced" system to reconstitute functional channel regulation. In fact, this strategy was recently employed to successfully demonstrate the minimal requisite components for skeletal muscle excitation-contraction coupling machinery (Perni et al. 2017). However, reconstitution of PKA regulation of $Ca_V1.2$ currents in heterologous systems has been a long-standing challenge. In HEK293 cells, when the canonical $Ca_V1.2$ subunits α_{1C}, β_{2B}, and $\alpha_2\delta$ alone are expressed, adrenergic agonists evoke a minimal change in peak currents, as is well established. Co-expression of Rad in addition to these core subunits reduces baseline current consistent with the inhibitory effect of Rad. Application of forskolin, however, markedly increased $Ca_V1.2$ currents, as well as evoked a hyperpolarizing shift in voltage-dependence of activation (Fig. 4B) (Liu et al. 2020). Of note, experiments were performed using perforated whole-cell patch clamp to preserve the normal intracellular milieu and signaling cascades. Similar effects were also independently observed in *Xenopus oocytes* (Katz et al. 2021). Sequence analysis of Rad revealed 14 distinct consensus PKA phosphorylation sites. Mass spectrometry analysis of Rad in forskolin-stimulated HEK293T cells and based on prior biochemical analysis, the potential phosphorylation sites were further narrowed to 4 sites: Ser^{25}, Ser^{38}, Ser^{272}, and Ser^{300}. Alanine-substitution of these potential phosphorylation sites (4SA-mutant) fully prevented forskolin/isoproterenol-induced increase in Ca^{2+} currents (Fig. 4B) (Liu et al. 2020). Furthermore, alanine-substitution of the two C-terminal residues (Ser^{272} and Ser^{300}) also sufficed to prevent PKA-dependent increase in $Ca_V1.2$ currents in heterologous systems. Taken together, these findings suggest that phosphorylation of key residues within Rad is both necessary and sufficient for functional reconstitution of β-adrenergic regulation (Liu et al. 2020).

Mechanistically, Rad can inhibit $Ca_V1.2$ by either interaction with the β_{2B} subunit or through direct interaction with the α_{1C} subunit. Previous studies have identified key interface residues responsible for Rad interaction with the $Ca_V\beta$ subunit including Arg^{208} and Leu^{235} in Rad and Asp^{244}, Asp^{320}, and Asp^{322} in the β_{2B} subunit (Béguin et al. 2007). Alanine-substitution of these interface residues in either Rad or in the β_{2B} subunit sufficed to prevent basal inhibition and forskolin-mediated change in $Ca_V1.2$ currents, suggesting that Rad-β subunit interaction is a critical determinant for PKA regulation of $Ca_V1.2$. How does PKA alter this interaction? In vitro analysis of Rad interaction is complex, as the C-terminus of Rad is membrane-associated resulting in instability in solution (Do Heo et al. 2006). As such, this region is often truncated for structural studies. Importantly, this region contains key residues critical for PKA modulation. Instead, a live-cell Forster resonance energy transfer (FRET) 2-hybrid assay has been used to measure association of the β subunit with full-length Rad tagged with a FRET donor and an acceptor, respectively (Liu et al. 2020; Papa et al. 2021). As single-cell measurement of FRET efficiency can be correlated with the concentration of the binding partners estimated based on their fluorescence intensities, this approach allows construction of a binding isotherm and determination of relative dissociation constants. At baseline, wild-type Rad associates robustly with the β_{2B} subunit. However, when PKA catalytic subunit is co-expressed, this interaction is markedly weakened (>10-fold). This reduction mirrors reduced biotinylation of Rad in the $Ca_V1.2$ neighborhood (Liu et al. 2020). By contrast, 4SA-mutant Rad, whose phosphorylation sites are disabled, showed strong binding to the β subunit even upon overexpression of the PKA catalytic subunit. A phosphorylation-dependent change in binding was also observed with other β subunits, hinting at generality of these mechanisms. Further supporting this notion, Rad co-expression also supported PKA-dependent upregulation of $Ca_V1.3$ and $Ca_V2.2$ currents. Altogether, these findings suggest that phosphorylation of Rad dislodges it from the β_{2B} subunit which relieves inhibition of the $Ca_V1.2$ channel (Liu et al. 2020).

9 Rad-Dependent Mechanism Recaptures Key Features of PKA Regulation of $Ca_V1.2$?

At the macroscopic level, Rad co-expression enables PKA regulation of $Ca_V1.2$ currents that coarsely matches modulation of endogenous channels in the heart. Specifically, forskolin stimulation results in up to fivefold increase in calcium current amplitude and hyperpolarizing shift in the voltage-dependence of activation (Fig. 4B) (Liu et al. 2020). Furthermore, the forskolin-induced fold increase in current magnitude was inversely proportional to the basal current density in HEK cells, paralleling similar observations in cardiomyocytes. This relationship likely stems from variable expression of Rad when heterologously expressed. That is, at low levels of Rad, only a fraction of channels are bound to Rad and thus basally inhibited (Miriyala et al. 2008). As a consequence, PKA phosphorylation and subsequent relief of inhibition yields only a muted increase in current density. By contrast, at intermediate high levels of Rad, a vast majority of channels are basally inhibited and the relief of inhibition by PKA yields a large increase in current density. This relation has also been observed in *Xenopus* oocytes (Katz et al. 2021). Notably, at still higher levels of Rad, the magnitude of current increase following PKA activation is actually diminished. One possibility is that given the overabundance of Rad, PKA activation may only phosphorylate a subfraction of Rad, while the non-phosphorylated Rad could still strongly inhibit the Ca_V channels. Alternatively, although phosphorylated Rad interacts poorly with the β subunit, it is possible that at high enough concentrations, it may still be inhibitory. Future studies may help resolve these ambiguities and dissect the precise molecular mechanisms underlying Rad/PKA modulation of $Ca_V1.2$ channels.

At the single channel level, co-expression of Rad with $Ca_V1.2$ potently inhibits channel openings. However, co-expression of PKA catalytic subunit along with Rad relieves this inhibition (Fig. 4C). Importantly, scrutiny of individual trials showed that when bound to Rad, channel exhibit either no openings corresponding to mode 0, or rare and brief openings akin to mode 0_a. When PKA catalytic subunit is co-expressed along with Rad, the fraction of blank sweeps decreases and a high-activity gating mode that qualitatively resemble mode 1 reemerges. The net effect is a ~ tenfold increase in the P_O. Importantly, this single channel signature of Rad-mediated PKA regulation of $Ca_V1.2$ bear striking resemblance to earlier reports of PKA-induced changes in gating of endogenous L-type channels in cardiomyocytes.

To delineate how Rad association with the β subunit alters the pore conformation, we probed the impact of mutations that uncouple the intracellular S6 gate with the β subunit (Papa et al. 2021). In particular, previous studies have shown that the distal S6 region is coupled to the intracellular I–II linker that harbors the β subunit through a contiguous helix. Altered conformation of the β subunit is thought to be transduced to the pore via this rigid linker. As such, we utilized a previously-established mutation that increases the flexibility of this linker by introducing three glycine residues (GGG-mutant). Although these channels still bind the β subunit, electrophysiological characterization in a transgenic mouse model revealed that these channels exhibit lower current density reflecting reduced P_O. Furthermore, this mutation also abolished forskolin-induced upregulation of $Ca_V1.2$ currents, suggesting that Rad modulation requires a rigid linker connecting the S6 gate with the β subunit. In all, a Rad-dependent mechanism largely recapitulates the signatures features of adrenergic regulation of $Ca_V1.2$ established over the past four decades.

10 Relevance of Rad Modulation in Cardiomyocytes

Although experiments in heterologous expression systems are certainly suggestive, conclusive evidence for a Rad-mediated mechanism for PKA modulation of $Ca_V1.2$ requires further corroboration in cardiomyocytes, particularly by disabling phosphorylation of key C-terminal residues of Rad. Nonetheless, several lines of evidence support a Rad-dependent mechanism as summarized in brief.

In mice, both whole-body and myocardial restricted deletion of Rad resulted in a marked basal increase in L-type Ca^{2+} current in ventricular cardiomyocytes as well as a hyperpolarizing shift in voltage-dependence of activation (Ahern et al. 2019; Manning et al. 2015). Importantly, isoproterenol stimulation yielded no further increase in Ca^{2+} currents. That is, Ca^{2+} current in the Rad null mice appears to mimic a constitutively β-adrenergic modulated conformation. These findings are consistent with the hypothesis that Rad serves as an endogenous inhibitor of $Ca_V1.2$, that is relieved upon adrenergic stimulation. In rat ventricular myocytes, adenoviral expression of Rad reduced basal currents and dramatically attenuated the β-adrenergic upregulation of Ca^{2+} currents (Wang et al. 2010). By comparison, shRNA-mediated downregulation of Rad enhanced basal $Ca_V1.2$ currents, and application of β-adrenergic agonist still evoked an increase in Ca^{2+} current amplitude to similar levels as control. In like manner in guinea pig ventricular myocytes, adenoviral overexpression of Rem resulted in a strong reduction in L-type Ca^{2+} currents (Puckerin et al. 2018). These channels were, however, recalcitrant to PKA upregulation, even as they could be partially enhanced using the L-type channel agonist BayK8644. The reduction in β-adrenergic response upon overexpression of either Rad or Rem may reflect the fact that PKA activation may only phosphorylate a subpopulation of Rad/Rem, while the overabundance of non-phosphorylated Rad/Rem can still inhibit Ca^{2+} current. In comparison to genetic ablation of Rad, shRNA-mediated downregulation may only partially reduce the ambient levels of Rad. Consequently, sufficient Rad may still be available to inhibit Ca^{2+} current and support adrenergic upregulation. In total, these findings largely support the role of Rad in mediating β-adrenergic regulation of $Ca_V1.2$ current. Future studies that mutate key Rad phosphorylation sites will be essential to conclusively demonstrate the physiological contribution of this mechanism to Ca^{2+} channel regulation, as well as to discern its importance to changes in Ca^{2+} transients and contractility following adrenergic stimulation.

11 Pathophysiological Implications

Given its central role in excitation-contraction coupling, dysregulation of $Ca_V1.2$ in cardiomyocytes is linked to various cardiac diseases including cardiac arrhythmias, heart failure, and sudden death (Zhang et al. 2016; Splawski et al. 2004; Dick et al. 2016; Hofmann et al. 2014; Goonasekera et al. 2012; Chen et al. 2002). Whole-cell recordings have revealed minor changes in Ca^{2+} current properties in heart failure (Hegyi et al. 2018; Chen et al. 2002). However, these changes are often variable depending on the species-specific differences, the specific model, or other factors. Interestingly, single channel recordings have revealed increased availability and increased P_O for single L-type Ca^{2+} channels in ventricular myocytes from human failing hearts (both dilated and ischemic cardiomyopathy) compared to nonfailing donor hearts (Schröder et al. 1998). Similarly, disruption of t-tubular structure has been shown to result in redistribution of functional $Ca_V1.2$ channels and subcellular region-specific differences in P_O (Wei et al. 2010; Song et al. 2006). Mechanisms underlying these pathophysiological changes are not fully understood, though several possibilities have emerged. First, aberrant alternative splicing of the $Ca_V1.2$ in heart failure results in increased inclusion of exon 9* that also upregulates channel activity (Papa et al. 2021). Second, as these pathological changes in channel properties mimic effects of β-adrenergic stimulation, alteration in phosphorylation status of the channel has also been proposed. Identification of a Rad-dependent mechanism for $Ca_V1.2$ modulation opens a new modulatory dimension for pathophysiological understanding. For instance, it is possible that increased $Ca_V1.2$ activity observed in heart failure may result from either increased phosphorylation or diminished expression of Rad.

The Rad-dependent mechanism also informs upon new avenues and molecular interfaces for modulation of β-adrenergic regulation of $Ca_V1.2$ in the heart and beyond. First, disruption of basal Rad association with the $Ca_V\beta$ subunit can upregulate Ca^{2+} entry into the heart and serve as an inotrope. Evidence supporting this notion emerges from Rad null mice that show enhanced heart output without structural remodeling or fibrosis, despite constitutively elevated Ca^{2+} entry (Ahern et al. 2019). Furthermore, Rad null mice have both reduced mortality and contractile dysfunction following myocardial infarction resulting from surgical ligation of the left anterior descending coronary artery (LAD ligation) (Manning et al. 2018). Alternatively, selectively preventing Rad phosphorylation or stabilizing Rad association with the β subunit could counteract sympathetic nervous system activation of cardiac Ca^{2+} entry and inotropy.

12 Conclusions

Despite its long and storied history, emerging work points to an elegant mechanism of adrenergic modulation of $Ca_V1.2$ in the heart: First, Rad is constitutively associated with a subpopulation of the $Ca_V1.2$ complex through interaction with the auxiliary β subunit that inhibits channel activity. Second, the binding of adrenaline or noradrenaline to the β-adrenergic receptor results in activation of adenylyl cyclase, conversion of ATP to cAMP, and subsequent activation of PKA. Third, PKA-dependent phosphorylation of Rad C-terminal residues Ser^{272} and Ser^{300} residues dislodges Rad from the auxiliary β subunit, resulting in a relief of channel inhibition (Fig. 4D). Future studies that specifically disable Rad phosphorylation will establish the relevance of this mechanism in vivo as well as inform upon the contribution of $Ca_V1.2$ modulation in the context of broader β-adrenergic reprogramming of various cardiac proteins. Furthermore, in-depth molecular understanding of Rad-Ca_V channel modulation may facilitate the development of new pharmacological agents that are therapeutically beneficial. In all, an exciting new era of discovery now looms on the horizon.

Acknowledgments This publication was supported by the following grants: NIH R01 HL140934, NIH R01 HL155377, NIH R01 HL146149, NIH R01 HL160089, NIH R01 HL121253 and NIH R01NS110672.

References

Ahern BM, Levitan BM, Veeranki S, Shah M, Ali N, Sebastian A, Wen S, Gong MC, Li J, Stelzer JE, Andres DA, Satin J (2019) Myocardial-restricted ablation of the GTPase RAD results in a pro-adaptive heart response in mice. J Biol Chem 294:10913–10927

Astrain B, Gabriela MN, Smith MJ (2022) Functional diversity in the RAS subfamily of small GTPases. Biochem Soc Trans 50:921–933

Bean BP, Nowycky MC, Tsien RW (1984) Beta-adrenergic modulation of calcium channels in frog ventricular heart cells. Nature 307:371–375

Béguin P, Nagashima K, Gonoi T, Shibasaki T, Takahashi K, Kashima Y, Ozaki N, Geering K, Iwanaga T, Seino S (2001) Regulation of Ca2+ channel expression at the cell surface by the small G-protein kir/Gem. Nature 411:701–706

Béguin P, Ng YJA, Krause C, Mahalakshmi RN, Ng MY, Hunziker W (2007) RGK small GTP-binding proteins interact with the nucleotide kinase domain of Ca2+−channel beta-subunits via an uncommon effector binding domain. J Biol Chem 282:11509–11520

Bers DM (2002) Cardiac excitation–contraction coupling. Nature 415:198–205

Brum G, Flockerzi V, Hofmann F, Osterrieder W, Trautwein W (1983) Injection of catalytic subunit of cAMP-dependent protein kinase into isolated cardiac myocytes. Pflugers Arch 398:147–154

Bunemann M, Gerhardstein BL, Gao T, Hosey MM (1999) Functional regulation of L-type calcium channels via protein kinase A-mediated phosphorylation of the beta(2) subunit. J Biol Chem 274:33851–33854

Catterall WA (2015) Regulation of cardiac calcium channels in the fight-or-flight response. Curr Mol Pharmacol 8:12–21

Chen X, Piacentino V, Furukawa S, Goldman B, Margulies KB, Houser SR (2002) L-type Ca^{2+} channel density and regulation are altered in failing human ventricular myocytes and recover after support with mechanical assist devices. Circ Res 91:517–524

Chien AJ, Carr KM, Shirokov RE, Rios E, Hosey MM (1996) Identification of palmitoylation sites within the L-type calcium channel beta2a subunit and effects on channel function. J Biol Chem 271:26465–26468

Correll RN, Pang C, Niedowicz DM, Finlin BS, Andres DA (2008) The RGK family of GTP-binding proteins: regulators of voltage-dependent calcium channels and cytoskeleton remodeling. Cell Signal 20:292–300

Curtis BM, Catterall WA (1985) Phosphorylation of the calcium antagonist receptor of the voltage-sensitive calcium channel by cAMP-dependent protein kinase. Proc Natl Acad Sci U S A 82: 2528–2532

De Jongh KS, Warner C, Colvin AA, Catterall WA (1991) Characterization of the two size forms of the alpha 1 subunit of skeletal muscle L-type calcium channels. Proc Natl Acad Sci U S A 88: 10778–10782

De Jongh KS, Murphy BJ, Colvin AA, Hell JW, Takahashi M, Catterall WA (1996) Specific phosphorylation of a site in the full-length form of the alpha 1 subunit of the cardiac L-type calcium channel by adenosine 3′,5′-cyclic monophosphate-dependent protein kinase. Biochemistry 35:10392–10402

Dick IE, Joshi-Mukherjee R, Yang W, Yue DT (2016) Arrhythmogenesis in Timothy syndrome is associated with defects in Ca2+−dependent inactivation. Nat Commun 7:10370

Do Heo W, Inoue T, Park WS, Kim ML, Park BO, Wandless TJ, Meyer T (2006) PI(3,4,5)P3 and PI (4,5)P2 lipids target proteins with polybasic clusters to the plasma membrane. Science 314: 1458–1461

Domes K, Ding J, Lemke T, Blaich A, Wegener JW, Brandmayr J, Moosmang S, Hofmann F (2011) Truncation of murine CaV1.2 at Asp-1904 results in heart failure after birth. J Biol Chem 286:33863–33871

Findeisen F, Minor DL Jr (2009) Disruption of the IS6-AID linker affects voltage-gated calcium channel inactivation and facilitation. J Gen Physiol 133:327–343

Finlin BS, Andres DA (1997) Rem is a new member of the rad- and Gem/Kir ras-related GTP-binding protein family repressed by lipopolysaccharide stimulation. J Biol Chem 272: 21982–21988

Finlin BS, Crump SM, Satin J, Andres DA (2003) Regulation of voltage-gated calcium channel activity by the rem and rad GTPases. Proc Natl Acad Sci U S A 100:14469–14474

Fu Y, Westenbroek RE, Yu FH, Clark JP 3rd, Marshall MR, Scheuer T, Catterall WA (2011) Deletion of the distal C terminus of CaV1.2 channels leads to loss of beta-adrenergic regulation and heart failure in vivo. J Biol Chem 286:12617–12626

Fuller MD, Emrick MA, Sadilek M, Scheuer T, Catterall WA (2010) Molecular mechanism of calcium channel regulation in the fight-or-flight response. Sci Signal 3:ra70

Ganesan AN, O'Rourke B, Maack C, Colecraft H, Sidor A, Johns DC (2005) Reverse engineering the L-type Ca2+ channel alpha1c subunit in adult cardiac myocytes using novel adenoviral vectors. Biochem Biophys Res Commun 329:749–754

Ganesan AN, Maack C, Johns DC, Sidor A, O'Rourke B (2006) Beta-adrenergic stimulation of L-type Ca2+ channels in cardiac myocytes requires the distal carboxyl terminus of alpha1C but not serine 1928. Circ Res 98:e11–e18

Gao T, Yatani A, Dell'Acqua ML, Sako H, Green SA, Dascal N, Scott JD, Hosey MM (1997) cAMP-dependent regulation of cardiac L-type Ca2+ channels requires membrane targeting of PKA and phosphorylation of channel subunits. Neuron 19:185–196

Gerhardstein BL, Puri TS, Chien AJ, Hosey MM (1999) Identification of the sites phosphorylated by cyclic AMP-dependent protein kinase on the beta 2 subunit of L-type voltage-dependent calcium channels. Biochemistry 38:10361–10370

Gerhardstein BL, Gao T, Bunemann M, Puri TS, Adair A, Ma H, Hosey MM (2000) Proteolytic processing of the C terminus of the alpha(1C) subunit of L-type calcium channels and the role of a proline-rich domain in membrane tethering of proteolytic fragments. J Biol Chem 275:8556–8563

Goonasekera SA, Hammer K, Auger-Messier M, Bodi I, Chen X, Zhang H, Reiken S, Elrod JW, Correll RN, York AJ, Sargent MA, Hofmann F, Moosmang S, Marks AR, Houser SR, Bers DM, Molkentin JD (2012) Decreased cardiac L-type Ca^{2+} channel activity induces hypertrophy and heart failure in mice. J Clin Invest 122:280–290

Greene D, Kang S, Kosenko A, Hoshi N (2012) Adrenergic regulation of HCN4 channel requires protein association with β2-adrenergic receptor. J Biol Chem 287:23690–23697

Haase H, Bartel S, Karczewski P, Morano I, Krause EG (1996) In-vivo phosphorylation of the cardiac L-type calcium channel beta-subunit in response to catecholamines. Mol Cell Biochem 163-164:99–106

Harvey RD, Hell JW (2013) CaV1.2 signaling complexes in the heart. J Mol Cell Cardiol 58:143–152

Hegyi B, Bossuyt J, Griffiths LG, Shimkunas R, Coulibaly Z, Jian Z, Grimsrud KN, Sondergaard CS, Ginsburg KS, Chiamvimonvat N, Belardinelli L, Varró A, Papp JG, Pollesello P, Levijoki J, Izu LT, Douglas Boyd W, Bányász T, Bers DM, Chen-Izu Y (2018) Complex electrophysiological remodeling in postinfarction ischemic heart failure. Proc Natl Acad Sci U S A 115: E3036–E3e44

Hell JW (2010) Beta-adrenergic regulation of the L-type Ca2+ channel Ca(V)1.2 by PKA rekindles excitement. Sci Signal 3:pe33

Hell JW, Westenbroek RE, Warner C, Ahlijanian MK, Prystay W, Gilbert MM, Snutch TP, Catterall WA (1993) Identification and differential subcellular localization of the neuronal class C and class D L-type calcium channel alpha 1 subunits. J Cell Biol 123:949–962

Herzig S, Patil P, Neumann J, Staschen C-M, Yue DT (1993) Mechanisms of beta-adrenergic stimulation of cardiac Ca2+ channels revealed by discrete-time Markov analysis of slow gating. Biophys J 65:1599–1612

Hess P, Lansman JB, Tsien RW (1984) Different modes of Ca channel gating behaviour favoured by dihydropyridine Ca agonists and antagonists. Nature 311:538–544

Hirano Y, Suzuki K, Yamawake N, Hiraoka M (1994) Multiple kinetic effects of beta-adrenergic stimulation on single cardiac L-type Ca channels. Am J Physiol 266:C1714–C1721

Hofmann F, Flockerzi V, Kahl S, Wegener JW (2014) L-type CaV1.2 calcium channels: from in vitro findings to in vivo function. Physiol Rev 94:303–326

Hulme JT, Konoki K, Lin TW, Gritsenko MA, Camp DG 2nd, Bigelow DJ, Catterall WA (2005) Sites of proteolytic processing and noncovalent association of the distal C-terminal domain of CaV1.1 channels in skeletal muscle. Proc Natl Acad Sci U S A 102:5274–5279

Hung V, Udeshi ND, Lam SS, Loh KH, Cox KJ, Pedram K, Carr SA, Ting AY (2016) Spatially resolved proteomic mapping in living cells with the engineered peroxidase APEX2. Nat Protoc 11:456–475

Ito DW, Hannigan KI, Ghosh D, Xu B, Del Villar SG, Xiang YK, Dickson EJ, Navedo MF, Dixon RE (2019) Beta-adrenergic-mediated dynamic augmentation of sarcolemmal CaV 1.2 clustering and co-operativity in ventricular myocytes. J Physiol 597:2139–2162

Jhun BS, O-Uchi J, Wang W, Ha CH, Zhao J, Kim JY, Wong C, Dirksen RT, Lopes CMB, Jin ZG (2012) Adrenergic signaling controls RGK-dependent trafficking of cardiac voltage-gated L-type Ca2+ channels through PKD1. Circ Res 110:59–70

Jork K, Kuschinsky G, Reuter H (1967) Influence of the extracellular calcium concentration on the action of noradrenalin, beta-adrenloytic agents and quinidine in isolated Guinea pig atria. Naunyn Schmiedebergs Arch Exp Pathol Pharmakol 258:59–68

Kamp TJ, Hell JW (2000) Regulation of cardiac L-type calcium channels by protein kinase A and protein kinase C. Circ Res 87:1095–1102

Katchman A, Yang L, Zakharov SI, Kushner J, Abrams J, Chen BX, Liu G, Pitt GS, Colecraft HM, Marx SO (2017) Proteolytic cleavage and PKA phosphorylation of alpha1C subunit are not required for adrenergic regulation of CaV1.2 in the heart. Proc Natl Acad Sci U S A 114:9194–9199

Katz M, Subramaniam S, Chomsky-Hecht O, Tsemakhovich V, Flockerzi V, Klussmann E, Hirsch JA, Weiss S, Dascal N (2021) Reconstitution of beta-adrenergic regulation of CaV1.2: rad-dependent and rad-independent protein kinase a mechanisms. Proc Natl Acad Sci U S A 118

Kentish JC, McCloskey DT, Layland J, Palmer S, Leiden JM, Martin AF, John Solaro R (2001) Phosphorylation of troponin I by protein kinase a accelerates relaxation and crossbridge cycle kinetics in mouse ventricular muscle. Circ Res 88:1059–1065

Lakatta EG, Maltsev VA, Vinogradova TM (2010) A coupled SYSTEM of intracellular Ca2+ clocks and surface membrane voltage clocks controls the timekeeping mechanism of the heart's pacemaker. Circ Res 106:659–673

Lefkowitz RJ, Rockman HA, Koch WJ (2000) Catecholamines, cardiac beta-adrenergic receptors, and heart failure. Circulation 101:1634–1637

Lemke T, Welling A, Christel CJ, Blaich A, Bernhard D, Lenhardt P, Hofmann F, Moosmang S (2008) Unchanged beta-adrenergic stimulation of cardiac L-type calcium channels in Ca v 1.2 phosphorylation site S1928A mutant mice. J Biol Chem 283:34738–34744

Li L, Desantiago J, Chu G, Kranias EG, Bers DM (2000) Phosphorylation of phospholamban and troponin I in beta-adrenergic-induced acceleration of cardiac relaxation. Am J Physiol Heart Circ Physiol 278:H769–H779

Lindemann JP, Jones LR, Hathaway DR, Henry BG, Watanabe AM (1983) Beta-adrenergic stimulation of phospholamban phosphorylation and Ca2+-ATPase activity in Guinea pig ventricles. J Biol Chem 258:464–471

Liu G, Papa A, Katchman AN, Zakharov SI, Roybal D, Hennessey JA, Kushner J, Yang L, Chen BX, Kushnir A, Dangas K, Gygi SP, Pitt GS, Colecraft HM, Ben-Johny M, Kalocsay M, Marx SO (2020) Mechanism of adrenergic CaV1.2 stimulation revealed by proximity proteomics. Nature 577:695–700

López-López JR, Shacklock PS, William Balke C, Wier WG (1995) Local calcium transients triggered by single L-type calcium channel currents in cardiac cells. Science 268:1042–1045

Lu F, Pu WT (2020) The architecture and function of cardiac dyads. Biophys Rev 12:1007–1017

Manning JR, Yin G, Kaminski CN, Magyar J, Feng HZ, Penn J, Sievert G, Thompson K, Jin JP, Andres DA, Satin J (2013) Rad GTPase deletion increases L-type calcium channel current leading to increased cardiac contraction. J Am Heart Assoc 2:e000459

Manning JR, Withers CN, Levitan B, Smith JD, Andres DA, Satin J (2015) Loss of rad-GTPase produces a novel adaptive cardiac phenotype resistant to systolic decline with aging. Am J Physiol Heart Circ Physiol 309:H1336–H1H45

Manning JR, Chelvarajan L, Levitan BM, Withers CN, Nagareddy PR, Haggerty CM, Fornwalt BK, Gao E, Tripathi H, Abdel-Latif A, Andres DA, Satin J (2018) Rad GTPase deletion attenuates post-ischemic cardiac dysfunction and remodeling. JACC Basic Transl Sci 3:83–96

Marks AR (2003) Calcium and the heart: a question of life and death. J Clin Investig 111:597–600

Marx SO, Reiken S, Hisamatsu Y, Jayaraman T, Burkhoff D, Rosemblit N, Marks AR (2000) PKA phosphorylation dissociates FKBP12.6 from the calcium release channel (ryanodine receptor): defective regulation in failing hearts. Cell 101:365–376

Marx SO, Kurokawa J, Reiken S, Motoike H, D'Armiento J, Marks AR, Kass RS (2002) Requirement of a macromolecular signaling complex for beta adrenergic receptor modulation of the KCNQ1-KCNE1 potassium channel. Science 295:496–499

Miriyala J, Nguyen T, Yue DT, Colecraft HM (2008) Role of CaVbeta subunits, and lack of functional reserve, in protein kinase A modulation of cardiac CaV1.2 channels. Circ Res 102: e54–e64

Mitterdorfer J, Froschmayr M, Grabner M, Moebius FF, Glossmann H, Striessnig J (1996) Identification of PK-A phosphorylation sites in the carboxyl terminus of L-type calcium channel alpha 1 subunits. Biochemistry 35:9400–9406

Morales D, Hermosilla T, Varela D (2019) Calcium-dependent inactivation controls cardiac L-type Ca2+ currents under β-adrenergic stimulation. J Gen Physiol 151:786–797

Moyers JS, Zhu J, Kahn CR (1998) Effects of phosphorylation on function of the rad GTPase. Biochem J 333(Pt 3):609–614

Naguro I, Nagao T, Adachi-Akahane S (2001) Ser(1901) of alpha(1C) subunit is required for the PKA-mediated enhancement of L-type Ca(2+) channel currents but not for the negative shift of activation. FEBS Lett 489:87–91

Nakada T, Kashihara T, Komatsu M, Kojima K, Takeshita T, Yamada M (2018) Physical interaction of junctophilin and the CaV1.1 C terminus is crucial for skeletal muscle contraction. Proc Natl Acad Sci U S A 115:4507–4512

Osterrieder W, Brum G, Hescheler J, Trautwein W, Flockerzi V, Hofmann F (1982) Injection of subunits of cyclic AMP-dependent protein kinase into cardiac myocytes modulates Ca2+ current. Nature 298:576–578

Papa A, Kushner J, Hennessey JA, Katchman AN, Zakharov SI, Chen BX, Yang L, Lu R, Leong S, Diaz J, Liu G, Roybal D, Liao X, Del Rivero Morfin PJ, Colecraft HM, Pitt GS, Clarke O, Topkara V, Ben-Johny M, Marx SO (2021) Adrenergic CaV1.2 activation via rad phosphorylation converges at alpha1C I-II loop. Circ Res 128:76–88

Perni S, Lavorato M, Beam KG (2017) De novo reconstitution reveals the proteins required for skeletal muscle voltage-induced Ca^{2+} release. Proc Natl Acad Sci 114:13822–13827

Poomvanicha M, Matthes J, Domes K, Patrucco E, Angermeier E, Laugwitz KL, Schneider T, Hofmann F (2017) Beta-adrenergic regulation of the heart expressing the Ser1700A/Thr1704A mutated Cav1.2 channel. J Mol Cell Cardiol 111:10–16

Puckerin AA, Chang DD, Shuja Z, Choudhury P, Scholz J, Colecraft HM (2018) Engineering selectivity into RGK GTPase inhibition of voltage-dependent calcium channels. Proc Natl Acad Sci U S A 115:12051–12056

Reiken S, Gaburjakova M, Guatimosim S, Gomez AM, D'Armiento J, Burkhoff D, Wang J, Vassort G, Lederer WJ, Marks AR (2003) Protein kinase A phosphorylation of the cardiac calcium release channel (ryanodine receptor) in normal and failing hearts. Role of phosphatases and response to isoproterenol. J Biol Chem 278:444–453

Reuter H (1974) Localization of beta adrenergic receptors, and effects of noradrenaline and cyclic nucleotides on action potentials, ionic currents and tension in mammalian cardiac muscle. J Physiol 242:429–451

Reuter H (1983) Calcium channel modulation by neurotransmitters, enzymes and drugs. Nature 301:569–574

Reuter H (1987) Calcium channel modulation by beta-adrenergic neurotransmitters in the heart. Experientia 43:1173–1175

Scamps F, Mayoux E, Charlemagne D, Vassort G (1990) Calcium current in single cells isolated from normal and hypertrophied rat heart. Effects of beta-adrenergic stimulation. Circ Res 67: 199–208

Schröder F, Handrock R, Beuckelmann DJ, Hirt S, Hullin R, Priebe L, Schwinger RHG, Weil J, Herzig S (1998) Increased availability and open probability of single L-type calcium channels from failing compared with nonfailing human ventricle. Circulation 98:969–976

Song LS, Sobie EA, McCulle S, Lederer WJ, Balke CW, Cheng H (2006) Orphaned ryanodine receptors in the failing heart. Proc Natl Acad Sci U S A 103:4305–4310

Splawski I, Timothy KW, Sharpe LM, Decher N, Kumar P, Bloise R, Napolitano C, Schwartz PJ, Joseph RM, Condouris K, Tager-Flusberg H, Priori SG, Sanguinetti MC, Keating MT (2004) Ca (V)1.2 calcium channel dysfunction causes a multisystem disorder including arrhythmia and autism. Cell 119:19–31

Taylor EW, Jordan D, Coote JH (1999) Central control of the cardiovascular and respiratory systems and their interactions in vertebrates. Physiol Rev 79:855–916

Torre E, Mesirca P, Mangoni ME (2021) Beta-1 adrenergic receptors modulate pacemaker activity of mouse sino-atrial myocytes through L-type Cav1.3 channels. EP Europace 23

Tsien RW, Bean BP, Hess P, Lansman JB, Nilius B, Nowycky MC (1986) Mechanisms of calcium channel modulation by beta-adrenergic agents and dihydropyridine calcium agonists. J Mol Cell Cardiol 18:691–710

Tsutsui K, Monfredi OJ, Sirenko-Tagirova SG, Maltseva LA, Bychkov R, Kim MS, Ziman BD, Tarasov KV, Tarasova YS, Zhang J, Wang M, Maltsev AV, Brennan JA, Efimov IR, Stern MD, Maltsev VA, Lakatta EG (2018) A coupled-clock system drives the automaticity of human sinoatrial nodal pacemaker cells. Sci Signal 11

van der Heyden MAG, Wijnhoven TJM, Opthof T (2005) Molecular aspects of adrenergic modulation of cardiac L-type Ca2+ channels. Cardiovasc Res 65:28–39

Van Petegem F, Clark KA, Chatelain FC, Minor DL Jr (2004) Structure of a complex between a voltage-gated calcium channel beta-subunit and an alpha-subunit domain. Nature 429:671–675

Villar SG, Del TL, Voelker MW, Reddy GR, Spooner HC, Navedo MF, Dickson EJ, Dixon RE (2021) β-Adrenergic control of sarcolemmal ca(V)1.2 abundance by small GTPase Rab proteins. Proc Natl Acad Sci U S A 118

Wang S-Q, Song L-S, Lakatta EG, Cheng H (2001) Ca2+ signalling between single L-type Ca2+ channels and ryanodine receptors in heart cells. Nature 410:592–596

Wang G, Zhu X, Xie W, Han P, Li K, Sun Z, Wang Y, Chen C, Song R, Cao C, Zhang J, Wu C, Liu J, Cheng H (2010) Rad as a novel regulator of excitation-contraction coupling and beta-adrenergic signaling in heart. Circ Res 106:317–327

Wei S, Guo A, Chen B, Kutschke W, Xie Y-P, Zimmerman K, Weiss RM, Anderson ME, Cheng H, Song L-S (2010) T-tubule remodeling during transition from hypertrophy to heart failure. Circ Res 107:520–531

Yang T, Colecraft HM (2013) Regulation of voltage-dependent calcium channels by RGK proteins. Biochim Biophys Acta 1828:1644–1654

Yang L, Katchman A, Kushner J, Kushnir A, Zakharov SI, Chen BX, Shuja Z, Subramanyam P, Liu G, Papa A, Roybal D, Pitt GS, Colecraft HM, Marx SO (2019) Cardiac CaV1.2 channels require beta subunits for beta-adrenergic-mediated modulation but not trafficking. J Clin Invest 129:647–658

Yanuar A, Sakurai S, Kitano K, Hakoshima T (2006) Crystal structure of human rad GTPase of the RGK-family. Genes Cells 11:961–968

Yu H, Yuan C, Westenbroek RE, Catterall WA (2018) The AKAP Cypher/Zasp contributes to beta-adrenergic/PKA stimulation of cardiac CaV1.2 calcium channels. J Gen Physiol 150:883–889

Yue DT, Herzig S, Marban E (1990) Beta-adrenergic stimulation of calcium channels occurs by potentiation of high-activity gating modes. Proc Natl Acad Sci U S A 87:753–757

Zhang X, Ai X, Nakayama H, Chen B, Harris DM, Tang M, Xie Y, Szeto C, Li Y, Li Y, Zhang H, Eckhart AD, Koch WJ, Molkentin JD, Chen X (2016) Persistent increases in Ca(2+) influx through Cav1.2 shortens action potential and causes Ca(2+) overload-induced afterdepolarizations and arrhythmias. Basic Res Cardiol 111:4

Modulation of $Ca_V1.2$ Channel Function by Interacting Proteins and Post-Translational Modifications: Implications in Cardiovascular Diseases and COVID-19

Kelvin Wei Zhern Loh, Zhenyu Hu, and Tuck Wah Soong

Contents

Kelvin Wei Zhern Loh and Zhenyu Hu contributed equally to this work.

K. W. Z. Loh · Z. Hu
Department of Physiology, Yong Loo Lin School of Medicine, National University of Singapore, Singapore, Singapore

Cardiovascular Diseases Translational Research Programme, National University of Singapore, Singapore, Singapore

T. W. Soong (✉)
Department of Physiology, Yong Loo Lin School of Medicine, National University of Singapore, Singapore, Singapore

Cardiovascular Diseases Translational Research Programme, National University of Singapore, Singapore, Singapore

Healthy Longevity Translational Research Programme, National University of Singapore, Singapore, Singapore
e-mail: phsstw@nus.edu.sg

J. Striessnig (ed.), *Voltage-gated Ca^{2+} Channels: Pharmacology, Modulation and their Role in Human Disease*, Handbook of Experimental Pharmacology 279,
https://doi.org/10.1007/164_2023_636

Abstract

$Ca_V1.2$ calcium channel is the primary conduit for Ca^{2+} influx into cardiac and smooth muscles that underscores its importance in the pathogenesis of hypertension, atherosclerosis, myocardial infarction, and heart failure. But, a few controversies still remain. Therefore, exploring new ways to modulate $Ca_V1.2$ channel activity will augment the arsenal of $Ca_V1.2$ channel-based therapeutics for treatment of cardiovascular diseases. Here, we will mainly introduce a couple of emerging $Ca_V1.2$ channel interacting proteins, such as Galectin-1 and Cereblon, and discuss their roles in hypertension and heart failure through fine-tuning $Ca_V1.2$ channel activity. Of current interest, we will also evaluate the implication of the role of $Ca_V1.2$ channel in SARS-CoV-2 infection and the potential treatments of COVID-19-related cardiovascular symptoms.

Keywords

$Ca_V1.2$ channel · COVID-19 · Galectin · Heart failure · Hypertension

1 Cardiovascular $Ca_V1.2$ Channel at a Glance

$Ca_V1.2$ channels are membrane protein complexes composed of a few subunits: pore-forming α_{1C} and the auxiliary $\alpha_2\delta$ and β subunits. This L-type calcium channel has been well accepted as the major pathway for the influx of Ca^{2+} to raise cytoplasmic Ca^{2+} to initiate contraction of smooth or cardiac muscles (Hofmann et al. 2014). The $\alpha_2\delta$ subunit in its auxiliary role facilitates voltage-dependent activation (Savalli et al. 2016) and channel trafficking to the plasma membrane without affecting the total pool of $Ca_V1.2$ channels in smooth muscle cells (Bannister et al. 2009). On the other hand, the β subunit has been reported to block E3 ligase RFP2-mediated $Ca_V1.2$ poly-ubiquitination and prevent the entry of $Ca_V1.2$ channels into the endoplasmic reticulum-associated protein degradation (ERAD) complex for degradation (Altier et al. 2011). Loss of β_3 subunit has been reported to significantly reduce angiotensin II-induced hypertension in mice (Kharade et al. 2013), and $\beta_2^{-/-}$ cardiomyocytes isolated from E9.5 pups displayed markedly diminished L-type Ca^{2+} currents (Weissgerber et al. 2006). Moreover, β_{2a} subunit over-expression led to larger L-type Ca^{2+} currents and cardiac hypertrophy in mice (Chen et al. 2011). These studies on the auxiliary subunits further emphasized the essential roles of $Ca_V1.2$ channels in cardiovascular diseases. Of note, the smooth muscle-specific knockout of $Ca_V1.2$ channels blunted the pressure-induced development of myogenic tone and reduced the arterial blood pressure in mice (Moosmang et al. 2003). In cardiac diseases, gain-of-function mutations in $Ca_V1.2$ channels (G402S and G406R) is associated with Timothy syndrome featured by a long QT interval and ventricular arrhythmia (Splawski et al. 2004, 2005); loss-of-function missense mutations (A39V and G490R) of $Ca_V1.2$ channels led to Brugada syndrome characterized by a short QT interval and sudden cardiac death (Antzelevitch et al. 2007). However, for structural heart diseases both cardiac-specific α_{1C} over-expression (Song et al. 2002) and heterozygous α_{1C} knockout (Goonasekera et al. 2012) were reported to induce cardiac hypertrophy, suggesting the role of $Ca_V1.2$ channels in structural heart disorders remains to be completely understood.

Given the essential role of $Ca_V1.2$ channels in smooth and cardiac muscle contractility, exploring the ways to modulate $Ca_V1.2$ channel activity will uncover more therapeutic targets for treatment of cardiovascular diseases. The modulation of $Ca_V1.2$ channel activity by alternative splicing, post-translational modifications, and interacting proteins has been extensively discussed (Loh et al. 2020; Hu et al. 2017; Liao et al. 2005). The major thrust of the past research is to provide more nuanced understanding of the roles of $Ca_V1.2$ channels in the pathogenesis of cardiovascular diseases, such as hypertension and heart failure, and to feature the potential targets for $Ca_V1.2$ channels-based therapeutics with less adverse effects by fine-tuning $Ca_V1.2$ channel function. Here, we will first discuss two of the recently reported $Ca_V1.2$ interacting proteins and their roles in vascular and cardiac diseases; second, we will highlight the post-translational modifications of $Ca_V1.2$ channels, such as sialylation and glycosylation, in SARS-CoV-2 infection.

2 Galectin-1

2.1 Galectin-1 Overview

Galectin-1 (Gal-1) is a small protein with a molecular weight of ~14 kDa. It is encoded by the *Lgals1* gene located on chromosome 22q12 (Camby et al. 2006). Gal-1 is biologically active as a monomer and also as a non-covalent homodimer that can be spontaneously dissociated at low concentrations (Kd ~ 7 μM) into the monomeric form (Camby et al. 2006; Cho and Cummings 1995a). Gal-1 lacks the recognizable signal sequences essential for conventional secretion and does not traffic along the classical endoplasmic reticulum/Golgi pathway (Hughes 1999). However, Gal-1 exists both intracellularly, within the cytosol and nucleus, and extracellularly (Liu and Rabinovich 2005; Cho and Cummings 1995b). Gal-1 can be secreted via a non-classical secretory pathway and it has been found in the extracellular matrices of various tissues (Cooper and Barondes 1990; Nickel 2005). Possibly, the Na^+/K^+-ATPase may play an important role in the secretory process (Nickel 2005), but the detailed mechanisms are still unknown.

As the first described member of the Galectin family of carbohydrate-binding proteins with an affinity for β-galactosides (Cooper and Barondes 1999), Gal-1 plays a number of roles in physiological and pathological situations: (1) promoting HIV1 infection via stronger adhesion of the virus to host cells (Cedeno-Laurent and Dimitroff 2012; Lanteri et al. 2003; Ouellet et al. 2005) and preventing inflammation in multiple sclerosis by deactivating M1 microglia and the downstream pro-inflammatory mediators (Starossom et al. 2012; Rabinovich and Toscano 2009); (2) mediating cell transformation via interacting with oncogenic Ras and acting as a soluble mediator of tumour cells to evade immune responses (Liu and Rabinovich 2005); (3) promoting axonal regeneration via selective interaction with the NRP-1/PlexinA4 receptor complex in injured neurons and subsequent interruption of the Sema3A-mediated inhibitory signals in spinal cord injury (Quinta et al. 2014; Ishibashi et al. 2007). With an increasing number of studies of Gal-1 in the cardiovascular system, Gal-1 is also considered to be involved in the pathogenesis of vascular remodelling-related cerebral ischemia and atherosclerosis (Roldan-Montero et al. 2022; Cheng et al. 2022), hypertension (Hu et al. 2018), and myocardial infarction (Ou et al. 2022). In this book chapter, we will focus on Gal-1-mediated regulation of $Ca_V1.2$ channels in smooth and cardiac muscles.

2.2 Galectin-1-Mediated $Ca_V1.2$ Regulation in Hypertension

Gal-1 is ubiquitously expressed in many tissues at low levels, but is abundant in skeletal and smooth muscles (Dias-Baruffi et al. 2010; Camby et al. 2001; Nio-Kobayashi 2016). Gal-1 was shown to modulate attachment, spreading, migration (Cho and Cummings 1995a), and even proliferation (Moiseeva et al. 2000) via interactions with extracellular matrix proteins and $\alpha_1\beta_1$ integrin in human vascular smooth muscle cells. Gal-1 serves as an adaptor between smooth muscle cells and the extracellular matrix. Of importance, Gal-1 deficiency induced higher pulmonary blood pressure after exposure to acute hypoxia (2 min, 0% O_2) (Case et al. 2007) and higher systolic blood pressure in adult male mice (Hu et al. 2018) and pregnant female mice (Freitag et al. 2013) than wild-type mice, suggesting that Gal-1 may be involved in the regulation of blood pressure. In addition, in early onset preeclampsia patients, usually with clinically recognizable features, such as hypertension and proteinuria, reduction of Gal-1 circulating levels in serum and protein expression levels in placental villi was identified (Freitag et al. 2013), which may result in the remodelling of maternal spiral arteries and abnormal uteroplacental arterial blood flow. Moreover, decreased circulating Gal-1 was also detected in patients with subsequent onset of gestational hypertension (Ohkuchi et al. 2014). Taken together, these results suggest that down-regulation of Gal-1 may correspond to the pathogenesis of hypertension.

One primary mechanism underlying Gal-1-mediated regulation of blood pressure is Gal-1-induced negative modulation of L-type $Ca_V1.2$ calcium channel through binding to the intracellular I-II loop and promoting proteasomal degradation of the $Ca_V1.2$ channels (Fig. 1) (Hu et al. 2018; Wang et al. 2011), the major pathway of Ca^{2+} influx into vascular smooth muscles (Hu et al. 2017).

Using a human aortic cDNA library, yeast 2-hybrid assays with $Ca_V1.2$ I-II loop as the bait uncovered Gal-1 as a potential $Ca_V1.2$-binding protein (Wang et al. 2011). As validated by GST pull-down assays and site-directed mutagenesis, Gal-1 was further found to bind to the acidic DEE motif within the C-terminus of exon 9 fragment in a splice variant-selective manner as it failed to interact with $Ca_V1.2$

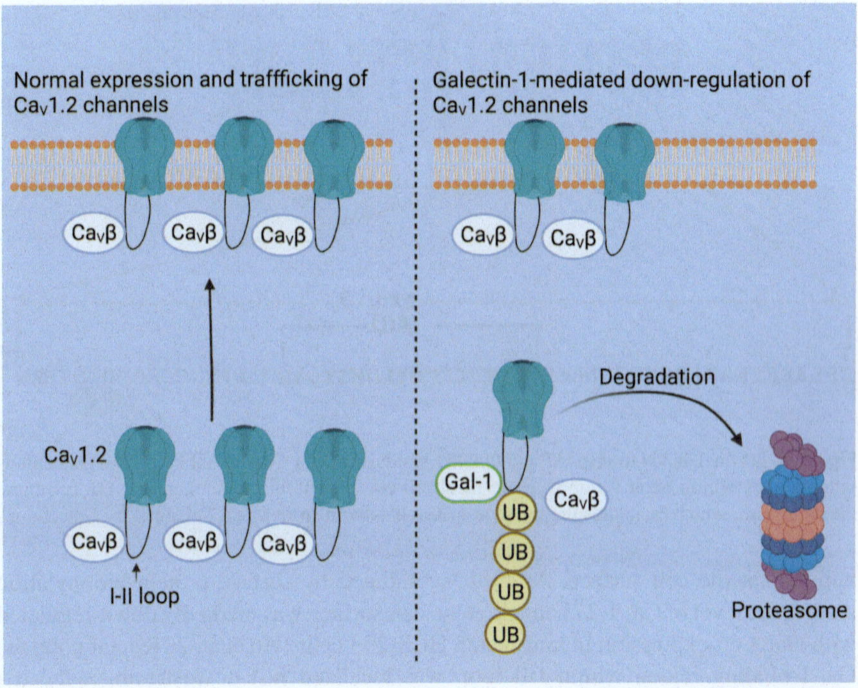

Fig. 1 Galectin-1-mediated negative regulation on $Ca_V1.2$ channels. Physiologically the $Ca_V\beta$ subunit-bound $Ca_V1.2$ channels are trafficked to cell surface as the functional gateways for Ca^{2+} influx. With binding of Galectin-1 to $Ca_V1.2$ channel, the $Ca_V\beta$ subunit is displaced, which subsequently leads to increased poly-ubiquitination and proteasomal degradation of $Ca_V1.2$ channel. Thus, Galectin-1 could be a novel target for hypertension through down-regulating $Ca_V1.2$ channel surface expression

I-II loop with exon 9*, an alternatively spliced exon downstream of exon 9 (Liao et al. 2007). The mechanisms underlying the selective binding of Gal-1 to exon 9*-exclusive I-II loop remain unclear, although in silico modelling suggested exon 9* increased the possibility of forming α-helix within the endoplasmic reticulum (ER) export signals, which was shown to be masked by Gal-1 and contributed to the reduced trafficking of $Ca_V1.2$ channels to the cell surface by Gal-1. It is noteworthy that the $Ca_V1.2$-binding sites within Gal-1 have still not identified. The specific binding residues within Gal-1 may facilitate the development of a minimal binding peptide that interacts with full-length $Ca_V1.2$ channels and mimics the function of a full-length Gal-1.

The functional studies using electrophysiological recordings showed that Gal-1 co-expression was able to significantly decrease $Ca_V1.2$ current density in transfected HEK 293 cells without affecting steady-state activation and inactivation properties (Hu et al. 2018; Wang et al. 2011). Moreover, Gal-1 also reduced the ON-gating currents representing the voltage sensor movement at reversal potential without ionic inward and outward currents, which could reflect the calcium channel

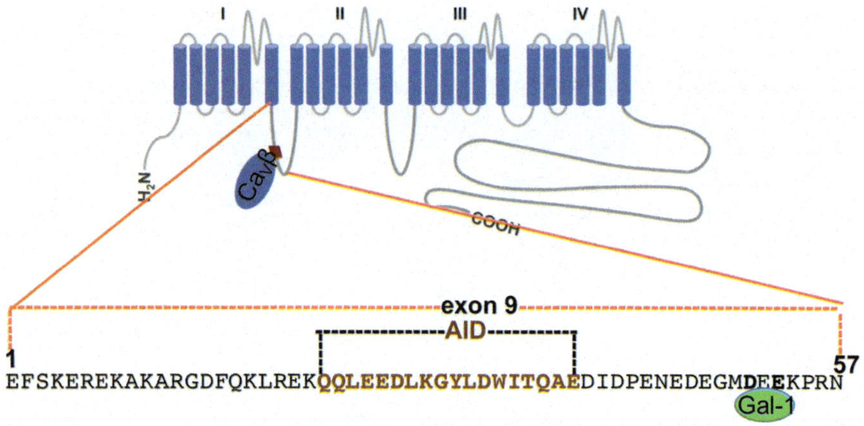

Fig. 2 Galectin-1 binds to Asp457 and Glu459 residues within $Ca_V1.2$ I-II loop. The Galectin-1-binding sites within exon 9 fragment are localized closely within the domain that $Ca_V\beta$ subunit interacts with, which facilitates the removal of $Ca_V\beta$ subunit from $Ca_V1.2$ channel by Galectin-1

numbers at the cell surface. As further validated by surface protein biotinylation assays, the level of $Ca_V1.2$ channels at the cell surface was markedly down-regulated with Gal-1 co-expression in transfected HEK 293 cells (Hu et al. 2018). Of note, the Gal-1-binding region within I-II loop was localized just downstream of the α_1-interacting domain (AID, Fig.2) that interacts with β subunit, an essential subunit that prevents the proteasomal degradation of L-type calcium channels (Altier et al. 2011). Therefore, it is reasonable to speculate that the Gal-1-mediated inhibitory effects on $Ca_V1.2$ channel function may be ascribed to the accelerated proteasomal degradation of $Ca_V1.2$ channels through interfering with $Ca_V1.2$-β interaction. This hypothesis was then substantiated by co-immunoprecipitations showing a dose-dependent reduction of β subunit bound to $Ca_V1.2$ channels by both Gal-1 co-expression in HEK 293 cells and purified Gal-1 protein (Hu et al. 2018). Subsequently, the disrupted $Ca_V1.2$-β interaction by Gal-1 exposed three lysines (K427, K435, and K460) to poly-ubiquitination, thus progressing the proteasomal degradation of $Ca_V1.2$ channels. In addition to accelerated degradation, Gal-1 also masked the ER export signals within exon 9 fragment of I-II loop (Hu et al. 2018), which contributed to the decreased level of $Ca_V1.2$ channels at the cell surface.

Knowing the molecular mechanisms that underlie the inhibitory regulation on $Ca_V1.2$ channels by Gal-1 strengthened the basis that Gal-1 would play a key role in blood pressure regulation in vivo. Indeed, by using the non-invasive tail-cuff measurement, the adult Gal-1 knockout mice were shown to have a significantly higher systolic blood pressure than wild-type mice (Fig.3) (Hu et al. 2018). This suggests that $Ca_V1.2$-Gal-1 protein interaction could be a novel target for hypertension treatment. Firstly, given the identified Gal-1-binding sites within the C-terminus of $Ca_V1.2$ exon 9 fragment, a short peptide Tat-e9c (**YGRKKRRQRRR**DIDPENEDEGMDEEKPRN) was rationally designed to disrupt the interactions between $Ca_V1.2$ channels and Gal-1, which increased the

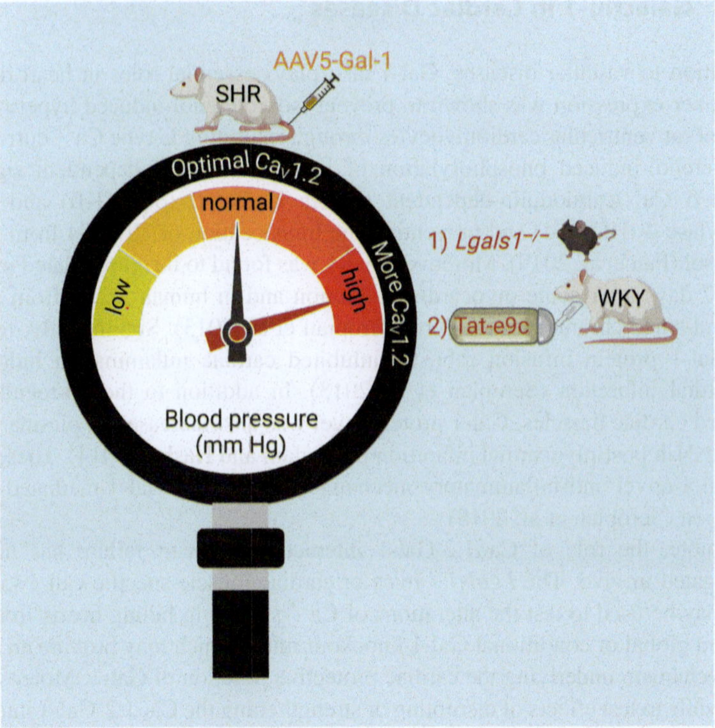

Fig. 3 Regulation of blood pressure by targeting the $Ca_V1.2$-Galectin-1 interaction in vivo. As measured by the non-invasive tail-cuff method, loss of Gal-1 or infusion of Tat-e9c peptide disrupting the $Ca_V1.2$-Galectin-1 interaction increases the blood pressure (BP) by up-regulating $Ca_V1.2$ channel surface expression in mice and WKY rats, respectively. In contrast, over-expression of smooth muscle Galectin-1 by injecting AAV5-Galectin-1 virus strongly reduces the blood pressure by down-regulating $Ca_V1.2$ channel surface expression in spontaneously hypertensive rats (SHR)

arterial contractility and myogenic tone of mesenteric arteries. Moreover, with infusion into jugular vein of WKY rats, it significantly increased the systolic blood pressure within 1 week (Fig.3) (Hu et al. 2018). These findings indicate that interfering with $Ca_V1.2$-Gal-1 interaction is able to increase the blood pressure in vivo. By contrast, AAV5-Gal-1 virus with a chimeric smooth muscle-specific enhancer/promoter (EnSM22α) was designed to selectively express exogenous Gal-1 in smooth muscles, which markedly reduced the blood pressure of spontaneously hypertensive rats for at least 1 month (Fig.3) (Hu et al. 2018).

In conclusion, Gal-1 has emerged as a novel target for hypertension treatment through promoting the proteasomal degradation of $Ca_V1.2$ channels and thereby fine-tuning the channel function. As Gal-1 selectively binds to $Ca_V1.2$ channel, and not the other three L-type calcium channels (Hu et al. 2018), this selectivity and associated fine-tuning of $Ca_V1.2$ channel activity will enable $Ca_V1.2$-Gal-1 interaction-targeted therapeutics to have less side effects.

2.3 Galectin-1 in Cardiac Diseases

In addition to vascular diseases, Gal-1 also plays essential roles in heart diseases. Gal-1 over-expression was shown to prevent isoproterenol-induced hypertrophy in neonatal rat ventricular cardiomyocytes through inhibiting L-type Ca^{2+} currents and isoproterenol-induced phosphorylation of intracellular Ca^{2+}-dependent signalling protein δ Ca^{2+}/calmodulin-dependent protein kinase II (δCaMKII) and histone deacetylase 4 (HDAC4) and preventing the translocation of HDAC4 from nucleus to cytosol (Fan et al. 2019). Moreover, Gal-1 was found to be up-regulated in mouse hearts 7 days after acute myocardial infarction and in human hearts from patients with end-stage chronic heart failure (Seropian et al. 2013). Significantly, recombinant Gal-1 protein infusion robustly inhibited cardiac inflammation induced by myocardial infarction (Seropian et al. 2013). In addition to the up-regulation in infarcted cardiac muscles, Gal-1 protein level was also increased in plasma of mice 4-h and 24-h post-myocardial infarction (Al-Salam and Hashmi 2014). These results indicate a novel anti-inflammatory mechanism underlying Gal-1-mediated cardiac protection (Seropian et al. 2018).

Of note, the role of $Ca_V1.2$-Gal-1 interaction in heart failure has not been investigated in vivo. The *Lgals1*$^{-/-}$ mice or cardiac muscle-specific Gal-1 knockout mice may be used to test the alterations of Ca^{2+} signals in failing hearts from wild-type and global or conditional Gal-1 knockout mice, which may provide an alternative mechanism underlying the cardiac protective function of Gal-1. Moreover, it is worthwhile to test effects of disrupting or strengthening the $Ca_V1.2$-Gal-1 interaction in cardiac function, which may enable the exploration and discovery of new therapeutics for heart diseases, coupled with less side effects than using the full-length Gal-1 protein that has multiple functions in various organs. The $Ca_V1.2$-binding sites within Gal-1 could be essential for the development of a therapeutic peptide that selectively targets $Ca_V1.2$-Gal-1 interaction in treatment of cardiovascular diseases.

2.4 Regulatory Signalling of Galectin-1

Given the important roles of Gal-1 in hypertension and myocardial infarction, it is essential to understand the signalling pathways up-stream of Gal-1 expression.

Hypoxia-inducible factor-1α (HIF-1α) was reported to up-regulate the mRNA and protein levels of Gal-1 in colorectal cancer cells (Zhao et al. 2010) with two hypoxia-responsive elements (HRE) located at −441 to −423 bp in the promoter of *Lgals1* gene. In addition, CCAAT/enhancer binding protein α (C/EBPα) was also reported to activate Gal-1 by binding to the −48 to −42 bp region of *Lgals1* promoter. Moreover, the chromatin immunoprecipitation (ChIP) assay further validated that the transcriptional activity of C/EBPα and HIF-1α on Gal-1 transcripts was synergistic (Zhao et al. 2011). Of note, controversy exists as to how HIF-1α plays in pulmonary hypertension. Although a few studies showed that increased HIF-1α expression level was associated with hypoxia-induced pulmonary arterial hypertension (Johns et al. 2016; Li et al. 2016; Jiang et al. 2015), HIF-1α was also down-regulated in isolated pulmonary smooth muscle cells from patients with idiopathic pulmonary hypertension (Barnes et al. 2017) or in pulmonary arteries from patients with pulmonary hypertension (Hu et al. 2018). Moreover, loss of smooth muscle HIF-1α increased both pulmonary arterial pressure (Kim et al. 2013) and systolic blood pressure (Huang et al. 2013) through up-regulating phosphorylation of myosin light chain and down-regulating peroxisome proliferator-activated receptor-γ-angiotensin II receptor type 1 axis, respectively. These findings are in line with the down-regulation of both HIF-1α and Gal-1 in hypertensive human pulmonary arteries (Hu et al. 2018), suggesting HIF-1α-mediated Gal-1 up-regulation in vivo. It is noteworthy that the specific roles of HIF-1α in different subtypes of pulmonary hypertension remain to be determined as, in addition to chronic hypoxia, many other factors were also reported to regulate HIF-1α expression or transcriptional activity, such as loss-of-function mutations of the von Hippel-Lindau tumour suppressor (VHL), a substrate recognition component of an E3 ubiquitin ligase, and a few vasomodulatory factors represented by nitric oxide and Endothelin 1 (Pullamsetti et al. 2020).

Gal-1 was shown to be highly expressed in kidneys of diabetic mouse, which was mainly regulated by activating enhancer binding protein 4 (AP4), a major transcription factor that activates Gal-1 by interacting with both the *Lgals1* promoter and Gal-1 protein under hyperglycaemia (Al-Obaidi et al. 2019). In addition, the transforming growth factor-β1 (TGF-β1) was reported to increase Gal-1 expression via activation of PI3K/Akt and p38 MAPK in human foetal lung fibroblasts (Lim et al. 2014). More importantly, TGF-β1 was significantly reduced in lung tissues and pulmonary arterial smooth muscle cells isolated from monocrotaline-induced rodent model of pulmonary arterial hypertension (Zakrzewicz et al. 2007). However, TGF-β1 was shown to be markedly up-regulated at the protein and mRNA levels in serums of African-American hypertensives (Suthanthiran et al. 2000) and in plasma from patients with essential hypertension (Derhaschnig et al. 2002). More studies will be required to further investigate whether reduced TGF-β1 in smooth muscles is correlated to hypertension development and whether increased TGF-β1 in serum can function as a biomarker for hypertension.

In summary, with Gal-1 as a key player in the pathogenesis of hypertension and myocardial infarction, understanding the regulation of Gal-1 by HIF-1α, C/EBPα, AP4, or TGF-β1 in smooth muscles and cardiac muscles may provide more alternatives for Galectin-targeted therapeutics for cardiovascular diseases.

2.5 Other Galectins

Including Gal-1, there are now a total of 15 mammalian galectin isoforms described, of which 12 have been detected in humans. All galectins are encoded by their respective *Lgals* gene and are identified by their carbohydrate-recognition domain (CRD). Because galectins are ubiquitously expressed in many tissues and cell types, these proteins are key players in regulating important cell signalling processes (Johannes et al. 2018).

Together with Gal-1, Gal-2, −5, -7, -10, -11, -13, -14, and -15 are categorized as prototype galectins because of similar structural characteristics (Nio-Kobayashi 2016). Here, each galectin monomer contains only the ~130 amino acid-long CRD. These galectins commonly dimerize. Similarly, Gal-4, -6, -8, -9, and -12 are categorized as tandem repeat galectins for expressing two distinct yet homologous CRDs joined by a short peptide linker within a monomer (Nio-Kobayashi 2016). Gal-3, on the other hand, is the only chimeric galectin and it expresses only one CRD tethered to an unusually long N-terminal tail consisting of repeat domains rich in proline and glycine residues (Nio-Kobayashi 2016; Suthahar et al. 2018). Furthermore, Gal-3 is the only galectin that is able to form an oligomer. However, the exact function of the Gal-3 oligomer remains to be elucidated.

The CRD is an important binding domain within galectins by which galectins are functionally active. As the name suggests, the CRD is responsible for the interaction between galectin and β-galactoside, as well as its derivatives (Rapoport et al. 2008). Through X-ray crystallography, crystal structures of a galectin CRDs revealed the presence of two globularly folded anti-parallel β-sheet sandwiches – five (F-face) and six (S-face) β-strands (Rini and Lobsanov 1999). It was later identified that the actual amino acids responsible for its carbohydrate-binding affinity were located within the S-face (Johannes et al. 2018). Thus, the S-face was designated the canonical binding site of galectins. Despite the conservation of the CRD between galectins, differences in amino acid sequences in parts of the CRD allow for substrate-specificity for each galectin, thereby preventing redundancy. Therefore, each galectin contributes to specific pathways for important biological processes.

2.5.1 Galectin Localization and Functions

Important to regulating galectins and their function is their subcellular localization. For example, Gal-3 has been detected in not just the nucleus and cytoplasm, but also in the extracellular matrix (Haudek et al. 2010; Ochieng et al. 2002). Moreover, the function of Gal-3 was reported to differ based on its subcellular localization relative to the cell. Therefore, this further increases the complexity of understanding the biology of galectin and their specific roles and functions in the various biological processes. A brief summary of this complexity using Gal-3 is provided below.

Gal-3 is one of the most widely studied galectin isoforms. As mentioned above, Gal-3 has been detected in the nucleus, cytoplasm, and extracellular matrix of the cell. Moreover, Gal-3 was reported to modulate different functions according to its subcellular localization. Within the nuclear structures and extracts, the expression of Gal-3, together with Gal-1, was observed in 3T3 mouse fibroblast and HeLa cell lines (Park et al. 2001; McLeod et al. 2018). In addition, within nuclear speckles, Gal-3 was observed to colocalize with known splicing factors for pre-mRNAs such as small nuclear ribonucleoproteins (snRNPs). Interestingly, when Gal-1/3 was removed, splicing activity was absent. The reintroduction of either Gal-1/3 to the nuclear extracts rescued this phenotype. Accordingly, galectins were later reported to be essential for the assembly of spliceosomes, whereby the absence of galectins resulted in diminished pre-mRNA splicing activity. Yet, it is interesting to note that in transgenic mouse models lacking either Gal-1 or -3, no significant changes in splicing activity and observable phenotypes were reported. This suggests that while Gal-1/3 may be important to regulate pre-mRNA splicing within the nuclear extracts, both could be redundant under physiological conditions. Nevertheless, it can be said that within the nuclear microenvironment, Gal-1/3 may provide some form of relief for chaperoning proper pre-mRNA splicing activities.

Gal-3 is also detected within the cytoplasm and may be involved in cellular apoptosis (Harazono et al. 2014; Yang et al. 1996). It was reported that under starvation condition, Gal-3 localizes to the cytoplasm in 3T3 mouse fibroblasts. Interestingly, a highly conserved NWGR motif has been found within the CRD of Gal-3 (Takenaka et al. 2004). This motif is homologous to the Bcl-2-homology (BH) domains of anti-apoptotic B-cell lymphoma (Bcl-2) protein family, therefore suggesting that Gal-3 may be involved in anti-apoptosis. Indeed, when treated with cisplatin, Gal-3 was detected at mitochondrial membranes following the release of cytochrome C. Furthermore, G > A substitution within NWGR of Gal-3 CRD resulted in an increase in apoptosis (Akahani et al. 1997). On the other hand, when stimulated with serum, Gal-3 was observed to translocate to the nucleus, possibly for pre-mRNA splicing activity. Therefore, these results strongly support the idea that subcellular localization is an important regulator of the biological activity of galectins.

Within the extracellular matrix, Gal-3 has been reported to modulate various inflammatory processes (Henderson and Sethi 2009). In skin healing, Gal-3 functions not just to recruit neutrophils, but also to activate them for the elimination of foreign debris and bacteria during the initial stages of inflammation. Subsequently, Gal-3 then facilitates the clearing of neutrophils by promoting the endocytosis of neutrophils by activated macrophages. Lastly, Gal-3 induces the alternative anti-inflammatory M2-polarized macrophage differentiation (MacKinnon et al. 2008). In $Gal\text{-}3^{-/-}$ mice, the absence of Gal-3 abrogated the M2 polarization of macrophages despite the presence of other inflammatory cytokines such as TNFα and interleukins (IL). From here, it can be seen extracellular Gal-3 functionally differs from that of nuclear and cytoplasmic Gal-3. However, it should also be noted that within the extracellular matrix itself, Gal-3 may be temporally regulated to modulate various processes during the different stages of inflammation.

Taken together, galectin biology is highly complex and is still poorly understood. While Gal-3 may function differently in various subcellular domains, the exact conditions for which Gal-3 is functionally active remain largely unknown. Could the starvation status of the cell affect Gal-3 nuclear translocation? Could Gal-1/3 help rescue pre-mRNA splicing activity in cultured cell lines lacking true physiological conditions and cellular interactions? Could the dysregulation of galectins expression and their activity be the underlying cause of known disease conditions? The recent studies have only begun to expound the importance of galectins in modulating various biological processes. Further work is required to better understand the regulation of galectins under specific disease conditions and cellular microenvironments.

2.5.2 Galectin-3 in Cardiovascular Diseases

Owing to the wide range of biological functions that galectins have been found to modulate, there is now greater recognition and appreciation for galectins in various diseases, notably cardiovascular diseases (CVD) (Hara et al. 2020; Besler et al. 2017). In the genome-wide association study (GWAS) involving the Prevention of Renal and Vascular End-stage Disease (PREVEND) Cohort, various single nucleotide polymorphisms (SNPs) of Gal-3 were identified by de Boer et al. in 2012 (de Boer et al. 2012). Of note, rs4644 and rs4652 were later further reported to be strongly associated with dilated cardiomyopathy in a group of Northern Han Chinese (Zhang et al. 2018). Apart from this, numerous reports have independently associated high levels of serum Gal-3 with CVD, largely through promoting fibrosis and tissue remodelling (McLeod et al. 2018; Calvier et al. 2016; Ho et al. 2012). This is consistent with studies reporting Gal-3 up-regulation in decompensated heart failure but not in compensated heart failure, thereby suggesting that Gal-3 does play a role in cardiac remodelling (Sharma et al. 2004). However, the underlying molecular mechanism still remains largely unknown. Nevertheless, Gal-3 is increasingly recognized as a potential serum biomarker for CVD such as heart failure, and further studies are ongoing to identify the molecular mechanisms through which Gal-3 may contribute to the progression of CVD.

3 Cereblon

Cereblon is a protein with 442 amino acids encoded by the *CRBN* gene containing 11 exons on chromosome 3 (Thakurta et al. 2014). It is expressed in many organs, such as brain, heart, and liver, and is mainly localized in the cytosol, plasma membrane, and nucleus (Kim et al. 2016). Cereblon is a substrate receptor of the Cullin-Ring ligase 4 E3 ubiquitin ligase complex that recognizes substrates for ubiquitination and degradation via ubiquitin-protease system (Sato et al. 2021). Cereblon was identified as the common factor contributing to the therapeutic effects of Thalidomide and its analogues, such as Lenalidomide and Pomalidomide, for the treatment of multiple myeloma by degrading casein kinase 1α (CK1α) and two C2H2 Zn-finger transcriptional factors, Ikaros (IKZF1) and Aiolos (IKZF3) (Chamberlain et al. 2019). In addition to anti-cancer activity, Cereblon also plays important roles in cardiac function as loss of Cereblon significantly reduced the infarcted area and fibrosis in left ventricles of mice subjected to myocardial infarction by negatively regulating AMP-activated protein kinase activity (Kim et al. 2014). Moreover, recently Cereblon was reported to bind to the N-terminus of $Ca_V1.2$ channels and enhance their ubiquitination and proteasomal degradation (Park et al. 2022). Loss of Cereblon significantly increased L-type calcium currents in isolated mouse cardiomyocytes, which may contribute to the enhanced cardiac contractility in global and cardiac-specific Cereblon knockout mice (Park et al. 2022). In contrast, intramyocardial administration of Cereblon adenovirus vector in the Cereblon knockout mice significantly reduced the ejection fraction by reducing Ca^{2+} currents in cardiomyocytes. These findings further substantiated the key roles of Ca^{2+} currents in cardiac function. As Galectin-1 has a similar function as compared to Cereblon in terms of promoting $Ca_V1.2$ channel degradation, it will be compelling to further investigate the changes of Ca^{2+} currents and cardiac contractility in both global and cardiac-specific Galectin-1 knockout mice.

4 $Ca_V1.2$, Sialylation, and COVID-19

Less widely known are the involvements of $Ca_V1.2$ channels in viral infection. In 1983, it was first reported that verapamil, a calcium channel blocker (CCB), was able to inhibit influenza virus replication in Madin-Darby canine kidney (MDCK) cells (Nugent and Shanley 1984). A second drug, chlorpromazine, was also found to exhibit the same anti-viral effects in cells. Together, Nugent and Shanley suggested that influenza virus replication can be inhibited by interfering with calmodulin-dependent intracellular activities, thereby affecting Ca^{2+} flux across the plasma membrane. Further studies by other groups revealed the importance of calcium channels and Ca^{2+} signalling in virus replication (Fujioka et al. 2013; Crespi and Alcock 2020). However, the mechanism by which calcium channels are involved remained unknown.

In 2018, a published study revealed the importance of sialylation on voltage-gated calcium channels in mediating virus entry and infection of mammalian cells (Fujioka et al. 2018). In their study, Fujioka et al. reported the hemagglutinin of influenza A virus (IAV) binds to $Ca_V1.2$, and the knockdown of $Ca_V1.2$ prevented IAV internalization. Next, the group identified asparagine residues at positions 1,436 and 1,487 to be critical for the association between $Ca_V1.2$ and IAV hemagglutinin, and IAV infection. Important to this function of $Ca_V1.2$ is the sialylation of the N1436 and N1487 residues. Indeed, the substitutions of asparagine to glutamine residues at N1436Q and N1487Q attenuated the binding between $Ca_V1.2$ and IAV hemagglutinin. While it is acknowledged that glycosylation is important for regulating protein folding and structure, it was noted that $Ca_V1.2_{N1436Q/N1487Q}$ showed similar levels of plasma membrane translocation as wild-type $Ca_V1.2$. Thus, this suggests that the reduced IAV infection levels were not due to a decrease in surface expression of $Ca_V1.2$ channels, but rather suggests the importance of glycosylation of asparagine residues in $Ca_V1.2$ in mediating IAV binding and entry. Going a step further, Fujioka et al. also reported that the CCB diltiazem was able to inhibit IAV infection both in vivo and ex vivo. The delivery of diltiazem into the nasal cavity of anesthetized mice significantly reduced the viral load within the nasal lavage fluid. This reduced viral load was also observed in human respiratory epithelial cell line BEAS-2B. Therefore, these results strongly suggest that sialylation of $Ca_V1.2$ asparagine residues is important to facilitate viral binding and entry and that the inhibition of $Ca_V1.2$ function is sufficient to abrogate virus infection.

In a more recent study, diltiazem was reported to inhibit SARS-CoV-2 binding and internalization in mouse lung (Wang et al. 2022). SARS-CoV-2 is the virus responsible for the ongoing COVID-19 pandemic. The lack of efficacious treatment methods coupled with the poor understanding of SARS-CoV-2 biology resulted in the loss of lives of many across the world. In this study, like in the IAV study, the authors reported that SARS-CoV-2 spike protein associated with $Ca_V1.2$ channel. To elucidate the importance of $Ca_V1.2$ in facilitating SARS-CoV-2 entry, RNA interference (RNAi) was employed to knock down $Ca_V1.2$ expression in Vero-E6 cells. *CACNA1C*-silenced cells yielded decreased viral titres as compared to control. Moreover, the use of cell lines of various origins showed that SARS-CoV-2 infection via $Ca_V1.2$ is largely specific towards the respiratory cell type (Vero-E6 and A549) as compared to HEK 293 cells. Similarly, the CCB diltiazem-treated cells also yielded significantly lower levels of viral RNA levels as compared to control cells. Furthermore, in diltiazem-treated mice, SARS-CoV-2 replication was observed to be significantly reduced in lung tissues, resulting in increased survivability. Results from this study supported that of an earlier study reporting the therapeutic potential of CCBs in COVID-19 treatment (Straus et al. 2021). Therefore, $Ca_V1.2$ represents a novel transmembrane protein that may be targeted for the treatment of SARS-CoV-2-infected patients.

Given that glycosylation of asparagine residues that are important to facilitate viral binding and entry are located within the extracellular portion of $Ca_V1.2$, this raises the question of whether galectins could be intermediary modulators of $Ca_V1.2$ and virus association. Like $Ca_V1.2$, various compartments of the viral protein such as the spike and envelope proteins are highly glycosylated (Gong et al. 2021; Ramírez Hernández et al. 2021; Petrović et al. 2021; Watanabe et al. 2020). Because of their innate ability to bind β-galactoside and its derivatives (Johannes et al. 2018), it is highly likely that galectins are involved in the binding and localization of viruses to cell surfaces. Should this occur, viruses will be brought into closer contact with $Ca_V1.2$, thereby resulting in increased opportunities for virus association and entry via $Ca_V1.2$ channels. Nevertheless, it should be noted that this is largely hypothetical and that further studies are needed to test this hypothesis.

In support of this hypothesis, intriguingly, is that Gal-3 is now proposed to be a potential biomarker and target for the treatment of COVID-19 (Cervantes-Alvarez et al. 2022; Caniglia et al. 2020). It has been reported that there is a high degree of homology between Gal-3 and the key domain of SARS-CoV-2S1 spike protein (Li 2016). As a result, anti-Gal-3 therapeutics are now being tested for the treatment of COVID-19 (Caniglia et al. 2020). Here, the authors proposed that Gal-3 inhibitors such as belapectin (GR-MD-02), and anti-Gal-3 antibody (TD139), could potentially bind to the homologous domain within the SARS-CoV-2S1 spike protein to prevent interaction with cell surface receptors, thereby reducing the likelihood of cell entry (Fig. 4). In addition, a second mechanism of action has been suggested. Because Gal-3 is known to modulate pro-inflammatory effects, Gal-3 inhibitors could potentially reduce Gal-3-mediated inflammation. Indeed, in a spinal cord injury model, Gal-3 inhibitors reduced pro-inflammatory cytokines such as TNFα and IL-6. However, it should be noted that such inhibitors are pan-Gal-3 inhibitors. In other words, given the ubiquitous expression and pleiomorphic effects of Gal-3, systemic inhibition of Gal-3 can potentially cause many side effects and thus be poorly tolerated by patients. Nonetheless, the aforementioned Gal-3 inhibitors are presently

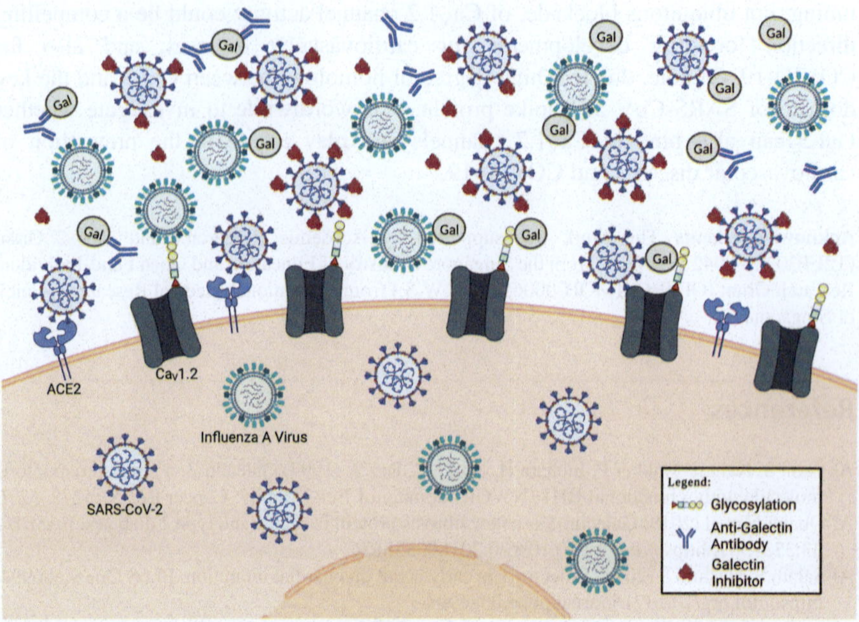

Fig. 4 Anti-galectin therapeutics for treatment of $Ca_V1.2$-mediated viral infection. Glycosylation present on the extracellular portion of $Ca_V1.2$ and virus spike proteins may facilitate galectin-mediated virus-binding and entry into the host cells. It is hypothesized that anti-galectin therapeutics may be administered to prevent virus-receptor binding, thereby reducing the likelihood of virus entry

under clinical trials with initial reports suggesting that the drugs are safe and well-tolerated. Therefore, although largely promising, it remains to be determined if anti-Gal-3 therapeutics are indeed a safe alternative for treatment of viral infections.

5 Concluding Remarks

$Ca_V1.2$ calcium channels have been extensively studied in the cardiovascular system as a primary pathway for Ca^{2+} influx to trigger muscle contraction. More importantly, research on $Ca_V1.2$ channel has been extended to COVID-19 disease as $Ca_V1.2$ channels directly interact with SARS-CoV-2 spike protein and facilitate virus entry. Thus, exploring novel ways to fine-tune $Ca_V1.2$ channel function could provide potential therapeutics for hypertension, heart failure, and even COVID-19 with less side effects compared to the complete blockade by calcium channel blockers. In this chapter, Galectin-1 and Cereblon binding to $Ca_V1.2$ I-II loop and N-terminus, respectively, are highlighted as negative modulators by promoting the channel degradation via the ubiquitin-proteasomal system. In addition, targeting Galectin-1 or Cereblon has been shown to protect against hypertension and heart failure, respectively, which further enlightens the concept that a selective fine-tuning, not ubiquitous blockade, of $Ca_V1.2$ channel activity could be a compelling direction for drug developments for cardiovascular diseases, and also for COVID-19. Of note, due to a high degree of homology between Gal-3 and the key domain of SARS-CoV-2S1 spike protein, it is worthwhile to investigate whether Gal-3 can also bind to $Ca_V1.2$ channels and play a role in the prevention of cardiovascular diseases and COVID-19.

Acknowledgements This work was supported by Academic Research Fund Tier 2 Grant (T2EP30221-0042 to T.W.S.) from the Singapore Ministry of Education and Open Fund-Individual Research Grant (OF-IRG) (MOH-000650 to T.W.S.) from the National Medical Research Council of Singapore.

References

Akahani S, Nangia-Makker P, Inohara H, Kim HR, Raz A (1997) Galectin-3: a novel antiapoptotic molecule with a functional BH1 (NWGR) domain of Bcl-2 family. Cancer Res 57:5272–5276

Al-Obaidi N et al (2019) Galectin-1 is a new fibrosis protein in type 1 and type 2 diabetes. FASEB J 33:373–387. https://doi.org/10.1096/fj.201800555RR

Al-Salam S, Hashmi S (2014) Galectin-1 in early acute myocardial infarction. PLoS One 9:e86994. https://doi.org/10.1371/journal.pone.0086994

Altier C et al (2011) The Cavbeta subunit prevents RFP2-mediated ubiquitination and proteasomal degradation of L-type channels. Nat Neurosci 14:173–180. https://doi.org/10.1038/nn.2712

Antzelevitch C et al (2007) Loss-of-function mutations in the cardiac calcium channel underlie a new clinical entity characterized by ST-segment elevation, short QT intervals, and sudden cardiac death. Circulation 115:442–449. https://doi.org/10.1161/CIRCULATIONAHA.106.668392

Bannister JP et al (2009) Smooth muscle cell alpha2delta-1 subunits are essential for vasoregulation by CaV1.2 channels. Circ Res 105:948–955. https://doi.org/10.1161/CIRCRESAHA.109.203620

Barnes EA, Chen CH, Sedan O, Cornfield DN (2017) Loss of smooth muscle cell hypoxia inducible factor-1alpha underlies increased vascular contractility in pulmonary hypertension. FASEB J 31:650–662. https://doi.org/10.1096/fj.201600557R

Besler C et al (2017) Plasma and cardiac Galectin-3 in patients with heart failure reflects both inflammation and fibrosis. Circ Heart Fail 10:e003804. https://doi.org/10.1161/CIRCHEARTFAILURE.116.003804

Calvier L et al (2016) Galectin-3 and aldosterone as potential tandem biomarkers in pulmonary arterial hypertension. Heart 102:390–396. https://doi.org/10.1136/heartjnl-2015-308365

Camby I et al (2001) Galectins are differentially expressed in supratentorial pilocytic astrocytomas, astrocytomas, anaplastic astrocytomas and glioblastomas, and significantly modulate tumor astrocyte migration. Brain Pathol 11:12–26

Camby I, Le Mercier M, Lefranc F, Kiss R (2006) Galectin-1: a small protein with major functions. Glycobiology 16:137R–157R., cwl025 [pii]. https://doi.org/10.1093/glycob/cwl025

Caniglia JL, Guda MR, Asuthkar S, Tsung AJ, Velpula KK (2020) A potential role for Galectin-3 inhibitors in the treatment of COVID-19. PeerJ 8:e9392. https://doi.org/10.7717/peerj.9392

Case D et al (2007) Mice deficient in galectin-1 exhibit attenuated physiological responses to chronic hypoxia-induced pulmonary hypertension. Am J Physiol Lung Cell Mol Physiol 292:L154–L164., 00192.2006 [pii]. https://doi.org/10.1152/ajplung.00192.2006

Cedeno-Laurent F, Dimitroff CJ (2012) Galectin-1 research in T cell immunity: past, present and future. Clin Immunol 142:107–116., S1521-6616(11)00280-4 [pii]. https://doi.org/10.1016/j.clim.2011.09.011

Cervantes-Alvarez E et al (2022) Galectin-3 as a potential prognostic biomarker of severe COVID-19 in SARS-CoV-2 infected patients. Sci Rep 12:1856. https://doi.org/10.1038/s41598-022-05968-4

Chamberlain PP et al (2019) Evolution of cereblon-mediated protein degradation as a therapeutic modality. ACS Med Chem Lett 10:1592–1602. https://doi.org/10.1021/acsmedchemlett.9b00425

Chen X et al (2011) Calcium influx through Cav1.2 is a proximal signal for pathological cardiomyocyte hypertrophy. J Mol Cell Cardiol 50:460–470. https://doi.org/10.1016/j.yjmcc.2010.11.012

Cheng YH et al (2022) Galectin-1 contributes to vascular remodeling and blood flow recovery after cerebral ischemia in mice. Transl Stroke Res 13:160–170. https://doi.org/10.1007/s12975-021-00913-5

Cho M, Cummings RD (1995a) Galectin-1, a beta-galactoside-binding lectin in Chinese hamster ovary cells. I. Physical and chemical characterization. J Biol Chem 270:5198–5206

Cho M, Cummings RD (1995b) Galectin-1, a beta-galactoside-binding lectin in Chinese hamster ovary cells. II. Localization and biosynthesis. J Biol Chem 270:5207–5212

Cooper DN, Barondes SH (1990) Evidence for export of a muscle lectin from cytosol to extracellular matrix and for a novel secretory mechanism. J Cell Biol 110:1681–1691

Cooper DN, Barondes SH (1999) God must love galectins; he made so many of them. Glycobiology 9:979–984., cwc107 [pii]

Crespi B, Alcock J (2020) Conflicts over calcium and the treatment of COVID-19. Evol Med Public Health 9:149–156. https://doi.org/10.1093/emph/eoaa046

de Boer RA et al (2012) A genome-wide association study of circulating Galectin-3. PLoS One 7: e47385. https://doi.org/10.1371/journal.pone.0047385

Derhaschnig U et al (2002) Increased levels of transforming growth factor-beta1 in essential hypertension. Am J Hypertens 15:207–211., S0895706101023275 [pii]

Dias-Baruffi M et al (2010) Differential expression of immunomodulatory galectin-1 in peripheral leukocytes and adult tissues and its cytosolic organization in striated muscle. Glycobiology 20: 507–520., cwp203 [pii]. https://doi.org/10.1093/glycob/cwp203

Fan J et al (2019) Galectin-1 attenuates cardiomyocyte hypertrophy through splice-variant specific modulation of CaV1.2 calcium channel. Biochim Biophys Acta Mol Basis Dis 1865:218–229. https://doi.org/10.1016/j.bbadis.2018.08.016

Freitag N et al (2013) Interfering with Gal-1-mediated angiogenesis contributes to the pathogenesis of preeclampsia. Proc Natl Acad Sci U S A 110:11451–11456., 1303707110 [pii]. https://doi.org/10.1073/pnas.1303707110

Fujioka Y et al (2013) A Ca2+-dependent signalling circuit regulates influenza a virus internalization and infection. Nat Commun 4:2763. https://doi.org/10.1038/ncomms3763

Fujioka Y et al (2018) A sialylated voltage-dependent Ca(2+) channel binds hemagglutinin and mediates influenza a virus entry into mammalian cells. Cell Host Microbe 23:809-818 e805. https://doi.org/10.1016/j.chom.2018.04.015

Gong Y, Qin S, Dai L, Tian Z (2021) The glycosylation in SARS-CoV-2 and its receptor ACE2. Signal Transduct Target Ther 6:396. https://doi.org/10.1038/s41392-021-00809-8

Goonasekera SA et al (2012) Decreased cardiac L-type Ca(2)(+) channel activity induces hypertrophy and heart failure in mice. J Clin Invest 122:280–290. https://doi.org/10.1172/JCI58227

Hara A et al (2020) Galectin-3: a potential prognostic and diagnostic marker for heart disease and detection of early stage pathology. Biomol Ther 10:1277. https://doi.org/10.3390/biom10091277

Harazono Y et al (2014) Galectin-3 leads to attenuation of apoptosis through Bax heterodimerization in human thyroid carcinoma cells. Oncotarget 5:9992–10001. https://doi.org/10.18632/oncotarget.2486

Haudek KC et al (2010) Dynamics of galectin-3 in the nucleus and cytoplasm. Biochim Biophys Acta Gen Subj 1800:181–189. https://doi.org/10.1016/j.bbagen.2009.07.005

Henderson NC, Sethi T (2009) The regulation of inflammation by galectin-3. Immunol Rev 230: 160–171. https://doi.org/10.1111/j.1600-065X.2009.00794.x

Ho JE et al (2012) Galectin-3, a marker of cardiac fibrosis, predicts incident heart failure in the community. J Am Coll Cardiol 60:1249–1256. https://doi.org/10.1016/j.jacc.2012.04.053

Hofmann F, Flockerzi V, Kahl S, Wegener JW (2014) L-type CaV1.2 calcium channels: from in vitro findings to in vivo function. Physiol Rev 94:303–326. https://doi.org/10.1152/physrev.00016.2013

Hu Z, Liang MC, Soong TW (2017) Alternative splicing of L-type CaV1.2 calcium channels: implications in cardiovascular diseases. Genes (Basel) 8. https://doi.org/10.3390/genes8120344

Hu Z et al (2018) Regulation of blood pressure by targeting CaV1.2-Galectin-1 protein interaction. Circulation 138:1431–1445. https://doi.org/10.1161/CIRCULATIONAHA.117.031231

Huang Y et al (2013) Hypoxia-inducible factor-1alpha in vascular smooth muscle regulates blood pressure homeostasis through a peroxisome proliferator-activated receptor-gamma-angiotensin II receptor type 1 axis. Hypertension 62:634–640. https://doi.org/10.1161/HYPERTENSIONAHA.111.00160

Hughes RC (1999) Secretion of the galectin family of mammalian carbohydrate-binding proteins. Biochim Biophys Acta 1473:172–185., S0304-4165(99)00177-4 [pii]

Ishibashi S et al (2007) Galectin-1 regulates neurogenesis in the subventricular zone and promotes functional recovery after stroke. Exp Neurol 207:302–313., S0014-4886(07)00260-9 [pii]. https://doi.org/10.1016/j.expneurol.2007.06.024

Jiang Y et al (2015) Increased SUMO-1 expression in response to hypoxia: interaction with HIF-1alpha in hypoxic pulmonary hypertension. Int J Mol Med 36:271–281. https://doi.org/10.3892/ijmm.2015.2209

Johannes L, Jacob R, Leffler H (2018) Galectins at a glance. J Cell Sci 131:jcs208884. https://doi.org/10.1242/jcs.208884

Johns RA et al (2016) Hypoxia-inducible factor 1alpha is a critical downstream mediator for hypoxia-induced mitogenic factor (FIZZ1/RELMalpha)-induced pulmonary hypertension. Arterioscler Thromb Vasc Biol 36:134–144., ATVBAHA.115.306710 [pii]. https://doi.org/10.1161/ATVBAHA.115.306710

Kharade SV et al (2013) The beta3 subunit contributes to vascular calcium channel upregulation and hypertension in angiotensin II-infused C57BL/6 mice. Hypertension 61:137–142. https://doi.org/10.1161/HYPERTENSIONAHA.112.197863

Kim YM et al (2013) Hypoxia-inducible factor-1alpha in pulmonary artery smooth muscle cells lowers vascular tone by decreasing myosin light chain phosphorylation. Circ Res 112:1230–1233. https://doi.org/10.1161/CIRCRESAHA.112.300646

Kim J, Lee KM, Park CS, Park WJ (2014) Ablation of cereblon attenuates myocardial ischemia-reperfusion injury. Biochem Biophys Res Commun 447:649–654. https://doi.org/10.1016/j.bbrc.2014.04.061

Kim HK et al (2016) Cereblon in health and disease. Pflugers Arch 468:1299–1309. https://doi.org/10.1007/s00424-016-1854-1

Lanteri M et al (2003) Altered T cell surface glycosylation in HIV-1 infection results in increased susceptibility to galectin-1-induced cell death. Glycobiology 13:909–918. https://doi.org/10.1093/glycob/cwg110. cwg110 [pii]

Li F (2016) Structure, function, and evolution of coronavirus spike proteins. Annu Rev Virol 3: 237–261. https://doi.org/10.1146/annurev-virology-110615-042301

Li Y et al (2016) Suppression of the expression of hypoxia-inducible factor-1alpha by RNA interference alleviates hypoxia-induced pulmonary hypertension in adult rats. Int J Mol Med. https://doi.org/10.3892/ijmm.2016.2773

Liao P, Yong TF, Liang MC, Yue DT, Soong TW (2005) Splicing for alternative structures of Cav1.2 Ca2+ channels in cardiac and smooth muscles. Cardiovasc Res 68:197–203. https://doi.org/10.1016/j.cardiores.2005.06.024

Liao P et al (2007) A smooth muscle Cav1.2 calcium channel splice variant underlies hyperpolarized window current and enhanced state-dependent inhibition by nifedipine. J Biol Chem 282:35133–35142. https://doi.org/10.1074/jbc.M705478200

Lim MJ et al (2014) Induction of galectin-1 by TGF-beta1 accelerates fibrosis through enhancing nuclear retention of Smad2. Exp Cell Res 326:125–135., S0014-4827(14)00234-1 [pii]. https://doi.org/10.1016/j.yexcr.2014.06.001

Liu FT, Rabinovich GA (2005) Galectins as modulators of tumour progression. Nat Rev Cancer 5: 29–41., nrc1527 [pii]. https://doi.org/10.1038/nrc1527

Loh KWZ, Liang MC, Soong TW, Hu Z (2020) Regulation of cardiovascular calcium channel activity by post-translational modifications or interacting proteins. Pflugers Arch 472:653–667. https://doi.org/10.1007/s00424-020-02398-x

MacKinnon AC et al (2008) Regulation of alternative macrophage activation by Galectin-3. J Immunol 180:2650. https://doi.org/10.4049/jimmunol.180.4.2650

McLeod K, Walker JT, Hamilton DW (2018) Galectin-3 regulation of wound healing and fibrotic processes: insights for chronic skin wound therapeutics. J Cell Commun Signal 12:281–287. https://doi.org/10.1007/s12079-018-0453-7

Moiseeva EP, Javed Q, Spring EL, de Bono DP (2000) Galectin 1 is involved in vascular smooth muscle cell proliferation. Cardiovasc Res 45:493–502., S0008-6363(99)00276-X [pii]

Moosmang S et al (2003) Dominant role of smooth muscle L-type calcium channel Cav1.2 for blood pressure regulation. EMBO J 22:6027–6034. https://doi.org/10.1093/emboj/cdg583

Nickel W (2005) Unconventional secretory routes: direct protein export across the plasma membrane of mammalian cells. Traffic 6:607–614., TRA302 [pii]. https://doi.org/10.1111/j.1600-0854.2005.00302.x

Nio-Kobayashi J (2016) Tissue- and cell-specific localization of galectins, beta-galactose-binding animal lectins, and their potential functions in health and disease. Anat Sci Int. https://doi.org/10.1007/s12565-016-0366-6. 10.1007/s12565-016-0366-6 [pii]

Nugent KM, Shanley JD (1984) Verapamil inhibits influenza a virus replication. Arch Virol 81: 163–170. https://doi.org/10.1007/BF01309305

Ochieng J, Furtak V, Lukyanov P (2002) Extracellular functions of galectin-3. Glycoconj J 19:527–535. https://doi.org/10.1023/B:GLYC.0000014082.99675.2f

Ohkuchi AHC, Nagayama S, Suzaki H, Takahashi K, Matsubara S, Suzuki M (2014) In: XIXth world congress for the study of hypertension in pregnancy 284-POS (New Orleans, Louisiana, USA)

Ou D et al (2022) Galectin-1 alleviates myocardial ischemia-reperfusion injury by reducing the inflammation and apoptosis of cardiomyocytes. Exp Ther Med 23:143. https://doi.org/10.3892/etm.2021.11066

Ouellet M et al (2005) Galectin-1 acts as a soluble host factor that promotes HIV-1 infectivity through stabilization of virus attachment to host cells. J Immunol 174:4120–4126., 174/7/4120 [pii]

Park JW, Voss PG, Grabski S, Wang JL, Patterson RJ (2001) Association of galectin-1 and galectin-3 with Gemin4 in complexes containing the SMN protein. Nucleic Acids Res 29:3595–3602. https://doi.org/10.1093/nar/29.17.3595

Park N et al (2022) Cereblon contributes to cardiac dysfunction by degrading Cav1.2alpha. Eur Heart J 43:1973–1989. https://doi.org/10.1093/eurheartj/ehac072

Petrović T, Lauc G, Trbojević-Akmačić I (2021) The importance of glycosylation in COVID-19 infection. Adv Exp Med Biol 1325:239–264. https://doi.org/10.1007/978-3-030-70115-4_12

Pullamsetti SS, Mamazhakypov A, Weissmann N, Seeger W, Savai R (2020) Hypoxia-inducible factor signaling in pulmonary hypertension. J Clin Invest 130:5638–5651. https://doi.org/10.1172/JCI137558

Quinta HR, Pasquini JM, Rabinovich GA, Pasquini LA (2014) Glycan-dependent binding of galectin-1 to neuropilin-1 promotes axonal regeneration after spinal cord injury. Cell Death Differ 21:941–955., cdd201414 [pii]. https://doi.org/10.1038/cdd.2014.14

Rabinovich GA, Toscano MA (2009) Turning 'sweet' on immunity: galectin-glycan interactions in immune tolerance and inflammation. Nat Rev Immunol 9:338–352., nri2536 [pii]. https://doi.org/10.1038/nri2536

Ramírez Hernández E et al (2021) The role of the SARS-CoV-2 S-protein glycosylation in the interaction of SARS-CoV-2/ACE2 and immunological responses. Viral Immunol 34:165–173. https://doi.org/10.1089/vim.2020.0174

Rapoport EM, Kurmyshkina OV, Bovin NV (2008) Mammalian galectins: structure, carbohydrate specificity, and functions. Biochemistry (Mosc) 73:393–405. https://doi.org/10.1134/S0006297908040032

Rini JM, Lobsanov YD (1999) New animal lectin structures. Curr Opin Struct Biol 9:578–584. https://doi.org/10.1016/S0959-440X(99)00008-1

Roldan-Montero R et al (2022) Galectin-1 prevents pathological vascular remodeling in atherosclerosis and abdominal aortic aneurysm. Sci Adv 8., eabm7322. https://doi.org/10.1126/sciadv.abm7322

Sato T, Ito T, Handa H (2021) Cereblon-based small-molecule compounds to control neural stem cell proliferation in regenerative medicine. Front Cell Dev Biol 9:629326. https://doi.org/10.3389/fcell.2021.629326

Savalli N et al (2016) The alpha2delta-1 subunit remodels CaV1.2 voltage sensors and allows Ca2+ influx at physiological membrane potentials. J Gen Physiol 148:147–159. https://doi.org/10.1085/jgp.201611586

Seropian IM et al (2013) Galectin-1 controls cardiac inflammation and ventricular remodeling during acute myocardial infarction. Am J Pathol 182:29–40. https://doi.org/10.1016/j.ajpath.2012.09.022

Seropian IM et al (2018) Galectin-1 as an emerging mediator of cardiovascular inflammation: mechanisms and therapeutic opportunities. Mediators Inflamm 2018:8696543. https://doi.org/10.1155/2018/8696543

Sharma UC et al (2004) Galectin-3 marks activated macrophages in failure-prone hypertrophied hearts and contributes to cardiac dysfunction. Circulation 110:3121–3128. https://doi.org/10.1161/01.Cir.0000147181.65298.4d

Song LS et al (2002) Ca(2+) signaling in cardiac myocytes overexpressing the alpha(1) subunit of L-type Ca(2+) channel. Circ Res 90:174–181. https://doi.org/10.1161/hh0202.103230

Splawski I et al (2004) Ca(V)1.2 calcium channel dysfunction causes a multisystem disorder including arrhythmia and autism. Cell 119:19–31. https://doi.org/10.1016/j.cell.2004.09.011

Splawski I et al (2005) Severe arrhythmia disorder caused by cardiac L-type calcium channel mutations. Proc Natl Acad Sci U S A 102:8089–8096.; discussion 8086–8088. https://doi.org/10.1073/pnas.0502506102

Starossom SC et al (2012) Galectin-1 deactivates classically activated microglia and protects from inflammation-induced neurodegeneration. Immunity 37:249–263., S1074-7613(12)00323-8 [pii]. https://doi.org/10.1016/j.immuni.2012.05.023

Straus MR et al (2021) Inhibitors of L-type calcium channels show therapeutic potential for treating SARS-CoV-2 infections by preventing virus entry and spread. ACS Infect Dis 7:2807–2815. https://doi.org/10.1021/acsinfecdis.1c00023

Suthahar N et al (2018) Galectin-3 activation and inhibition in heart failure and cardiovascular disease: an update. Theranostics 8:593–609. https://doi.org/10.7150/thno.22196

Suthanthiran M et al (2000) Transforming growth factor-beta 1 hyperexpression in African-American hypertensives: a novel mediator of hypertension and/or target organ damage. Proc Natl Acad Sci U S A 97:3479–3484. https://doi.org/10.1073/pnas.050420897. 050420897 [pii]

Takenaka Y et al (2004) Nuclear export of phosphorylated Galectin-3 regulates its antiapoptotic activity in response to chemotherapeutic drugs. Mol Cell Biol 24:4395–4406. https://doi.org/10.1128/MCB.24.10.4395-4406.2004

Thakurta A et al (2014) Absence of mutations in cereblon (CRBN) and DNA damage-binding protein 1 (DDB1) genes and significance for IMiD therapy. Leukemia 28:1129–1131. https://doi.org/10.1038/leu.2013.315

Wang J et al (2011) Splice variant specific modulation of CaV1.2 calcium channel by galectin-1 regulates arterial constriction. Circ Res 109:1250–1258., CIRCRESAHA.111.248849 [pii]. https://doi.org/10.1161/CIRCRESAHA.111.248849
Wang X et al (2022) Diltiazem inhibits SARS-CoV-2 cell attachment and internalization and decreases the viral infection in mouse lung. PLoS Pathog 18:e1010343. https://doi.org/10.1371/journal.ppat.1010343
Watanabe Y, Allen Joel D, Wrapp D, McLellan Jason S, Crispin M (2020) Site-specific glycan analysis of the SARS-CoV-2 spike. Science 369:330–333. https://doi.org/10.1126/science.abb9983
Weissgerber P et al (2006) Reduced cardiac L-type Ca2+ current in Ca(V)beta2−/− embryos impairs cardiac development and contraction with secondary defects in vascular maturation. Circ Res 99:749–757. https://doi.org/10.1161/01.RES.0000243978.15182.c1
Yang RY, Hsu DK, Liu FT (1996) Expression of galectin-3 modulates T-cell growth and apoptosis. Proc Natl Acad Sci U S A 93:6737–6742. https://doi.org/10.1073/pnas.93.13.6737
Zakrzewicz A et al (2007) The transforming growth factor-beta/Smad2,3 signalling axis is impaired in experimental pulmonary hypertension. Eur Respir J 29:1094–1104., 09031936.00138206 [pii]. https://doi.org/10.1183/09031936.00138206
Zhang Y et al (2018) Influence of LGALS3 gene polymorphisms on susceptibility and prognosis of dilated cardiomyopathy in a northern Han Chinese population. Gene 642:293–298. https://doi.org/10.1016/j.gene.2017.11.026
Zhao XY et al (2010) Hypoxia inducible factor-1 mediates expression of galectin-1: the potential role in migration/invasion of colorectal cancer cells. Carcinogenesis 31:1367–1375., bgq116 [pii]. https://doi.org/10.1093/carcin/bgq116
Zhao XY, Zhao KW, Jiang Y, Zhao M, Chen GQ (2011) Synergistic induction of galectin-1 by CCAAT/enhancer binding protein alpha and hypoxia-inducible factor 1alpha and its role in differentiation of acute myeloid leukemic cells. J Biol Chem 286:36808–36819., M111.247262 [pii]. https://doi.org/10.1074/jbc.M111.247262

Part II

Pharmacological Approaches

The Role of Striatal Cav1.3 Calcium Channels in Therapeutics for Parkinson's Disease

Margaret E. Caulfield, Fredric P. Manfredsson, and Kathy Steece-Collier

Contents

M. E. Caulfield
Department of Translational Neuroscience, College of Human Medicine, Michigan State University, Grand Rapids, MI, USA

F. P. Manfredsson
Parkinson's Disease Research Unit, Department of Translational Neuroscience, Barrow Neurological Institute, Phoenix, AZ, USA

K. Steece-Collier (✉)
Department of Translational Neuroscience, College of Human Medicine, Michigan State University, Grand Rapids, MI, USA

Hauenstein Neuroscience Center, Mercy Health Saint Mary's, Grand Rapids, MI, USA
e-mail: collie68@msu.edu

J. Striessnig (ed.), *Voltage-gated Ca^{2+} Channels: Pharmacology, Modulation and their Role in Human Disease*, Handbook of Experimental Pharmacology 279,
https://doi.org/10.1007/164_2022_629

Abstract

Parkinson's disease (PD) is a relentlessly progressive neurodegenerative disorder with typical motor symptoms that include rigidity, tremor, and akinesia/bradykinesia, in addition to a host of non-motor symptoms. Motor symptoms are caused by progressive and selective degeneration of dopamine (DA) neurons in the SN pars compacta (SNpc) and the accompanying loss of striatal DA innervation from these neurons. With the exception of monogenic forms of PD, the etiology of idiopathic PD remains unknown. While there are a number of symptomatic treatment options available to individuals with PD, these therapies do not work uniformly well in all patients, and eventually most are plagued with waning efficacy and significant side-effect liability with disease progression. The incidence of PD increases with aging, and as such the expected burden of this disease will continue to escalate as our aging population increases (Dorsey et al. Neurology 68:384–386, 2007). The daunting personal and socioeconomic burden has pressed scientists and clinicians to find improved symptomatic treatment options devoid side-effect liability and meaningful disease-modifying therapies. Federal and private sources have supported clinical investigations over the past two-plus decades; however, no trial has yet been successful in finding an effective therapy to slow progression of PD, and there is currently just one FDA approved drug to treat the antiparkinsonian side-effect known as levodopa-induced dyskinesia (LID) that impacts approximately 90% of all individuals with PD. In this review, we present biological rationale and experimental evidence on the potential therapeutic role of the L-type voltage-gated Cav1.3 calcium (Ca^{2+}) channels in two distinct brain regions, with two distinct mechanisms of action, in impacting the lives of individuals with PD. Our primary emphasis will be on the role of Cav1.3 channels in the striatum and the compelling evidence of their involvement in LID side-effect liability. We also briefly discuss the role of these same Ca^{2+} channels in the SNpc and the longstanding interest in Cav1.3 in this brain region in halting or delaying progression of PD.

Keywords

Cav1.3 calcium channels · Levodopa-induced dyskinesias · Medium spiny neurons · Parkinson's disease · Striatum · Substantia nigra

1 Voltage-Gated Calcium Channels and Channelopathies in the Central Nervous System (CNS)

1.1 Voltage-Gated Calcium Channels and the Cav1 Subfamily

Calcium (Ca^{2+}) ions are critical intracellular messengers that drive a multitude of physiologic functions throughout the body, including muscle contraction, hormone release, cardiac pacemaking, neuronal plasticity, and neurotransmitter release (Callens et al. 2022). Voltage-gated calcium channels (VGCCs) mediate Ca^{2+} entry into electrically excitable cells in response to membrane depolarization (e.g., (Kang et al. 2013b)). This is achieved by mediating Ca^{2+} ion influx, or by direct conformational coupling to intracellular Ca^{2+} release channels (Koschak 2010). VGCCs are subdivided into various subfamilies based on their physiological function and include L-type (Cav1 family), T-type (Cav3 family), and the P/Q-type, R-type and N-type VGCCs (Cav2 family) (Berger and Bartsch 2014; Catterall et al. 2005; Yeow et al. 2021). All VGCCs function as an assembly of five distinct channel subunits (α1, α2, β, γ, δ). The α1 subunit is the central pore-forming unit, determining most of the biophysical and pharmacological properties of the respective VGCCs. The auxiliary subunits, alternative splicing variants of the α1 subunit, and associated regulatory proteins are responsible for the proper fine-tuning of the channel characteristics (Berger and Bartsch 2014; Catterall et al. 2005; Yeow et al. 2021). There is extensive research that focuses on the impact of these fine-tuning mechanisms. For example, it was recently reported that regulation of neuronal excitability involved in learning and memory by Cav1.3 L-type Ca^{2+} channels is impacted by alternative splicing both within the C-terminus of their pore-forming α1-subunits, as well as outside the C-terminal, a novel finding regarding channel fine-tuning via stabilization of negative activation and the inactivation gating properties (Hofer et al. 2021).

The members of the L-type VGCC (LTCC) subfamily (Cav1.1, Cav1.2, Cav1.3, and Cav1.4) have well-defined anatomical expression patterns. Cav1.1 and Cav1.4 channels are highly restricted to the skeletal muscle (Cav1.1) and retinal photoreceptors, ganglion cells, and leukocytes (Cav1.4), respectively (Sven Moosmang 2009a, b). Cav1.2 and Cav1.3 channels are detected in a variety of organs including the heart, smooth muscle, pancreas, adrenal glands as well as a variety of brain regions. In many tissues, including neurons, Cav1.2 and Cav1.3 are present within the same cell, suggesting that the two channels possess unique cellular functions (Berger and Bartsch 2014; Striessnig et al. 2006). Cav1 subfamily member proteins are encoded by the human genes *CACNA1S* (Cav1.1), *CACNA1C* (Cav1.2), *CACNA1D* (Cav1.3), and *CACNA1F* (Cav1.4) and are called "L-type" because of their long-lasting barium ion (Ba^{2+}) conductance, which is attributed to the slow inactivation kinetics of LTCCs (~500 ms). Unlike the other VGCCs, LTCCs are the only group that can bind to various 1,4-dihydropyridines (DHP) and are sometimes referred to as dihydropyridine (DHP) receptors (Berger and Bartsch 2014; Yeow et al. 2021).

1.2 CaV1 Subfamily Channelopathies

While the focus of this review is to provide insight into the therapeutic potential of Cav1.3 channels in Parkinson's disease (PD), we first provide here a brief highlight of the current understanding of the roles of Cav1 subfamily LTCCs in other disorders.

CaV1 subfamily channel abnormalities have been associated with a wide variety of clinical conditions. Alterations in the CACNA1S gene (CaV1.1) contribute to a range of skeletal muscle disorders, including hypokalemic periodic paralysis, malignant hyperthermia sensitivity, myopathies, and myotonic dystrophy (Flucher 2020). Variations in the CACNA1C gene (CaV1.2) have been linked to conditions involving the heart, endocrine system, and nervous system such as Timothy Syndrome (TS), neuropsychiatric disorders (e.g., bipolar disorder, major depressive disorder, anxiety, schizophrenia, etc.), and cardiovascular diseases (e.g., Brugada Syndrome) (Berger and Bartsch 2014). CACNA1D (Cav1.3) variants contributes to Sinoatrial Node Dysfunction and Deafness Syndrome, and investigations are ongoing into the specific role of Cav1.3 in neurodegeneration (discussed in detail below), neuropsychiatric disorders, and age-dependent cognitive decline (Berger and Bartsch 2014; Ortner et al. 2020). Finally, alterations in the CACNA1F gene (CaV1.4) have been shown to produce incomplete congenital stationary night blindness type 2 and X-linked cone-rod dystrophy type 3 (Koschak et al. 2021).

In the mammalian CNS, CaV1.2 and Cav1.3 have a broad and overlapping expression profile (Hell et al. 1993; Sinnegger-Brauns et al. 2009) and are present in specific brain areas implicated in learning and memory, mood and anxiety (e.g., hippocampus, amygdala, prefrontal cortex) (Hell et al. 1993; Ludwig et al. 1997). These channels have been identified to be essential for normal brain development, connectivity, and function (Ebert and Greenberg 2013; Kabir et al. 2016; Zamponi et al. 2015) and are critical mediators of experience-dependent plasticity in the brain (Kabir et al. 2017). Further, neuropsychiatric genetic risk variants in CACNA1C (Bhat et al. 2012; Heyes et al. 2015), and CACNA1D (Ortner and Striessnig 2016; Pinggera et al. 2015) alter levels and function of Cav1.2 and Cav1.3 channels, respectively, with consequences on neural processing and connectivity (Gurung and Prata 2015; Kabir et al. 2017; Romme et al. 2017). Specifically, in humans, CACNA1C has emerged as one of the most prominent candidate risk genes for neuropsychiatric disorders such as bipolar disorder, schizophrenia, major depressive disorder, autism spectrum disorder, and attention deficit hyperactivity disorder. Separately, CACNA1D has been found to be associated with bipolar disorder and autism spectrum disorder, as well as cocaine dependence, a comorbid feature associated with psychiatric disorders (Kabir et al. 2017). Although the mechanisms of action of Cav1.2 and Cav1.3 in each distinct disorder remains to be elucidated, decades of studies suggest that selective modulation of Cav1.2 and Cav1.3 channels could be a promising therapeutic strategy for the treatment of mood and addictive behavior disorders. In addition, dysregulated Ca^{2+} homeostasis is a central theme in neurodegenerative diseases ranging from Wolfram syndrome to Alzheimer's disease to the topic of this review, PD (Callens et al. 2022; Kang et al. 2012; Liss and Striessnig 2019).

2 Physiologic Functions of Cav1.3 Ca^{2+} Channels in Basal Ganglia

While Cav1.3 channels have prominent functions in many brain regions, they play different roles within various neuron populations, and have potential to be therapeutic targets for a variety of maladies. The focus of this review is the rationale and potential therapeutic utility of this target in PD. We will focus principally on Cav1.3 Ca^{2+} channels in the striatum and the role that these channels are thought to play in side-effect development of the antiparkinsonian medication, levodopa. We will also discuss, in a more brief fashion, the longstanding interest in the role of Cav1.3 Ca^{2+} channels in the substantia nigra (SN), their role in neurodegeneration, and how they may be another valuable but distinct therapeutic target in PD. Both the striatum and the SN are part of a grouping of nuclei known as the basal ganglia.

2.1 The Basal Ganglia

The word "basal ganglia" derives in part from the older term, "ganglia" which was once commonly used by neuroanatomists to describe an anatomical grouping of neurons. This term, when referring to a grouping of neurons in the brain, has been largely replaced with the word "nucleus." "Basal" is an *adjective* meaning at, or forming the bottom of something. Thus, the basal ganglia is a collection of nuclei located at the bottom, or at the base of the telencephalon.

The nuclei that make up the basal ganglia are the caudate and putamen, which together are often referred to as the *striatum*; the globus pallidus (GP), which has two distinct nuclei called the GP internal segment (GPi) and the GP external segment (GPe); the subthalamic nucleus (STN); and the SN, which also has two distinct regions called the SN pars compacta (SNpc) and SN pars reticulata (SNpr) (Fig. 1). The SNpc contains the dopamine (DA) neurons that degenerate in PD.

This constellation of nuclei is critical to goal-directed (procedural) motor learning and eventual automatic execution of learned motor plans (routine/"habit" behaviors). Within the basal ganglia, the striatum is considered the center for modulation and execution of these behaviors (e.g., (Redgrave et al. 2010)). The first functional model of the basal ganglia was described by DeLong and Georgopoulos in the early 1980s based on anatomical and physiological observations in primates (DeLong 1990) (Fig. 1).

2.2 The Striatum, Medium Spiny Neurons, and Cav1.3 Channels

While there have been many updates to the original functional model of the basal ganglia, it is still held that cortical motor areas send excitatory glutamatergic input topographically to the striatum, where these afferent inputs make synaptic connections with gamma-aminobutyric acid (GABA) containing spiny projection neurons known commonly as medium spiny neurons (MSNs) (Figs. 1 and 2). MSNs

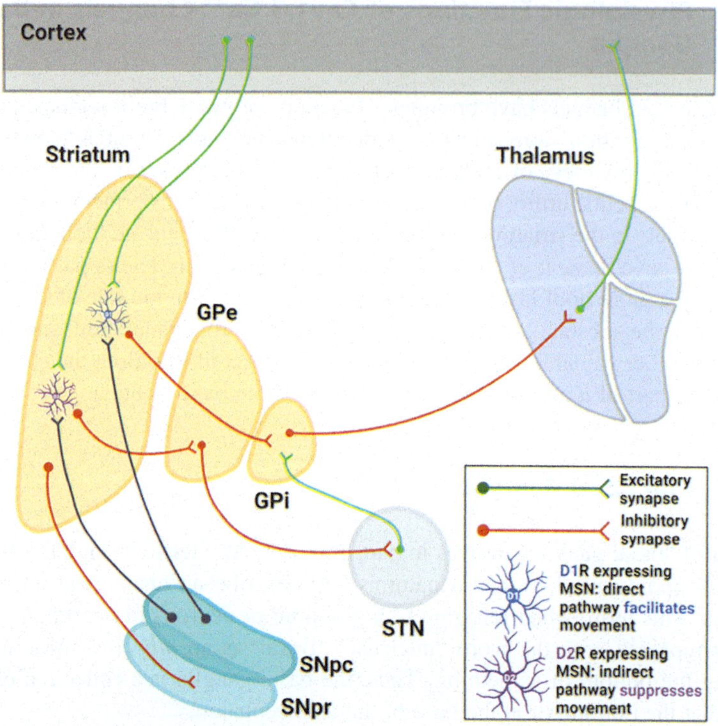

Fig. 1 **Model of Basal Ganglia Nuclei and Circuitry**. Broad regions of cortex send excitatory glutamatergic input to the striatum making afferent synapses with D1 or D2 receptor positive MSNs forming the monosynaptic pathway "direct pathway" or the polysynaptic pathway "indirect pathway," respectively. Dopaminergic SNpc neurons send ascending axons (gray lines) to also make appositions with striatal MSNs. *Abbreviations: MSN* medium spiny neuron, *GPi* globus pallidus internus, *GPe* globus pallidus externus, *STN* subthalamic nucleus, *SNpc* substantia nigra pars compacta, *SNpr* substantia nigra pars reticulata. Green lines indicate excitatory glutamatergic projections. Red lines indicate inhibitory GABAergic projections. Created in BioRender.com

constitute approximately 95–98% of all neurons in the striatum (Matamales et al. 2009). They are anatomically defined as having an extensive dendritic arbor that is densely studded with spiny protuberances called *dendritic spines* (Fig. 2). While there are several other important brains regions that provide input to MSNs, dendritic spines are critical targets of convergent cortical glutamate and SNpc DA input, which are the principle inputs to this class of striatal neuron. Normal extrapyramidal motor behavior depends on normal DA/glutamate interaction at these spines. Indeed, descending cortical glutamate afferents preferentially synapse onto the heads of dendritic spines and ascending afferent DA input from the SNpc preferentially makes appositions on the neck of the same dendritic spine (Figs. 2b and 3a) (Bolam et al. 2000), with DA critically modulating the strength of input at corticostriatal synapses (Surmeier et al. 2014; Zhai et al. 2018). Specifically, there are two populations of MSNs and activity of these two populations of projection

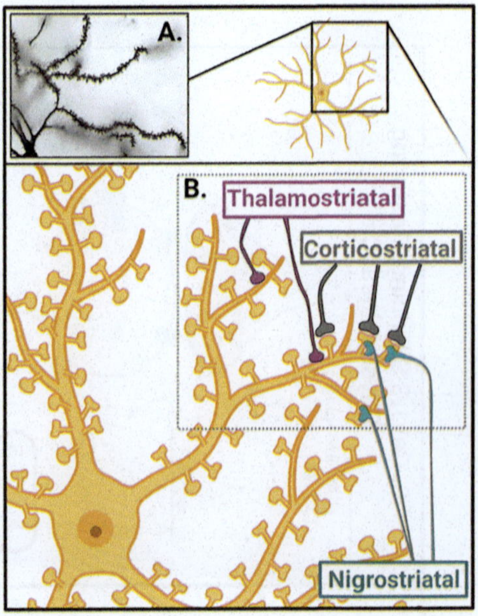

Fig. 2 Striatal MSN morphology and synaptic connections. (**a**) Micrograph of a Golgi-Cox impregnated striatal MSN. The cell soma is located at the bottom left with extending dendrites and extensive dendritic spines. (**b**) Schematic representation of MSN cell soma, dendrites, and dendritic spines with thalamo-striatal, corticostriatal, and nigrostriatal connections onto the dendrite, spine head, and neck of the spine, respectively (Bolam et al. 2000). Created in BioRender.com

neurons is modulated differentially by DA based on distinct expression of D1 versus D2 DA receptors (Gerfen and Surmeier 2011). MSNs activity is conveyed to the basal ganglia output nuclei (GPi and SNpr) through either the monosynaptic "direct pathway" with MSNs of this pathway expressing D1 receptors that enhance long-term potentiation at corticostriatal synapses, or the polysynaptic "indirect pathway" with MSNs giving rise to this pathway expressing D2 receptors that enhance long-term depression (Fig. 1) (Bateup et al. 2010; Wall et al. 2013).

In PD, SNpc DA neurons degenerate and there is an accompanying decrease in levels of striatal DA. It has long been known that nigrostriatal dopaminergic deafferentation is associated with transsynaptic atrophy of post-synaptic cells, namely a marked loss of dendritic spines and retraction of dendrites on MSNs (Ingham et al. 1989, 1993; McNeill et al. 1988) (Fig. 3a, b). While the loss of dendritic spines on striatal MSNs secondary to loss of DA has been known for since 1988 (McNeill et al. 1988), the mechanism underlying this was only revealed in 2006. Specifically, in a landmark report by Day et al. (2006), it was revealed that striatal DA depletion triggers dysregulation of intraspinous Cav1.3 L-type Ca^{2+} channels on MSNs that leads to dendritic spine retraction (Fig. 3b).

Cav1.3 channels, as with other VGCCs, contain a pore-forming α1D subunit and these specific channels are opportunistically positioned in dendritic spines (Fig. 3) to modulate these cytoarchitectural elements and their synapses (Day et al. 2006). In MSNs, these Ca^{2+} channels are found near the post-synaptic density region of glutamatergic synapses located at the head of the dendritic spines. In addition to regulating synaptic strength of corticostriatal input, as introduced above, DA plays a key role in regulating Cav1.3 channel activity. Specifically, D2 receptors reduce

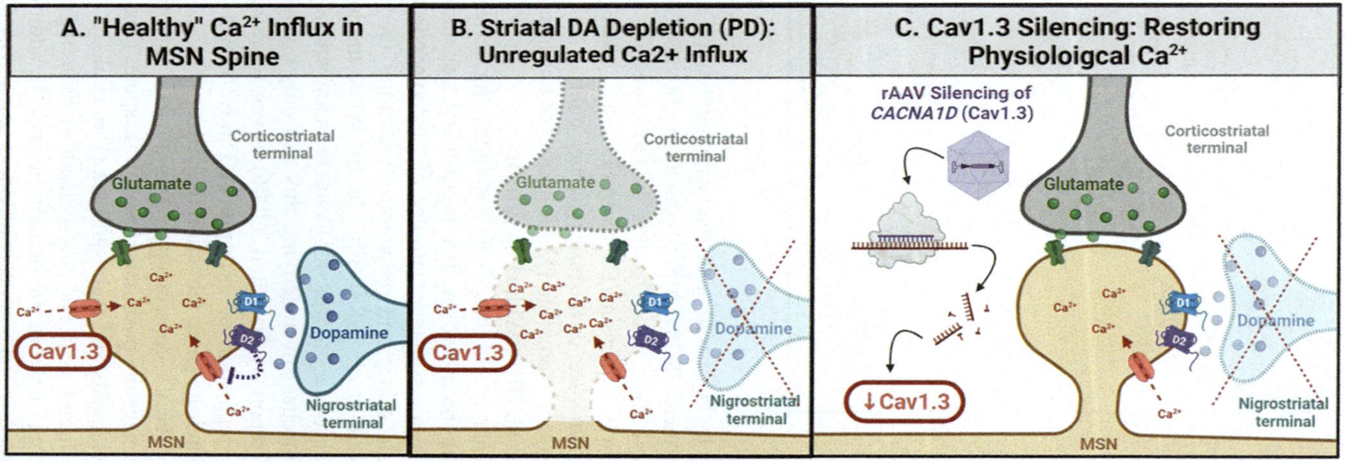

Fig. 3 **Cav1.3 and Ca^{2+} influx in striatal MSNs**. (**a**) Under physiological conditions a single corticostriatal afferent synapses preferentially onto the head of a MSN dendritic spine, with ascending SNpc DA terminals making appositions onto the neck of the same spine. Intraspinous Cav1.3 channels in these neurons regulate Ca^{2+} influx into the dendritic spines from the extracellular space. Under homeostatic conditions, Cav1.3 channel function is positively and negatively regulated by the dopamine D1 (blue) and D2 (purple) receptors, respectively, maintaining "normal" spine density and synaptic plasticity. (**b**) In the presence of SNpc DA neuron death in PD and loss of striatal DA tone, the inhibition of Cav1.3 function through D2 receptors is lost/diminished, resulting in increased intracellular Ca^{2+}, spine retraction, and loss of the corresponding cortical input depicted here by dashed lines and lighter color indicating loss of indicated structures (Ingham et al. 1989, 1993; McNeill et al. 1988; Day et al. 2006; Zhang et al. 2013). (**c**) Cav1.3 channel antagonism prevents loss of dendritic spines despite severe DA depletion (Schuster et al. 2009; Soderstrom et al. 2010). It was hypothesized that under conditions of optimized Cav1.3 silencing achieved with rAAV-Cav1.3-shRNA that "normalized" intracellular Ca^{2+} would result, as depicted here, in the maintenance of normal spine density and corticostriatal synapses despite severe loss of nigrostriatal DA input, and in this cytoarchitectural environment, DA replacement with levodopa would prevent induction of LID (Steece-Collier et al. 2019). *shRNA* = short hairpin ribonucleic acid. Created in BioRender.com

channel opening through a Ca^{2+}-dependent protein phosphatase mechanism (Hernandez-Lopez et al. 2000) and D1 receptors enhance Cav1.3 channel opening through a protein kinase A mechanism (Surmeier et al. 1995).

In the presence of DA depletion and reduced inhibitory action of D2 receptors, Cav1.3 channels are considered to be "disinhibited," promoting increases in intraspinous Ca^{2+} through an increase in Ca^{2+} transients via readily opened Cav1.3 channels (Day et al. 2006). The consequence of this scenario is retraction of dendritic spines and pruning of their respective presynaptic glutamatergic profiles (Day et al. 2006; Zhang et al. 2013) (Fig. 3b). There remains controversy as to whether striatal DA depletion results in spine loss specifically in MSNs of D2 receptor containing indirect pathway MSNs or whether it occurs in MSNs of both indirect and direct pathway MSNs (for review (Bastide et al. 2015)).

Regardless, there is ample evidence that striatal dendritic spine loss, through Cav1.3 mechanisms, and the associated synaptopathology sets the stage for development of the all-too-common side-effect of antiparkinsonian medication known as levodopa-induced dyskinesia (LID) (Bastide et al. 2015; Schuster et al. 2009; Soderstrom et al. 2010; Steece-Collier et al. 2019; Zhang et al. 2013). Accordingly, *Cav1.3 channels in striatal MSNs are a potential target whose modulation may benefit individuals with PD.* The details and therapeutic implications of striatal Cav1.3-associated pathology are discussed below.

2.3 Substantia Nigra, Tonically Active Dopamine Neurons, and Cav1.3 Channels

In addition to Cav1.3 channels playing an important role in the physiology and pathology of striatal MSNs, these voltage-gated channels in SNpc DA neurons play a distinct and critical role in the physiology and pathology of these neurons. Specifically, the nigral DA neurons that provide input to striatal MSNs have a unique physiology involving autonomous pacemaking activity that, while being influenced by a host of ion channels, transporters and receptors (Liss and Striessnig 2019), is prominently modulated by the engagement of Cav1.3 L-type Ca^{2+} channels (Fig. 4a) (e.g., (Surmeier et al. 2012; Surmeier and Schumacker 2013)). Autonomous pacemaking activity results in tonic release of DA in the striatum (Fig. 4b), although phasic increases occur in the presence of rewarding stimuli and decreases in response to aversive signaling (Gerfen and Surmeier 2011; Graybiel 1995). While these patterns of DA release are necessary for voluntary goal-directed behavior, maintaining physiological ion gradients associated with autonomous excitability of these tonically active neurons results in significant energetic burden in addition to biological challenges associated with elevated intracellular Ca^{2+} fluxes. Specifically, while elevated intracellular Ca^{2+} levels support the normal physiology of these cells, it can lead to neurodegeneration through complex and diverse mechanisms including activation of Ca^{2+}-dependent apoptotic enzymes in a population of neurons that have only modest Ca^{2+} buffering capacity (Foehring et al. 2009; Mouatt-Prigent et al. 1994). To counter toxic effects of continuous influx of Ca^{2+} in SNpc, there is a high

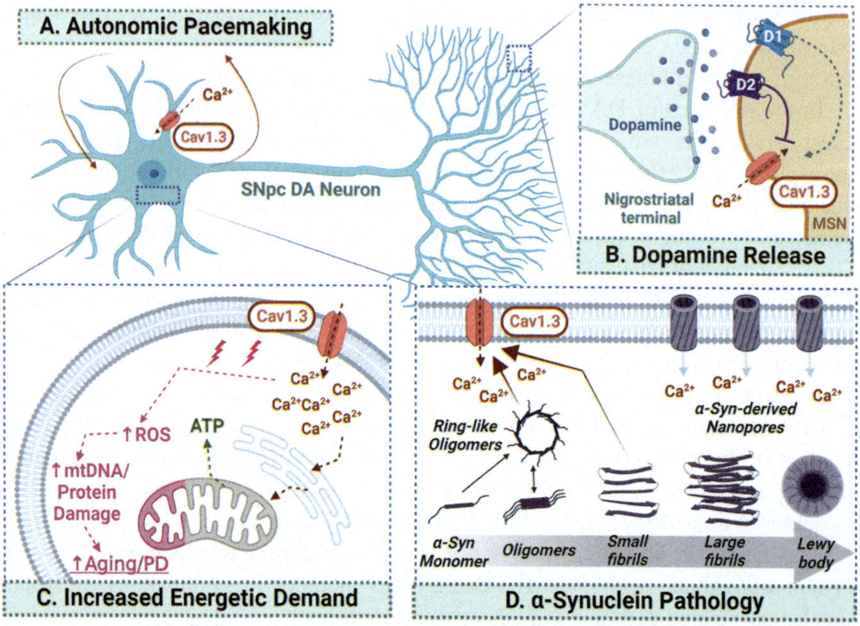

Fig. 4 Cav1.3 in SNpc DA neurons, autonomous pacemaking activity, and Ca^{2+} influx: Contributions to increased vulnerability in aging & PD. (**a**, **b**) Neurons of the SNpc have autonomous pacemaking activity that results in tonic release of DA in the striatum. (**c**) There is a high energetic demand in these autonomously active neurons to extrude Ca^{2+} ions by adenosine triphosphate (ATP)-dependent processes mediated by mitochondrial oxidative phosphorylation that results in generation of superoxide and reactive oxygen species (ROS). Additional environmental or genetic insults (pink lightning bolts) can further augment generation of mitochondrial oxidant stress rendering these neurons vulnerability to neurodegeneration in aging and PD, e.g., (Bolam and Pissadaki 2012; Guzman et al. 2010; Kang et al. 2013a; Surmeier et al. 2012). (**d**) Elevated cytosolic Ca^{2+} in SNpc induces oligomerization and aggregation of α-synuclein (α-syn) (Follet et al. 2013; Lowe et al. 2004; Nath et al. 2011), and in turn, α-syn through its regulation of Ca^{2+} entry pathways is thought to promote dysregulation of Ca^{2+} homeostasis. In addition, it was hypothesized more than two decades ago that PD pathogenesis may involve membrane permeabilization by protofibrillar α-syn to form pore-like structures recently referred to "α-syn nanopores" that promote elevation of intracellular Ca^{2+} (e.g., Volles and Lansbury 2002; Ghio et al. 2019; Parres-Gold et al. 2020). Created in BioRender.com

energetic demand in these autonomously active neurons to extrude Ca^{2+} ions by adenosine triphosphate (ATP)-dependent mechanisms. These processes are mediated by mitochondrial oxidative phosphorylation, a process that results in generation of superoxide and reactive oxygen species (ROS) and their by-products (Fig. 4c). The energetic burden is further compounded by the fact that SNpc neurons have extensively large, complex axonal arbors where it is estimated that a single DA neuron influences 75,000 striatal neurons via *en passant* appositions of their long and highly branched axons (Bolam and Pissadaki 2012). The excessively high metabolic demand in SNpc DA neurons has been implicated in the generation of mitochondrial oxidant stress leading to their vulnerability and neurodegeneration in aging and PD (Bolam and Pissadaki 2012; Catoni et al. 2019; Guzman et al. 2010; Kang et al. 2013a; Surmeier et al. 2012, 2017a, b).

There is an additional pathogenic mechanism of elevated intracellular Ca^{2+} related to α-synuclein (α-syn), a presynaptic protein that forms toxic oligomers and abnormal cytoplasmic aggregates in Lewy body disorders including PD. Specifically, a variety of evidence suggests that elevated cytosolic Ca^{2+} in SNpc induces pathogenic oligomerization and aggregation of cytosolic α-syn (Follett et al. 2013; Lowe et al. 2004; Nath et al. 2011); and that α-syn through its regulation of Ca^{2+} entry pathways may promote dysregulation of Ca^{2+} homeostasis, potentially resulting in a feedforward cycle of pathogenesis (Fig. 4d) (Hettiarachchi et al. 2009). While some of the effects of elevated intracellular Ca^{2+} on α-syn aggregation and toxicity have been shown to be blocked by CaV1.2/1.3 channel antagonists (Hettiarachchi et al. 2009), other evidence suggests that PD pathogenesis involving Ca^{2+} dysregulation may also be related to membrane permeabilization by protofibrillar α-syn which can form pore-like structures (Ghio et al. 2019; Parres-Gold et al. 2020; Tosatto et al. 2012; Tsigelny et al. 2012; Volles and Lansbury 2002) (Fig. 4d) resistant to modification by L-type Ca^{2+} channel blockers (Furukawa et al. 2006). Taken together an abundance of data suggest that *Cav1.3 channels in SNpc neurons present another target whose modulation may benefit individuals with PD.*

3 Therapeutic Striatal Cav1.3 Targeting for Motor Complications in PD

The role of Cav1.3 channels to the contribution of total Ca^{2+} currents, their physiological roles, and significance as a drug target for diseases like PD was largely unknown just two decades ago (Platzer et al. 2000). The unique therapeutic potential of this molecular target in PD, in two distinct brain regions, specifically for the motor complication known as LID and neuroprotection is discussed in the following sections, with emphasis on the role of Cav1.3 in LID liability.

3.1 Levodopa Treatment in PD

As reviewed by noted PD clinician Dr. Stanley Fahn (2015), it was not until 150 years after James Parkinson first described in 1817 the illness that came to bear his name, that an effective therapy for this disorder was introduced. This therapy, introduced by George Cotzias in 1967, was a high-dosage levodopa. Levodopa is the synthetic precursor to DA which, unlike DA, is readily taken up into the brain and converted to DA in neurons with necessary synthetic enzymes (Daubner et al. 2011; Sellnow et al. 2019). It took many years to define/refine levodopa dosing and to realize the necessity of co-administering a peripheral decarboxylase inhibitor to prevent peripheral degradation thereby: 1) enhancing the amount of levodopa that got into the brain, and 2) preventing the often intolerable side-effects of nausea, anorexia, tachycardia, irritability/hostility, and sleep disturbances. These painstaking efforts by clinicians in the 1960s and 1970s resulted in a revolutionary new treatment for PD, which remains the gold-standard for treating PD (Fahn 2015; Fahn and Poewe 2015; Lees et al. 2015; Poewe and Antonini 2015).

3.2 Motor Complications of Levodopa: An Unmet Clinical Need

Despite the outstanding benefit of this revolutionary treatment, long-term levodopa therapy is associated with motor complications including the development of LID, abnormal involuntary movements including chorea and dystonia that can negatively impact quality of life of those afflicted. While there are varying accounts of the occurrence, LID are estimated to occur in approximately 50% of individuals with PD after 3–5 years of treatment with the incidence escalating to approximately 90% after 10–15 years (Hauser et al. 2017; Huot et al. 2013). Despite the fact that nearly 90% of all individuals treated with levodopa will develop LID, there is currently only one FDA approved drug to treat dyskinesia in PD. Extended release amantadine (Gocovri®) was approved in 2017 as an oral, extended-release drug for the treatment of LID in individuals with PD. Results from two clinical trials have provided a mixed picture of the antidyskinetic efficacy of this drug, showing a modest effects in some patients (~61% of patients in the amantadine groups versus 47% in the placebo group met or exceeded the 8-point primary endpoint threshold), however nearly 20% of participants in the higher, most effective dose had to drop out because of significant adverse side-effects (Rascol et al. 2022).

An alternative approach for managing LID is deep brain stimulation (DBS). DBS is a neurosurgical procedure that involves the placement of a stimulating electrode, which sends electrical impulses to specific nuclei of the basal ganglia. STN DBS has been the primary surgical treatment of choice for motor benefit in individuals with advanced PD, however the GPi and pedunculopontine nucleus (PPN) are other common targets (Chen et al. 2021; Lin et al. 2022). It was once considered that GPi DBS could have a direct circuitry-related impact in reducing LID whereas dyskinesia reduction following STN DBS was thought to be related to the post-operative reduction in dose of levodopa (Follett 2004). However, more recent data suggests that bilateral STN DBS can improve dyskinesias beyond what would be expected from levodopa reduction alone (Mossner et al. 2019). While DBS can be an effective treatment for LID and it is the primary *surgical* procedure employed for motor fluctuations in PD, it is still an invasive method and less than 10% of individuals with PD receive this treatment. As such, LID remain an unmet clinical need in PD.

3.3 Striatal Cav1.3 Channels Antagonism for LID: The Initial Promise

Central to the biology of LID are changes in synaptic plasticity associated with striatal MSNs. As introduced above, the pathognomonic loss of striatal DA in PD results in dysregulation/disinhibition of striatal Cav1.3 channels leading to synaptic pathology that appears to be involved in LID. Specifically, in PD and in animal models of PD, it is well established that striatal DA depletion results in loss of dendritic spines on MSNs; an aberrant feature that is accompanied by the secondary loss of glutamatergic synapses from corticostriatal projections (Figs. 3b and 5b) (Day et al. 2006; Fieblinger and Cenci 2015; Fieblinger et al. 2014; Ingham et al. 1989; McNeill et al. 1988; Nishijima et al. 2014; Soderstrom et al. 2010; Zhang et al. 2013). Introduction of levodopa in this environment results in the restoration of dendritic spines (see (Bastide et al. 2015) for details of this complex phenomenon) and re-establishment of glutamate input, but the pulsatile nature of oral administration results in an aberrant pattern of apparent "mis-wiring" (Fig. 5c) (Fieblinger et al. 2014; Nishijima et al. 2014; Suarez et al. 2014; Zhang et al. 2013). As introduced above, in the "normal" striatum, DA projections provide input principally onto the neck of dendritic spines with cortical glutamate projections generally providing a *single synapse* onto the head of these same spines (Figs. 3a and 5a; for review (Zhang et al. 2013)). However, in parkinsonian rats administered levodopa that develop LID, ultrastructural data have revealed atypical characteristics of afferent

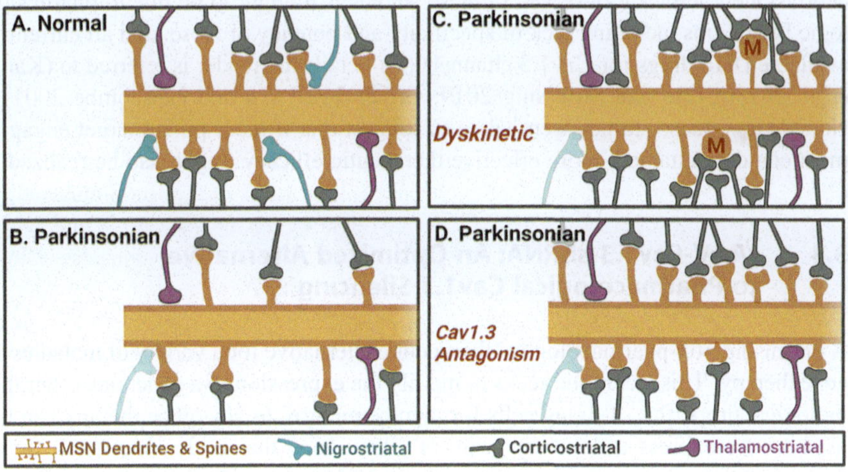

Fig. 5 **Synaptic profiles of striatal MSNs** under physiological conditions (**a**) and in the presence of DA depletion (**b**) where there is marked loss of dendritic spines and their corticostriatal glutamatergic inputs. (**c**) In the presence of DA-depletion related synaptopathology, levodopa administration and LID development result in restoration of spine and corticostriatal input, however in an aberrant pattern of "mis-wiring." "M" = mushroomed spine. (**d**) Despite a loss of striatal DA, Cav1 channel antagonists can prevent loss of spines (Schuster et al. 2009; Soderstrom et al. 2010), and presumably corticostriatal input. Created in BioRender.com

cortical glutamate projections including *multisynaptic* contacts onto individual spines and adaptive morphological changes of these spinous targets (i.e., increased number of mushroom spines) on the MSNs (Fig. 5c) (Zhang et al. 2013).

Since the initiating event of spine retraction in the presence of DA depletion has been linked to overactivity of the voltage-gated L-type Cav1.3 calcium channels located on dendritic spines (Day et al. 2006), it was hypothesized that if it were possible to use Cav1.3 channel antagonism to prevent the initial spine loss and the subsequent LID-associated synaptopathology in the parkinsonian brain, levodopa-mediated DA replacement would allow for motor benefit in the absence of LID. Indeed, the idea was that silencing Cav1.3 channels in the presence of severe DA depletion would theoretically allow for maintenance of "normal" synaptic profiles (Fig. 5a, d), thereby preventing the resultant aberrant re-wiring of glutamate inputs in the presence of DA replacement with levodopa, and significantly diminish liability for LID. The first attempts to test this hypothesis were undertaken using CaV1 channel antagonism with DHP drugs (i.e., nimodipine, isradipine). Specifically, a study from our laboratory (Soderstrom et al. 2010) and one lead by Schuster et al. (2009) used subcutaneous, slow-release pellets containing nimodipine (Soderstrom et al. 2010) or isradipine (Schuster et al. 2009) in unilaterally parkinsonian rats and demonstrated that the expression of LID produced by low-dose (6 mg/kg, (Schuster et al. 2009)) or high-dose levodopa (12.5 mg/kg, (Soderstrom et al. 2010)) was indeed abated. However, while these initial data suggested that Cav1.3 may have potential as an antidyskinetic target, the efficacy in the presence of pharmacological silencing was only *partial and lost over time*. The limited and transient antidyskinetic efficacy observed in these studies is thought to be due to pharmacologic limitations including lack of specificity and potency of these, and all currently available DHP drugs for Cav1.3 channels (for details the reader is referred to (Kang et al. 2013b; Liss and Striessnig 2019; Ortner 2021; Xu and Lipscombe 2001)). Nonetheless, these studies were taken to suggest that if more potent target engagement could be attained, more effective therapeutic efficacy might also be realized.

3.4 rAAV-Cav1.3-shRNA: An Optimized Alternative to Pharmacological Cav1.3 Silencing

A promising non-pharmacological therapeutic alternative for a variety of maladies is gene therapy. This technique seeks to modify the expression of a gene, or to alter the biological properties of living cells for therapeutic use. *In vivo* gene therapy can be used to overexpress a deficient gene, or silence one that is dysfunctional. Gene therapy can involve viral vectors or plasmid DNA genetically engineered to carry therapeutic DNA into cells (Sandoval et al. 2021), gene editing using techniques like CRISPR-Cas9 (Rahman et al. 2022), or cellular gene therapy where, for example, a patient's own cells can be biopsied, genetically modified to produce a gene product of interest, and grafted back into the individual (Frederiksen et al. 2020; Paolini Sguazzi et al. 2021).

Based on the significant limitations of currently available pharmacological Cav1.3 channel antagonists, we sought to use a viral vector gene therapy approach to provide unequivocal target validation of the utility of Cav1.3 channel silencing as an antidyskinetic therapy. We employed recombinant adeno-associated virus (rAAV)-mediated short hairpin RNA (shRNA) to provide continuous, high potency and isoform selective, mRNA-level silencing of striatal Cav1.3 channels in a series of studies where we examined the capacity and limitations of this approach to prevent induction and/or reverse existing LID in a rat model of PD (Steece-Collier et al. 2019).

As we recently reported, we first tested the hypothesis that rAAV-shRNA-mediated silencing of striatal CaV1.3 channels prior to the introduction of levodopa would prevent development and provide sustained amelioration of LID. In these "LID prevention" studies, adult male Sprague Dawley rats received unilateral injections of either rAAV-Cav1.3-shRNA or the control rAAV-Scramble(Scr)-shRNA into the dorsolateral striatum (Steece-Collier et al. 2019) (Fig. 6a). Vector surgery preceded by one week 6-hydroxydopamine (6-OHDA) surgery used to induce unilateral parkinsonism. Three weeks following induction of parkinsonism, four weeks post-vector, rats began receiving daily levodopa administered using a dose escalation paradigm, with doses ranging from low (6 mg/kg) to moderate (9 mg/g) to high (12 mg/kg) and, finally, to what we refer to as "extreme" (18 mg/kg) (Fig. 6b). Each dose was given daily (M-Fr) for 2 weeks and LID behaviors were rated on days 1, 6, and 10 of each dose. The results from this study were compelling. Specifically, in rats receiving the control rAAV-Scr-shRNA vector there was a typical escalation in LID severity over time that remained stable and severe with increasing doses of levodopa (Fig. 6c, black line). In striking contrast, rats receiving the rAAV-Cav1.3-shRNA vector show a significant dampening in the severity of LID that progressed over time into a near complete absence of dyskinetic behavior in all subjects (Fig. 6c, d, red lines) (Steece-Collier et al. 2019). This robust antidyskinetic effect was maintained even in the presence of continuous, escalating doses of daily levodopa over 2 months.

While being able to provide near complete protection against LID induction is highly valuable, it is perhaps more clinically relevant to be able to reverse existing dyskinesias. Thus, we next examined the ability to diminish already established, severe LID using this gene therapy approach. The ability to reverse what is often considered to be a steadfast phenomenon was found previously to be completely *un*affected with pharmacological Cav1 silencing, even against low-dose levodopa (Schuster et al. 2009). We hypothesized that if more potent target engagement could be attained in parkinsonian rats with established LID behaviors, diminishing Cav1.3-mediated dysfunctional calcium signaling in an environment of established "mis-wiring" would lessen but not completely reverse LID. We also reasoned that if preventing the initial spine loss and the secondary aberrant re-wiring of cortical glutamate inputs was necessary to impact LID expression, that reducing CaV1.3 channel expression *after* LID induction may not be capable of reversing this behavior. In our "LID reversibility" experiment, rats were first rendered unilaterally parkinsonian as described above and three weeks later began receiving high-dose

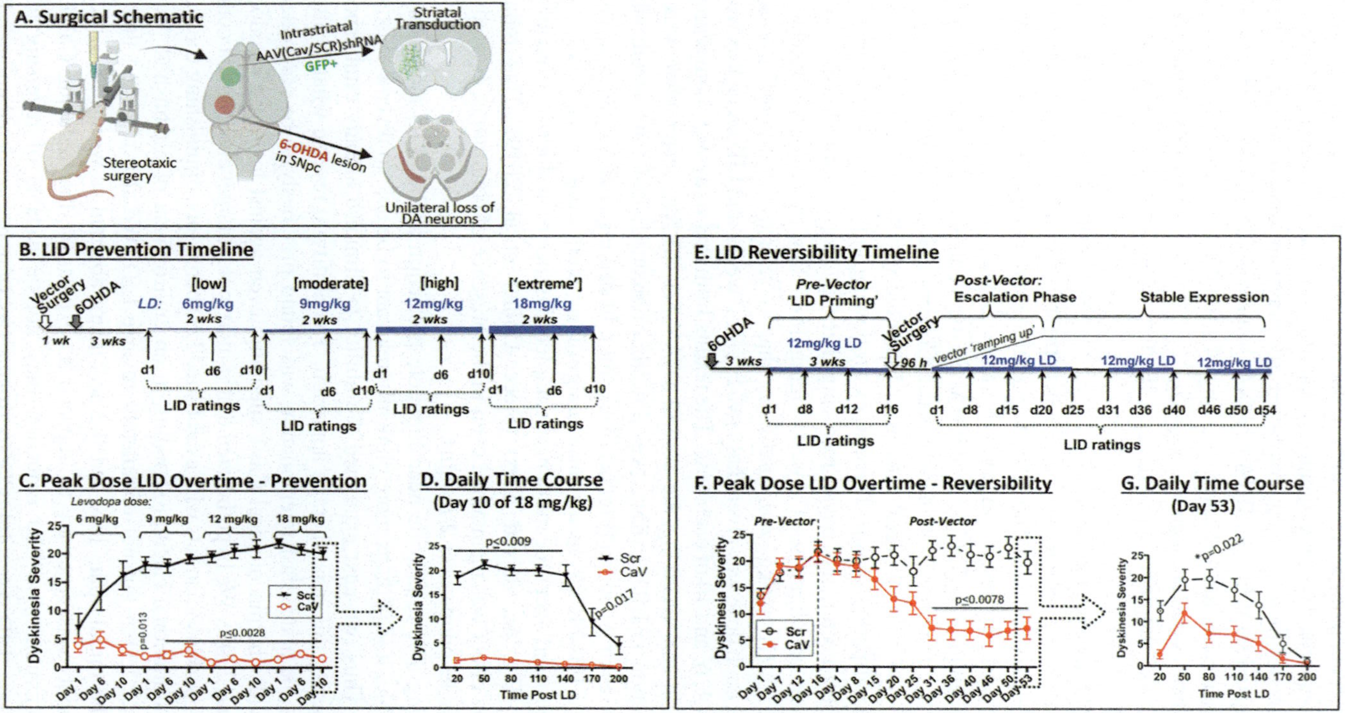

Fig. 6 Impact of rAAV-Cav1.3/SCR-shRNA on LID in parkinsonian rats. (**a**) Schematic representation of stereotaxic surgery strategy including nigral 6-OHDA lesioning and intrastriatal rAAV-shRNA injection. (**b**) Treatment timeline for the "LID prevention" study where rats received intrastriatal vector one week prior to nigral lesioning. Levodopa was administered daily (M-Fr) using a dose escalation paradigm as detailed in the text. (**c**) Impact of vector on peak-dose LID severity (80 min post-levodopa) across time and doses. (**d**) Daily time course of LID behavior ranging from 20 to 200 min post-levodopa *on the final*

Fig. 6 (continued) day of levodopa treatment (Day 10 of 18 mg/kg levodopa). Statistics: Friedman and Kruskal–Wallis tests with Dunn's multiple-comparison post hoc as shown in graphs. *Scr*, rAAV-Scrambled-shRNA ($n = 7$); *CaV*, rAAV-CaV1.3-shRNA ($n = 10$). (**e**) Treatment timeline for "LID reversibility" study where rats were first rendered parkinsonian with intranigral 6-OHDA followed by 3 weeks of daily high dose (12 mg/kg) levodopa to establish stable, severe LID behavior (F, "pre-vector"). Rats then received intrastriatal vector and LID were evaluated over 53 days. (**f**) Impact of vector on peak-dose LID severity (80 min post-levodopa) across the time course of the experiment. (**g**) Daily time course of LID behavior ranging from 20 to 200 min post-levodopa *on the final day of levodopa treatment* (Day 53 post-vector). Data represents the mean ± SEM. Statistics: Kruskal–Wallis tests with Dunn's multiple-comparison post-hoc test as shown in graphs. *LD*, levodopa; *Scr*, rAAV-Scrambled-shRNA ($n = 12$); *CaV*, rAAV-CaV1.3-shRNA ($n = 11$) (Steece-Collier et al. 2019). Data modified with permission (Steece-Collier et al. 2019). "A" created in BioRender.com

daily levodopa (12 mg/kg, M-Fr) (Fig. 6e). Once significant and stable LID behavior was established, unilaterally parkinsonian rats received either rAAV-Cav1.3-shRNA or the control rAAV-Scr-shRNA vector into the dorsolateral DA depleted striatum. In contrast to previous studies using pharmacological approaches, we observed that mRNA level Cav1.3 silencing in parkinsonian rats with already established severe LID could significantly, albeit partially by ~62%, ameliorate these aberrant behaviors (Fig. 5f, g). Future studies examining the ability of earlier intervention when levodopa dose and LID severity are more modest are underway in our labs.

Of critical importance in developing clinically relevant strategies to reduce LID is maintaining motor benefit of levodopa/DA replacement strategies. Our recent data using viral vector silencing of striatal Cav1.3 channels demonstrates that despite a potent impact on LID amelioration, this approach does not appear to impair general motor behavioral responses following levodopa. Specifically, in a limited series of initial investigations, we quantified motoric responses (i.e., exploratory rearing and rotational asymmetry) under baseline, drug-free conditions, and in response to low-dose (6mg/kg) levodopa. These motor behaviors were examined using a modified cylinder task (Schallert 2006) where rats were put into a clear plexiglass cylinder. Under normal circumstances rodents will rear and explore the novel environment of the cylinder. Unilaterally lesioned parkinsonian rats will generally show less exploratory behavior compared to intact control subjects, however this can be enhanced in the presence of DA replacement therapies. In addition, when exploring, unilaterally parkinsonian rodents tend to ambulate in the direction contralateral to the lesion, which is also generally increased in response to DA releasing agents like levodopa. In the rats from both studies discussed above (i.e., LID prevention and LID reversibility), there was no significant difference in either rearing or rotational motor behavior under baseline, drug-free conditions (Fig. 7a, b, "- LD") between the two viral vector treatment groups. However, in the presence of low-dose levodopa (6 mg/kg) only the parkinsonian rats receiving intrastriatal rAAV-Cav1.3-shRNA but not those receiving rAAV-Scr-shRNA showed a significant increase in exploratory rearing behavior and contralateral rotational response (Fig. 7a, b, "+ LD"). These studies suggest that normalizing/suppressing overactivity of Cav1.3 channels in the DA-depleted striatum may allow for enhanced motor response to levodopa. The reader also is referred to the videos in (Steece-Collier et al. 2019) to see additional evidence demonstrating marked motor enhancement, in the absence of dyskinesias, after levodopa administration in parkinsonian rats with gene-level striatal Cav1.3 silencing.

These data provide compelling evidence that knockdown of striatal Cav1.3 can entirely ameliorate the expression of LID when applied as a "prevention" therapy even when doses of levodopa are escalated over time, and that effective silencing of this molecular target can significantly reduce the severity of existing LID behavior in parkinsonian rats maintained on high-dose levodopa by mechanism(s) that remain to be elucidated. Importantly, these proof-of-principle studies suggest a *dual benefit of striatal Cav1.3 silencing in PD* by not only providing for substantial dyskinesia abatement but potentially also providing for enhanced motor benefit of DA agonist therapy. As we discuss in the following section, there is a third attribute of PD that

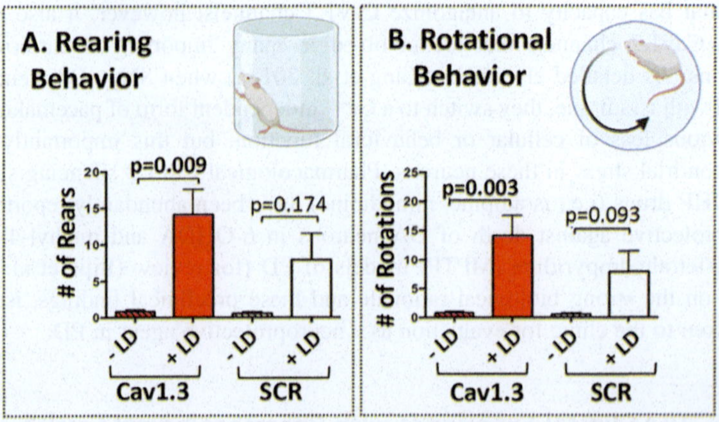

Fig. 7 Motor response to low dose levodopa in parkinsonian rats receiving rAAV-Cav1.3/ SCR sh-RNA. Data represent total number of exploratory rears (**a**) or contralateral ambulatory rotations (**b**) over a 5-min test period prior to administration of levodopa (-LD) or 50 min after 6 mg/kg of levodopa (+LD). Behavioral responses shown here are from the "LID prevention" study (Steece-Collier et al. 2019; modified with permission). Statistics: 2-way ANOVA with post-hoc Sidak's multiple-comparisons. *LD* levodopa, *SCR* rAAV-Scrambled-shRNA, *CaV* rAAV-CaV1.3-shRNA. Portions created in BioRender.com

may benefit from silencing of this same Cav1.3 molecular target, albeit in a different basal ganglia nuclei.

4 Therapeutic Nigral Cav1.3 Targeting for Neuroprotection in PD

While symptomatic therapies exist for individuals with PD, there is currently a lack of neuroprotective therapies to slow or halt progression of a disease that impacts more than ten million individuals worldwide. Great optimism has surrounded Cav1.3 channel antagonism in SNpc neurons as a much needed neuroprotective therapy for PD. This optimism was unfortunately dampened following results from a recent clinical trial (STEADY-PD III Trial) in individuals with PD that employed the voltage-gated Ca^{2+} channel CaV1 antagonist drug isradipine (Parkinson Study Group 2020).

4.1 Nigral Cav1.3 Channels Antagonism for Neuroprotection: The Initial Promise

As introduced above, autonomous pacemaking activity and Ca^{2+} influx mediated by Cav1.3 channels in SNpc DA neurons are thought to contribute to increased vulnerability of these neurons in aging & PD. Isradipine is a highly brain penetrant

DHP that has capacity to antagonize Cav1.3 channels; however, it also potently blocks Cav1.2 channels that are involved in many important in cardiovascular functions. As detailed elsewhere (Kang et al. 2013a) when SNpc DA neurons are treated with isradipine, they switch to a Ca^{2+}-independent form of pacemaking, with no obvious loss of cellular or behavioral function, but this importantly lowers mitochondrial stress in these neurons. Pharmacological Cav1.3 silencing strategies with DHP drugs (i.e., isradipine, nimodipine) have been abundantly reported to be neuroprotective against death of DA neurons in 6-OHDA and methyl-4-phenyl-1,2,3,6-tetrahydropyridine (MPTP) models of PD (for review (Ilijic et al. 2011)). Based on the strong biological rationale and these preclinical findings, isradipine was taken to the clinic for evaluation as a neuroprotective agent in PD.

4.2 The Clinical Experience with Pharmacological Cav1.3 Silencing

Two clinical trials, STEADY-PD II and STEADY-PD III, were entered into with great hope of advancing a path to neuroprotective therapy development. In these trials, the CaV-1 L-type Ca^{2+} channel antagonist isradipine was employed. STEADY-PD II, aimed at examining safety, tolerability, and dose selection, used varying doses of extended release isradipine (DynaCirc ER; 5, 10, 20 mg) given *once daily*. STEADY-PD III, aimed at examining the rate of clinical progression of PD, used immediate release isradipine given intermittently *2x daily* at a lower more clinically tolerable dose (5 mg).

The initial report for the phase III trial was that long-term treatment failed to find a change in either the primary or secondary outcomes, showing no evidence of neuroprotection in early-stage PD (Parkinson Study Group 2020). However, follow-up analysis modeling the plasma pharmacokinetics and the associations between drug exposure and longitudinal clinical outcome measures revealed that higher isradipine plasma exposure modestly decreased cumulative levodopa dose requirement and the time for initiation of antiparkinsonian treatment in the phase III patients (Venuto et al. 2021). This re-evaluation has been taken to suggest that if appropriate target engagement could be attained, CaV1.3 antagonism could have capacity to slow PD progression.

Indeed, it has been speculated that a lack of adequate target engagement with a clinically tolerated dose may be responsible for the negative clinical trial outcome of STEADY-PD III (Maiti and Perlmutter 2020). While this is a feasible explanation, an alternative scenario is possible. Specifically, while there is abundant evidence that Cav1.3 channel antagonism with isradipine or nimodipine can prevent oxidative and mitochondrial toxin-induced death of SNpc DA neurons following 6OHDA or MPTP in animal models, there is a lack of evidence documenting the ability of Cav1.3 antagonism to impact α-syn aggregation and/or toxicity *in vivo*. Given that α-syn oligomerization and aggregation is currently considered a central factor in the death of SNpc DA neurons in PD, studies are needed to elucidate whether the lack of/suboptimal efficacy of isradipine in STEADY-III was due to the pharmacological limitations associated with this drug (Kang et al. 2013a, b) that might be readily addressed with new generations of drugs or gene therapy approaches, and/or whether this is because of the inability of SNpc Cav1.3 silencing to impact the α-syn-induced pathology central to PD (Fig. 4d). It is noteworthy to mention here that in addition to Cav1.3 LTCC, there is novel evidence suggesting that R- and T-type Ca^{2+} channels (RTCC, TTCC), which are also found in SNpc DA neurons, may hold promise as neuroprotective targets in PD (Ortner 2021). While selective and clinically suitable TTCC inhibitors exist, like Cav1.3 LTCC, there are no selective RTCC inhibitors (for review (Ortner 2021)). Given that adequate target engagement will be essential regardless of the Ca^{2+} channel target phenotype or intracellular pathogen(s), we posit that the evidence that gene level silencing with rAAV-CaV1.3-shRNA showed superior "therapeutic potency" compared to pharmacological Cav1.3 antagonists for LID (Schuster et al. 2009 and Soderstrom et al. 2010 *versus* Steece-Collier et al. 2019) suggests that advances in genetic medicine may pave the way for the next-generation technologies that will ensure adequate target engagement and meaningful neuroprotection in PD.

5 Advantages and Disadvantages of Gene Therapy

In the simplest terms, gene therapy involves the introduction of nucleic acid-based molecules into a patient to treat a given disease. Although gene therapy is still a relatively cutting-edge therapeutic strategy, its roots reach back as far as Fredrick Griffith's 1928 transformation experiment in which he explored the transfer of genetic material between bacteria (Wirth et al. 2013).

5.1 Gene Therapy: Clinical Challenges and Advances

In general, gene therapy can be classified into two categories, those that affect only somatic cells and those that are inherited through germline cells. Not surprisingly, ethical and legal parameters dictate that gene therapy in humans can only be performed in somatic cells (Wirth et al. 2013). Gene therapies can be further classified by the class of treated disease, vector type, mechanistic strategy, and delivery method (Anguela and High 2019; Iqubal et al. 2020). The first gene therapy clinical study at the National Institutes of Health began in 1990, investigating a rare immunodeficiency disorder known as Adenosine Deaminase Deficiency (Anguela and High 2019). In 2008, Cerepro®, a gene-based treatment for malignant brain tumors, became the first adenoviral vector to complete a phase III clinical trial (Sandmair et al. 2000). More recently, the AAV-based gene therapy LUXTURNAR® was the first of its kind in the USA to gain approval in 2017 after a randomized controlled Phase III study produced convincing evidence regarding its efficacy in inherited retinal dystrophy with bi-allelic mutations in RPE65 (Anguela and High 2019; Maguire et al. 2019; Russell et al. 2017). LUXTURNA®, a prescription gene therapy product, was seminal in its utilization of AAV technology, paving the way for AAV as a clinically relevant technology (Maguire et al. 2019).

There is significant promise for gene therapy within the CNS. Delivery could occur either by remote/systemic delivery, where the product would need to cross the blood-brain barrier (BBB) after non-invasive administration, or direct delivery via stereotaxic injection into a particular site within CNS (O'Connor and Boulis 2015). Undoubtedly, viral immunogenicity and toxicity are of particular concern for all gene therapies, but significant strides have been made in the engineering of viral vectors to reduce such complications (Bulaklak and Gersbach 2020). In 1999, the tragic death of renowned scientific pioneer, 18-year-old Jesse Gelsinger, following his involvement in a gene therapy safety trial for an X-linked genetic disease (ornithine transcarbamylase deficiency) raised ethical concerns about the feasibility of gene therapy in human (Liang and Mackey 2010; Wilson 2010; Wirth and Yla-Herttuala 2006). He was the first person publicly identified as dying from toxic immunogenicity and multiorgan failure, which occurred four days after treatment (Wirth and Yla-Herttuala 2006). Moreover, in recent tragic events in the Audentes AT132 clinical trial for X-linked myotubular myopathy, two patients perished, likely due to a combination of liver disease and the extremely high dose of AAV delivered (Biotech 2020). This has led to the field re-evaluating delivery paradigms, and it is likely that CNS gene therapy will involve targeted delivery to avoid both immunological and dosing risks. As such, decades of clinical and preclinical investigations, filled with successes and failures, have laid the foundation for significant advances in the next generation of technologies. Efforts and advances have been directed at means to circumvent immune obstacles by modifying AAV capsids to evade pre-existing neutralizing antibodies and development of new methods for temporarily clearing of antibodies from circulation together with advances in approaches of avoiding adaptive immunity to vectors (for review (Bulaklak and Gersbach 2020)).

5.2 Gene Therapy for PD

Current pharmacologic and surgical therapies for individuals with PD include dopamine agonists, monoamine (MAO)-B and catechol-o-methyl transferase (COMT) inhibitors, focused ultrasound neuroablative surgery, and deep brain stimulation (DBS). Although PD was first described in 1817 by English physician James Parkinson as a "shaking palsy", and many treatments have been tried over the past 150 years, there remains no "perfect" therapy. Medication related side-effects such as LIDs, waning therapeutic efficacy, and cognitive and emotional deterioration in late-stage PD following DBS (Fasano et al. 2015) continue to drive researchers and clinicians to hunt for additional/novel treatment options for individuals with PD. Interestingly, PD represents a prime candidate for directed gene therapy considering the centrally located damage in the substantia nigra and afferent nuclei.

One of the first series of gene transfer studies conducted in late-stage PD patients involved viral vector deliver of the growth factor neurturin (NRTN) delivered stereotaxically to the putamen and/or SN (Bartus et al. 2011; Herzog et al. 2013; Marks et al. 2010). NRTN belongs to the glial cell line-derived neurotrophic factor (GDNF) family, which regulates the survival and function of neurons (Kotzbauer et al. 1996). This trophic family of factors has received a great deal of attention in PD since it was established to potently support the vitality and neurite outgrowth of SNpc DA neurons (Gasmi et al. 2007; Kordower et al. 2006). Despite several clinical trials with AAV2-NRTN (CRE-120), meaningful clinical effects were not realized, however a great deal was learned about targeting and dose. Further, it was realized that given the significant degenerative state of DA neurons in late-stage PD, administration of trophic factor gene therapies might require administration early in the disease to preserve DA neurons and their connections that are lost in late-stage PD (Bartus et al. 2013; Manfredsson et al. 2020).

Another early gene therapy approach that showed some efficacy in PD and advanced to a Phase II trial was one involving infusion of AAV2-glutamic acid decarboxylase (GAD) delivered to the STN. GAD is the enzyme that catalyzes synthesis of the inhibitory neurotransmitter GABA. The loss of SNpc DA neurons in PD alters striato-pallidal circuitry resulting in decreased GABA input to the STN nucleus, rendering it "disinhibited." In this Phase II double-blind, randomized control study, LeWitt et al. (2011) used gene transfer to modulate/increase GABA in the STN of late-stage PD patients. Although the AAV2-GAD was shown to be effective compared to the sham control treatment, the improvements were not greater than standard PD treatment and the study was terminated.

Gene therapy research continues to examine various targets and define optimal delivery and patient profiles. For example, a unique approach with the idea that a single target might be insufficient to address a complex neurodegenerative condition involved ProSavin, a gene therapy utilizing a tricistronic vector system. This approach was designed to introduce three DA synthetic enzymes, AADC (aromatic-l-amino acid decarboxylase), TH (tyrosine hydroxylase), and GCH (GTP cyclohydroxylase I, a critical TH cofactor) within a single lentiviral viral vector that was injected bilaterally into the putamen (Azzouz et al. 2002; Palfi et al. 2014). After the success of this tricistronic vector system in rats (Azzouz et al. 2002) and non-human primates (Jarraya et al. 2009), investigation with ProSavin proceeded as an open-label Phase 1 clinical trial whose primary goal was to establish the safety of the treatments in humans (Palfi et al. 2014). While the study provided encouraging safety data, the clinical efficacy was modest. These findings were attributed to imprecise administration of the ProSavin, highlighting an additional need for gene therapy, specifically, the need for real-time MRI-guided injections.

Gene therapy is currently one of the most exciting areas of biotechnology (Bulaklak and Gersbach 2020). After decades of research, there are great expectations for the future ability of gene therapy technologies to impact a variety of maladies in humans, including PD. Presently, there are 12 trials involving PD and gene therapy listed on clinicaltrials.gov, although five are listed as "terminated." The overarching goal of most of these trials is to enhance striatal DA. However, DA replacement is just the beginning of the potential for gene modification in PD. There are currently no clinical trials directly targeting the debilitating side-effect of LID or aimed at providing neuroprotection in PD. With proper vector design and delivery to limit negative side effects, preclinical assessment of safety and efficacy in multiple models, as well as phased clinical trials, gene therapy could represent a promising future for individuals with PD, potentially impacting symptomatic benefit, reducing side-effect liability, and delaying or halting disease progression. We posit that *gene therapy interventions to silence Cav1.3 in striatal and nigral neurons* hold potential to impact all three aspects of PD, a disease that has knowingly plagued our aging population for centuries.

6 Conclusion: Cav1.3 Silencing Providing a Trifecta of Therapeutic Benefit in PD

The combinatorial outcomes of preclinical pharmacological and gene-therapy-mediated studies provides strong impetus for moving forward with continued development of genetic approaches to target Cav1.3 channels in PD. There is a strong likelihood that the ability to overcome pharmacological limitations observed in the treatment of LID using a continuous and precise gene-therapy approach will also translate into disease-modifying strategies. Thus, gene-therapy-mediated targeting of these channels may represent a "trifecta" in the treatment of PD-retarding or halting disease progression, improving L-DOPA response, whilst avoiding LID. With increasing skillsets and comfort with neurosurgical techniques, paired with continuous vector development, there remain few roadblocks for translating this approach into clinical reality.

Acknowledgements We would like to acknowledge the assistance of Devon Eldred and Rachel Harmon for their contributions on researching literature on channelopathies (Eldred) and the gene therapy (Harmon) and providing information used in the writing of this manuscript.

Author Contributions KSC and MEC conceived and wrote the text, and designed and composed the figures. FPM has provided intellectual input to all studies discussed in this review and provided extensive editing of this manuscript.

References

Anguela XM, High KA (2019) Entering the modern era of gene therapy. Annu Rev Med 70:273–288

Azzouz M, Martin-Rendon E, Barber RD, Mitrophanous KA, Carter EE, Rohll JB, Kingsman SM, Kingsman AJ, Mazarakis ND (2002) Multicistronic lentiviral vector-mediated striatal gene transfer of aromatic L-amino acid decarboxylase, tyrosine hydroxylase, and GTP cyclohydrolase I induces sustained transgene expression, dopamine production, and functional improvement in a rat model of Parkinson's disease. J Neurosci 22:10302–10312

Bartus RT, Brown L, Wilson A, Kruegel B, Siffert J, Johnson EM Jr, Kordower JH, Herzog CD (2011) Properly scaled and targeted AAV2-NRTN (neurturin) to the substantia nigra is safe, effective and causes no weight loss: support for nigral targeting in Parkinson's disease. Neurobiol Dis 44:38–52

Bartus RT, Baumann TL, Brown L, Kruegel BR, Ostrove JM, Herzog CD (2013) Advancing neurotrophic factors as treatments for age-related neurodegenerative diseases: developing and demonstrating "clinical proof-of-concept" for AAV-neurturin (CERE-120) in Parkinson's disease. Neurobiol Aging 34:35–61

Bastide MF, Meissner WG, Picconi B, Fasano S, Fernagut PO, Feyder M, Francardo V, Alcacer C, Ding Y, Brambilla R, Fisone G, Jon Stoessl A, Bourdenx M, Engeln M, Navailles S, De Deurwaerdere P, Ko WK, Simola N, Morelli M, Groc L, Rodriguez MC, Gurevich EV, Quik M, Morari M, Mellone M, Gardoni F, Tronci E, Guehl D, Tison F, Crossman AR, Kang UJ, Steece-Collier K, Fox S, Carta M, Angela Cenci M, Bezard E (2015) Pathophysiology of L-dopa-induced motor and non-motor complications in Parkinson's disease. Prog Neurobiol 132: 96–168

Bateup HS, Santini E, Shen W, Birnbaum S, Valjent E, Surmeier DJ, Fisone G, Nestler EJ, Greengard P (2010) Distinct subclasses of medium spiny neurons differentially regulate striatal motor behaviors. Proc Natl Acad Sci U S A 107:14845–14850

Berger SM, Bartsch D (2014) The role of L-type voltage-gated calcium channels Cav1.2 and Cav1.3 in normal and pathological brain function. Cell Tissue Res 357:463–476

Bhat S, Dao DT, Terrillion CE, Arad M, Smith RJ, Soldatov NM, Gould TD (2012) CACNA1C (Cav1.2) in the pathophysiology of psychiatric disease. Prog Neurobiol 99:1–14

Biotech NIB-N (2020) High-dose AAV gene therapy deaths. Nat Biotechnol 38:910

Bolam JP, Pissadaki EK (2012) Living on the edge with too many mouths to feed: why dopamine neurons die. Mov Disord 27:1478–1483

Bolam JP, Hanley JJ, Booth PA, Bevan MD (2000) Synaptic organisation of the basal ganglia. J Anat 196(Pt 4):527–542

Bulaklak K, Gersbach CA (2020) The once and future gene therapy. Nat Commun 11:5820

Callens M, Loncke J, Bultynck G (2022) Dysregulated Ca(2+) homeostasis as a central theme in neurodegeneration: lessons from Alzheimer's disease and Wolfram syndrome. Cells 11

Catoni C, Cali T, Brini M (2019) Calcium, dopamine and neuronal calcium sensor 1: their contribution to Parkinson's disease. Front Mol Neurosci 12:55

Catterall WA, Goldin AL, Waxman SG (2005) International Union of Pharmacology. XLVII. Nomenclature and structure-function relationships of voltage-gated sodium channels. Pharmacol Rev 57:397–409

Chen T, Lin F, Cai G (2021) Comparison of the efficacy of deep brain stimulation in different targets in improving gait in Parkinson's disease: a systematic review and Bayesian network meta-analysis. Front Hum Neurosci 15:749722

Daubner SC, Le T, Wang S (2011) Tyrosine hydroxylase and regulation of dopamine synthesis. Arch Biochem Biophys 508:1–12

Day M, Wang Z, Ding J, An X, Ingham CA, Shering AF, Wokosin D, Ilijic E, Sun Z, Sampson AR, Mugnaini E, Deutch AY, Sesack SR, Arbuthnott GW, Surmeier DJ (2006) Selective elimination of glutamatergic synapses on striatopallidal neurons in Parkinson disease models. Nat Neurosci 9:251–259

DeLong MR (1990) Primate models of movement disorders of basal ganglia origin. Trends Neurosci 13:281–285

Ebert DH, Greenberg ME (2013) Activity-dependent neuronal signalling and autism spectrum disorder. Nature 493:327–337

Fahn S (2015) The medical treatment of Parkinson disease from James Parkinson to George Cotzias. Mov Disord 30:4–18

Fahn S, Poewe W (2015) Levodopa: 50 years of a revolutionary drug for Parkinson disease. Mov Disord 30:1–3

Fasano A, Aquino CC, Krauss JK, Honey CR, Bloem BR (2015) Axial disability and deep brain stimulation in patients with Parkinson disease. Nat Rev Neurol 11:98–110

Fieblinger T, Cenci MA (2015) Zooming in on the small: the plasticity of striatal dendritic spines in L-DOPA-induced dyskinesia. Mov Disord 30:484–493

Fieblinger T, Graves SM, Sebel LE, Alcacer C, Plotkin JL, Gertler TS, Chan CS, Heiman M, Greengard P, Cenci MA, Surmeier DJ (2014) Cell type-specific plasticity of striatal projection neurons in parkinsonism and L-DOPA-induced dyskinesia. Nat Commun 5:5316

Flucher BE (2020) Skeletal muscle CaV1.1 channelopathies. Pflugers Arch 472:739–754

Foehring RC, Zhang XF, Lee JC, Callaway JC (2009) Endogenous calcium buffering capacity of substantia nigral dopamine neurons. J Neurophysiol 102:2326–2333

Follett KA (2004) Comparison of pallidal and subthalamic deep brain stimulation for the treatment of levodopa-induced dyskinesias. Neurosurg Focus 17:E3

Follett J, Darlow B, Wong MB, Goodwin J, Pountney DL (2013) Potassium depolarization and raised calcium induces alpha-synuclein aggregates. Neurotox Res 23:378–392

Frederiksen HR, Doehn U, Tveden-Nyborg P, Freude KK (2020) Non-immunogenic induced pluripotent stem cells, a promising way forward for allogenic transplantations for neurological disorders. Front Genome Ed 2:623717

Furukawa K, Matsuzaki-Kobayashi M, Hasegawa T, Kikuchi A, Sugeno N, Itoyama Y, Wang Y, Yao PJ, Bushlin I, Takeda A (2006) Plasma membrane ion permeability induced by mutant alpha-synuclein contributes to the degeneration of neural cells. J Neurochem 97:1071–1077

Gasmi M, Brandon EP, Herzog CD, Wilson A, Bishop KM, Hofer EK, Cunningham JJ, Printz MA, Kordower JH, Bartus RT (2007) AAV2-mediated delivery of human neurturin to the rat nigrostriatal system: long-term efficacy and tolerability of CERE-120 for Parkinson's disease. Neurobiol Dis 27:67–76

Gerfen CR, Surmeier DJ (2011) Modulation of striatal projection systems by dopamine. Annu Rev Neurosci 34:441–466

Ghio S, Camilleri A, Caruana M, Ruf VC, Schmidt F, Leonov A, Ryazanov S, Griesinger C, Cauchi RJ, Kamp F, Giese A, Vassallo N (2019) Cardiolipin promotes pore-forming activity of alpha-synuclein oligomers in mitochondrial membranes. ACS Chem Nerosci 10:3815–3829

Graybiel AM (1995) Building action repertoires: memory and learning functions of the basal ganglia. Curr Opin Neurobiol 5:733–741

Gurung R, Prata DP (2015) What is the impact of genome-wide supported risk variants for schizophrenia and bipolar disorder on brain structure and function? A systematic review. Psychol Med 45:2461–2480

Guzman JN, Sanchez-Padilla J, Wokosin D, Kondapalli J, Ilijic E, Schumacker PT, Surmeier DJ (2010) Oxidant stress evoked by pacemaking in dopaminergic neurons is attenuated by DJ-1. Nature 468:696–700

Hauser RA, Pahwa R, Tanner CM, Oertel W, Isaacson SH, Johnson R, Felt L, Stempien MJ (2017) ADS-5102 (Amantadine) extended-release capsules for levodopa-induced dyskinesia in Parkinson's disease (EASE LID 2 Study): interim results of an open-label safety study. J Parkinsons Dis 7:511–522

Hell JW, Westenbroek RE, Warner C, Ahlijanian MK, Prystay W, Gilbert MM, Snutch TP, Catterall WA (1993) Identification and differential subcellular localization of the neuronal class C and class D L-type calcium channel alpha 1 subunits. J Cell Biol 123:949–962

Hernandez-Lopez S, Tkatch T, Perez-Garci E, Galarraga E, Bargas J, Hamm H, Surmeier DJ (2000) D2 dopamine receptors in striatal medium spiny neurons reduce L-type Ca2+ currents and excitability via a novel PLC[beta]1-IP3-calcineurin-signaling cascade. J Neurosci Off J Soc Neurosci 20:8987–8995

Herzog CD, Brown L, Kruegel BR, Wilson A, Tansey MG, Gage FH, Johnson EM Jr, Bartus RT (2013) Enhanced neurotrophic distribution, cell signaling and neuroprotection following substantia nigral versus striatal delivery of AAV2-NRTN (CERE-120). Neurobiol Dis 58:38–48

Hettiarachchi NT, Parker A, Dallas ML, Pennington K, Hung CC, Pearson HA, Boyle JP, Robinson P, Peers C (2009) alpha-Synuclein modulation of Ca2+ signaling in human neuroblastoma (SH-SY5Y) cells. J Neurochem 111:1192–1201

Heyes S, Pratt WS, Rees E, Dahimene S, Ferron L, Owen MJ, Dolphin AC (2015) Genetic disruption of voltage-gated calcium channels in psychiatric and neurological disorders. Prog Neurobiol 134:36–54

Hofer NT, Pinggera A, Nikonishyna YV, Tuluc P, Fritz EM, Obermair GJ, Striessnig J (2021) Stabilization of negative activation voltages of Cav1.3 L-Type Ca(2+)-channels by alternative splicing. Channels (Austin) 15:38–52

Huot P, Johnston TH, Koprich JB, Fox SH, Brotchie JM (2013) The pharmacology of L-DOPA-induced dyskinesia in Parkinson's disease. Pharmacol Rev 65:171–222

Ilijic E, Guzman JN, Surmeier DJ (2011) The L-type channel antagonist isradipine is neuroprotective in a mouse model of Parkinson's disease. Neurobiol Dis 43:364–371

Ingham CA, Hood SH, Arbuthnott GW (1989) Spine density on neostriatal neurones changes with 6-hydroxydopamine lesions and with age. Brain Res 503:334–338

Ingham CA, Hood SH, van Maldegem B, Weenink A, Arbuthnott GW (1993) Morphological changes in the rat neostriatum after unilateral 6-hydroxydopamine injections into the nigrostriatal pathway. Exp Brain Res 93:17–27

Iqubal A, Iqubal MK, Khan A, Ali J, Baboota S, Haque SE (2020) Gene therapy, a novel therapeutic tool for neurological disorders: current progress, challenges and future prospective. Curr Gene Ther 20:184–194

Jarraya B, Boulet S, Ralph GS, Jan C, Bonvento G, Azzouz M, Miskin JE, Shin M, Delzescaux T, Drouot X, Herard AS, Day DM, Brouillet E, Kingsman SM, Hantraye P, Mitrophanous KA, Mazarakis ND, Palfi S (2009) Dopamine gene therapy for Parkinson's disease in a nonhuman primate without associated dyskinesia. Sci Transl Med 1:2ra4

Kabir ZD, Lee AS, Rajadhyaksha AM (2016) L-type Ca(2+) channels in mood, cognition and addiction: integrating human and rodent studies with a focus on behavioural endophenotypes. J Physiol 594:5823–5837

Kabir ZD, Martinez-Rivera A, Rajadhyaksha AM (2017) From gene to behavior: L-type calcium channel mechanisms underlying neuropsychiatric symptoms. Neurotherapeutics 14:588–613

Kang S, Cooper G, Dunne SF, Dusel B, Luan CH, Surmeier DJ, Silverman RB (2012) CaV1.3-selective L-type calcium channel antagonists as potential new therapeutics for Parkinson's disease. Nat Commun 3:1146

Kang S, Cooper G, Dunne SF, Luan CH, James Surmeier D, Silverman RB (2013a) Antagonism of L-type Ca2+ channels CaV1.3 and CaV1.2 by 1,4-dihydropyrimidines and 4H-pyrans as dihydropyridine mimics. Bioorg Med Chem 21:4365–4373

Kang S, Cooper G, Dunne SF, Luan CH, Surmeier DJ, Silverman RB (2013b) Structure-activity relationship of N,N'-disubstituted pyrimidinetriones as Ca(V)1.3 calcium channel-selective antagonists for Parkinson's disease. J Med Chem 56:4786–4797

Kordower JH, Herzog CD, Dass B, Bakay RA, Stansell J 3rd, Gasmi M, Bartus RT (2006) Delivery of neurturin by AAV2 (CERE-120)-mediated gene transfer provides structural and functional neuroprotection and neurorestoration in MPTP-treated monkeys. Ann Neurol 60:706–715

Koschak A (2010) Impact of gating modulation in CaV1.3 L-type calcium channels. Channels (Austin) 4:523–525

Koschak A, Fernandez-Quintero ML, Heigl T, Ruzza M, Seitter H, Zanetti L (2021) Cav1.4 dysfunction and congenital stationary night blindness type 2. Pflugers Arch 473:1437–1454

Kotzbauer PT, Lampe PA, Heuckeroth RO, Golden JP, Creedon DJ, Johnson EM Jr, Milbrandt J (1996) Neurturin, a relative of glial-cell-line-derived neurotrophic factor. Nature 384:467–470

Lees AJ, Tolosa E, Olanow CW (2015) Four pioneers of L-dopa treatment: Arvid Carlsson, Oleh Hornykiewicz, George Cotzias, and Melvin Yahr. Mov Disord 30:19–36

LeWitt PA, Rezai AR, Leehey MA, Ojemann SG, Flaherty AW, Eskandar EN, Kostyk SK, Thomas K, Sarkar A, Siddiqui MS, Tatter SB, Schwalb JM, Poston KL, Henderson JM, Kurlan RM, Richard IH, Van Meter L, Sapan CV, During MJ, Kaplitt MG, Feigin A (2011) AAV2-GAD gene therapy for advanced Parkinson's disease: a double-blind, sham-surgery controlled, randomised trial. Lancet Neurol 10:309–319

Liang BA, Mackey T (2010) Confronting conflict: addressing institutional conflicts of interest in academic medical centers. Am J Law Med 36:136–187

Lin Z, Zhang C, Li D, Sun B (2022) Preoperative levodopa response and deep brain stimulation effects on motor outcomes in Parkinson's disease: a systematic review. Mov Disord Clin Pract 9: 140–155

Liss B, Striessnig J (2019) The potential of L-type calcium channels as a drug target for neuroprotective therapy in Parkinson's disease. Annu Rev Pharmacol Toxicol 59:263–289

Lowe R, Pountney DL, Jensen PH, Gai WP, Voelcker NH (2004) Calcium(II) selectively induces alpha-synuclein annular oligomers via interaction with the C-terminal domain. Protein Sci 13: 3245–3252

Ludwig A, Flockerzi V, Hofmann F (1997) Regional expression and cellular localization of the alpha1 and beta subunit of high voltage-activated calcium channels in rat brain. J Neurosci 17: 1339–1349

Maguire AM, Russell S, Wellman JA, Chung DC, Yu ZF, Tillman A, Wittes J, Pappas J, Elci O, Marshall KA, McCague S, Reichert H, Davis M, Simonelli F, Leroy BP, Wright JF, High KA, Bennett J (2019) Efficacy, safety, and durability of voretigene neparvovec-rzyl in RPE65 mutation-associated inherited retinal dystrophy: results of phase 1 and 3 trials. Ophthalmology 126:1273–1285

Maiti B, Perlmutter JS (2020) A clinical trial of isradipine: what went wrong? Ann Intern Med 172: 625–626

Manfredsson FP, Polinski NK, Subramanian T, Boulis N, Wakeman DR, Mandel RJ (2020) The future of GDNF in Parkinson's disease. Front Aging Neurosci 12:593572

Marks WJ Jr, Bartus RT, Siffert J, Davis CS, Lozano A, Boulis N, Vitek J, Stacy M, Turner D, Verhagen L, Bakay R, Watts R, Guthrie B, Jankovic J, Simpson R, Tagliati M, Alterman R, Stern M, Baltuch G, Starr PA, Larson PS, Ostrem JL, Nutt J, Kieburtz K, Kordower JH, Olanow CW (2010) Gene delivery of AAV2-neurturin for Parkinson's disease: a double-blind, randomised, controlled trial. Lancet Neurol 9:1164–1172

Matamales M, Bertran-Gonzalez J, Salomon L, Degos B, Deniau JM, Valjent E, Herve D, Girault JA (2009) Striatal medium-sized spiny neurons: identification by nuclear staining and study of neuronal subpopulations in BAC transgenic mice. PLoS One 4:e4770

McNeill TH, Brown SA, Rafols JA, Shoulson I (1988) Atrophy of medium spiny I striatal dendrites in advanced Parkinson's disease. Brain Res 455:148–152

Mossner JM, Patil PG, Chou KL (2019) Subthalamic nucleus deep brain stimulation improves dyskinesias in Parkinson's disease beyond levodopa reduction. J Neural Transm (Vienna) 126: 1479–1483

Mouatt-Prigent A, Agid Y, Hirsch EC (1994) Does the calcium binding protein calretinin protect dopaminergic neurons against degeneration in Parkinson's disease? Brain Res 668:62–70

Nath S, Goodwin J, Engelborghs Y, Pountney DL (2011) Raised calcium promotes alpha-synuclein aggregate formation. Mol Cell Neurosci 46:516–526

Nishijima H, Suzuki S, Kon T, Funamizu Y, Ueno T, Haga R, Suzuki C, Arai A, Kimura T, Suzuki C, Meguro R, Miki Y, Yamada J, Migita K, Ichinohe N, Ueno S, Baba M, Tomiyama M (2014) Morphologic changes of dendritic spines of striatal neurons in the levodopa-induced dyskinesia model. Mov Disord 29:336–343

O'Connor DM, Boulis NM (2015) Gene therapy for neurodegenerative diseases. Trends Mol Med 21:504–512

Ortner NJ (2021) Voltage-gated Ca(2+) channels in dopaminergic substantia nigra neurons: therapeutic targets for neuroprotection in Parkinson's disease? Front Synaptic Neurosci 13: 636103

Ortner NJ, Striessnig J (2016) L-type calcium channels as drug targets in CNS disorders. Channels (Austin) 10:7–13

Ortner NJ, Kaserer T, Copeland JN, Striessnig J (2020) De novo CACNA1D Ca(2+) channelopathies: clinical phenotypes and molecular mechanism. Pflugers Arch 472:755–773

Palfi S, Gurruchaga JM, Ralph GS, Lepetit H, Lavisse S, Buttery PC, Watts C, Miskin J, Kelleher M, Deeley S, Iwamuro H, Lefaucheur JP, Thiriez C, Fenelon G, Lucas C, Brugieres P, Gabriel I, Abhay K, Drouot X, Tani N, Kas A, Ghaleh B, Le Corvoisier P,

Dolphin P, Breen DP, Mason S, Guzman NV, Mazarakis ND, Radcliffe PA, Harrop R, Kingsman SM, Rascol O, Naylor S, Barker RA, Hantraye P, Remy P, Cesaro P, Mitrophanous KA (2014) Long-term safety and tolerability of ProSavin, a lentiviral vector-based gene therapy for Parkinson's disease: a dose escalation, open-label, phase 1/2 trial. Lancet 383:1138–1146

Paolini Sguazzi G, Muto V, Tartaglia M, Bertini E, Compagnucci C (2021) Induced pluripotent stem cells (iPSCs) and gene therapy: a new era for the treatment of neurological diseases. Int J Mol Sci 22

Parkinson Study Group STEADY-PD III Investigators (2020) Isradipine versus placebo in early Parkinson disease: a randomized trial. Ann Intern Med 172:591–598

Parres-Gold J, Chieng A, Wong Su S, Wang Y (2020) Real-time characterization of cell membrane disruption by alpha-synuclein oligomers in live SH-SY5Y neuroblastoma cells. ACS Chem Nerosci 11:2528–2534

Pinggera A, Lieb A, Benedetti B, Lampert M, Monteleone S, Liedl KR, Tuluc P, Striessnig J (2015) CACNA1D de novo mutations in autism spectrum disorders activate Cav1.3 L-type calcium channels. Biol Psychiatry 77:816–822

Platzer J, Engel J, Schrott-Fischer A, Stephan K, Bova S, Chen H, Zheng H, Striessnig J (2000) Congenital deafness and sinoatrial node dysfunction in mice lacking class D L-type Ca2+ channels. Cell 102:89–97

Poewe W, Antonini A (2015) Novel formulations and modes of delivery of levodopa. Mov Disord 30:114–120

Rahman MU, Bilal M, Shah JA, Kaushik A, Teissedre PL, Kujawska M (2022) CRISPR-Cas9-based technology and its relevance to gene editing in Parkinson's disease. Pharmaceutics 14

Rascol O, Tonges L, deVries T, Jaros M, Quartel A, Jacobs D, ALLAY-LID I and ALLAY-LID II Study Groups (2022) Immediate-release/extended-release amantadine (OS320) to treat Parkinson's disease with levodopa-induced dyskinesia: analysis of the randomized, controlled ALLAY-LID studies. Parkinsonism Relat Disord 96:65–73

Redgrave P, Rodriguez M, Smith Y, Rodriguez-Oroz MC, Lehericy S, Bergman H, Agid Y, DeLong MR, Obeso JA (2010) Goal-directed and habitual control in the basal ganglia: implications for Parkinson's disease. Nat Rev Neurosci 11:760–772

Romme IA, de Reus MA, Ophoff RA, Kahn RS, van den Heuvel MP (2017) Connectome disconnectivity and cortical gene expression in patients with schizophrenia. Biol Psychiatry 81:495–502

Russell S, Bennett J, Wellman JA, Chung DC, Yu ZF, Tillman A, Wittes J, Pappas J, Elci O, McCague S, Cross D, Marshall KA, Walshire J, Kehoe TL, Reichert H, Davis M, Raffini L, George LA, Hudson FP, Dingfield L, Zhu X, Haller JA, Sohn EH, Mahajan VB, Pfeifer W, Weckmann M, Johnson C, Gewaily D, Drack A, Stone E, Wachtel K, Simonelli F, Leroy BP, Wright JF, High KA, Maguire AM (2017) Efficacy and safety of voretigene neparvovec (AAV2-hRPE65v2) in patients with RPE65-mediated inherited retinal dystrophy: a randomised, controlled, open-label, phase 3 trial. Lancet 390:849–860

Sandmair AM, Loimas S, Puranen P, Immonen A, Kossila M, Puranen M, Hurskainen H, Tyynela K, Turunen M, Vanninen R, Lehtolainen P, Paljarvi L, Johansson R, Vapalahti M, Yla-Herttuala S (2000) Thymidine kinase gene therapy for human malignant glioma, using replication-deficient retroviruses or adenoviruses. Hum Gene Ther 11:2197–2205

Sandoval IM, Marmion DJ, Meyers KT, Manfredsson FP (2021) Gene therapy to modulate alpha-synuclein in synucleinopathies. J Parkinsons Dis 11:S189–S197

Schallert T (2006) Behavioral tests for preclinical intervention assessment. NeuroRx 3:497–504

Schuster S, Doudnikoff E, Rylander D, Berthet A, Aubert I, Ittrich C, Bloch B, Cenci MA, Surmeier DJ, Hengerer B, Bezard E (2009) Antagonizing L-type Ca2+ channel reduces development of abnormal involuntary movement in the rat model of L-3,4-dihydroxyphenylalanine-induced dyskinesia. Biol Psychiatry 65:518–526

Sellnow RC, Newman JH, Chambers N, West AR, Steece-Collier K, Sandoval IM, Benskey MJ, Bishop C, Manfredsson FP (2019) Regulation of dopamine neurotransmission from

serotonergic neurons by ectopic expression of the dopamine D2 autoreceptor blocks levodopa-induced dyskinesia. Acta Neuropathol Commun 7:8
Sinnegger-Brauns MJ, Huber IG, Koschak A, Wild C, Obermair GJ, Einzinger U, Hoda JC, Sartori SB, Striessnig J (2009) Expression and 1,4-dihydropyridine-binding properties of brain L-type calcium channel isoforms. Mol Pharmacol 75:407–414
Soderstrom KE, O'Malley JA, Levine ND, Sortwell CE, Collier TJ, Steece-Collier K (2010) Impact of dendritic spine preservation in medium spiny neurons on dopamine graft efficacy and the expression of dyskinesias in parkinsonian rats. Eur J Neurosci 31:478–490
Steece-Collier K, Stancati JA, Collier NJ, Sandoval IM, Mercado NM, Sortwell CE, Collier TJ, Manfredsson FP (2019) Genetic silencing of striatal CaV1.3 prevents and ameliorates levodopa dyskinesia. Mov Disord 34:697–707
Striessnig J, Koschak A, Sinnegger-Brauns MJ, Hetzenauer A, Nguyen NK, Busquet P, Pelster G, Singewald N (2006) Role of voltage-gated L-type Ca2+ channel isoforms for brain function. Biochem Soc Trans 34:903–909
Suarez LM, Solis O, Carames JM, Taravini IR, Solis JM, Murer MG, Moratalla R (2014) L-DOPA treatment selectively restores spine density in dopamine receptor D2-expressing projection neurons in dyskinetic mice. Biol Psychiatry 75:711–722
Surmeier DJ, Schumacker PT (2013) Calcium, bioenergetics, and neuronal vulnerability in Parkinson's disease. J Biol Chem 288:10736–10741
Surmeier DJ, Bargas J, Hemmings HC Jr, Nairn AC, Greengard P (1995) Modulation of calcium currents by a D1 dopaminergic protein kinase/phosphatase cascade in rat neostriatal neurons. Neuron 14:385–397
Surmeier DJ, Guzman JN, Sanchez J, Schumacker PT (2012) Physiological phenotype and vulnerability in Parkinson's disease. Cold Spring Harb Perspect Med 2:a009290
Surmeier DJ, Graves SM, Shen W (2014) Dopaminergic modulation of striatal networks in health and Parkinson's disease. Curr Opin Neurobiol 29:109–117
Surmeier DJ, Halliday GM, Simuni T (2017a) Calcium, mitochondrial dysfunction and slowing the progression of Parkinson's disease. Exp Neurol 298:202–209
Surmeier DJ, Schumacker PT, Guzman JD, Ilijic E, Yang B, Zampese E (2017b) Calcium and Parkinson's disease. Biochem Biophys Res Commun 483:1013–1019
Sven Moosmang FH (2009a) Cav1.1 voltage-gated calcium channel. In: Enna DBBSJ (ed) xPharm: the comprehensive pharmacology reference. Elsevier, pp 1–4
Sven Moosmang FH (2009b) Cav1.4 voltage-gated calcium channel. In: Enna DBBSJ (ed) xPharm: the comprehensive pharmacology reference. Elsevier, pp 1–7
Tosatto L, Andrighetti AO, Plotegher N, Antonini V, Tessari I, Ricci L, Bubacco L, Dalla Serra M (2012) Alpha-synuclein pore forming activity upon membrane association. Biochim Biophys Acta 1818:2876–2883
Tsigelny IF, Sharikov Y, Wrasidlo W, Gonzalez T, Desplats PA, Crews L, Spencer B, Masliah E (2012) Role of alpha-synuclein penetration into the membrane in the mechanisms of oligomer pore formation. FEBS J 279:1000–1013
Venuto CS, Yang L, Javidnia M, Oakes D, James Surmeier D, Simuni T (2021) Isradipine plasma pharmacokinetics and exposure-response in early Parkinson's disease. Ann Clin Transl Neurol 8(3):603–612
Volles MJ, Lansbury PT Jr (2002) Vesicle permeabilization by protofibrillar alpha-synuclein is sensitive to Parkinson's disease-linked mutations and occurs by a pore-like mechanism. Biochemistry 41(14):4595–4602
Wall NR, De La Parra M, Callaway EM, Kreitzer AC (2013) Differential innervation of direct- and indirect-pathway striatal projection neurons. Neuron 79:347–360
Wilson RF (2010) The death of Jesse Gelsinger: new evidence of the influence of money and prestige in human research. Am J Law Med 36:295–325
Wirth T, Yla-Herttuala S (2006) Gene technology based therapies in the brain. Adv Tech Stand Neurosurg 31:3–32
Wirth T, Parker N, Yla-Herttuala S (2013) History of gene therapy. Gene 525:162–169

Xu W, Lipscombe D (2001) Neuronal Ca(V)1.3alpha(1) L-type channels activate at relatively hyperpolarized membrane potentials and are incompletely inhibited by dihydropyridines. J Neurosci Off J Soc Neurosci 21:5944–5951

Yeow SQZ, Loh KWZ, Soong TW (2021) Calcium channel splice variants and their effects in brain and cardiovascular function. Adv Exp Med Biol 1349:67–86

Zamponi GW, Striessnig J, Koschak A, Dolphin AC (2015) The physiology, pathology, and pharmacology of voltage-gated calcium channels and their future therapeutic potential. Pharmacol Rev 67:821–870

Zhai S, Tanimura A, Graves SM, Shen W, Surmeier DJ (2018) Striatal synapses, circuits, and Parkinson's disease. Curr Opin Neurobiol 48:9–16

Zhang Y, Meredith GE, Mendoza-Elias N, Rademacher DJ, Tseng KY, Steece-Collier K (2013) Aberrant restoration of spines and their synapses in L-DOPA-induced dyskinesia: involvement of corticostriatal but not thalamostriatal synapses. J Neurosci Off J Soc Neurosci 33:11655–11667

Design and Applications of Genetically-Encoded Voltage-Dependent Calcium Channel Inhibitors

Ariana C. Gavin and Henry M. Colecraft

Contents

Abstract

Ca^{2+} influx through high-voltage-gated Ca^{2+} channels (HVGCCs; Ca_V1/Ca_V2) is an exceptionally powerful and versatile signal that controls numerous cell and physiological functions including neurotransmission, muscle contraction, and regulation of gene expression. The impressive ability of a singular signal, Ca^{2+} influx, to have such a plethora of functional outcomes is enabled by: molecular diversity of HVGCC pore-forming α_1 and auxiliary subunits; organization of HVGCCs with extrinsic modulatory and effector protein to form discrete macro-molecular complexes with unique properties; distinctive distribution of HVGCCs

A. C. Gavin
Department of Molecular Pharmacology and Therapeutics, Columbia University Irving Medical Center, New York, NY, USA

H. M. Colecraft (✉)
Department of Molecular Pharmacology and Therapeutics, Columbia University Irving Medical Center, New York, NY, USA

Department of Physiology and Cellular Biophysics, Columbia University Irving Medical Center, New York, NY, USA
e-mail: hc2405@cumc.columbia.edu

J. Striessnig (ed.), *Voltage-gated Ca^{2+} Channels: Pharmacology, Modulation and their Role in Human Disease*, Handbook of Experimental Pharmacology 279,
https://doi.org/10.1007/164_2023_656

into separate subcellular compartments; and varying expression profiles of HVGCC isoforms among different tissues and organs. The capacity to block HVGCCs with selectivity and specificity with respect to the different levels of their organization is critical for fully understanding the scope of functional consequences of Ca^{2+} influx through them, and is also important for realizing their full potential as therapeutic targets. In this review, we discuss the gaps in the current landscape of small-molecule HVGCC blockers and how these may be addressed with designer genetically-encoded Ca^{2+} channel inhibitors (GECCIs) that draw inspiration from physiological protein inhibitors of HVGCCs.

1 Introduction

Ca^{2+} entry into cells through voltage-gated calcium channels (VGCCs) is an obligatory signal for many essential biological responses. In excitable cells including neurons, cardiac myocytes, and pancreatic β cells, distinct surface membrane VGCCs that are closed at resting membrane potentials open in response to membrane depolarization encoded in action potentials. The stochastic millisecond time-scale openings of VGCCs provide a water-filled pathway through which Ca^{2+} ions rush into the cell down a steep electrochemical gradient. The inflowing Ca^{2+} binds diverse Ca^{2+}-dependent regulatory and effector proteins to initiate essential physiological effects including (but not limited to) synaptic transmission in neurons, excitation-contraction coupling in cardiac myocytes, smooth muscle contraction, hormonal release from endocrine cells, and excitation-transcription coupling (Zamponi et al. 2015).

VGCCs are molecularly diverse and are traditionally classified into two main branches: 1) low-voltage-gated calcium channels (LVGCCs; $Ca_V3.1$–$Ca_V3.3$) which open at a threshold voltage of ~−70 mV and 2) high-voltage-gated calcium channels (HVGCCs; $Ca_V1.1$–$Ca_V1.4$, $Ca_V2.1$–$Ca_V2.3$) that have intermediate (~−55 mV; $Ca_V1.3$ and $Ca_V2.3$) and high (~−40 mV) thresholds for activation, respectively. The ten members of the VGCC family are comprised of distinct pore-forming α_1 subunits ($Ca_V1.1$ – *CACNA1S*; $Ca_V1.2$ – *CACNA1C*; $Ca_V1.3$ – *CACNA1D*; $Ca_V1.4$ – *CACNA1F*; $Ca_V2.1$ – *CACNA1A*; $Ca_V2.2$ – *CACNA1B*; $Ca_V2.3$ – *CACNA1E*; $Ca_V3.1$ – *CACNA1H*; $Ca_V3.2$ – *CACNA1H*; and $Ca_V3.3$ – *CACNA1I*) that share a similar topology with four homologous domains (DI–DIV) each with six transmembrane segments (S1–S6) (Catterall et al. 2005). DI – DIV are stitched together by intracellular loops of varying lengths (I-II, II-III, and III-IV loops), with cytosolic N- and C-termini preceding DI and following DIV, respectively. The S1–S4 modules of DI–DIV form non-equivalent voltage-sensors while S5–S6 segments come together to form the selectivity filter and pore. Expression of $Ca_V3.1$ (α_{1G}), $Ca_V3.2$ (α_{1H}), and $Ca_V3.3$ (α_{1I}) pore-forming α_1 subunits in heterologous cells is sufficient to produce robust LVGCC currents. By contrast, similar experiments with HVGCCs in mammalian cells typically yield meager currents mostly because the pore-forming α_1

subunits are retained in intracellular compartments. HVGCCs are multi-subunit complexes wherein α_1 subunits are associated with auxiliary proteins – $Ca_V\beta$, $\alpha_2\delta$, γ (in some cases) and calmodulin (CaM). There are four distinctive $Ca_V\beta$ ($Ca_V\beta_1$–$Ca_V\beta_4$) and $\alpha_2\delta$ ($\alpha_2\delta$-1–$\alpha_2\delta$-4) isoforms, respectively. The molecular diversity of HVGCCs is further enriched by alternative splicing that occurs in the α_1, $Ca_V\beta$, and $\alpha_2\delta$ subunits. $Ca_V\beta$s are typically cytosolic proteins (or membrane-associated in the case of β_{2a} and β_{2e}) and bind with nM affinity to a conserved α_1-interaction domain (AID) present in HVGCC α_1-subunit I-II loops. The α_1/$Ca_V\beta$ interaction promotes HVGCC α_1 subunit targeting to the cell surface; shifts the voltage dependence of channel activation in the hyperpolarizing direction; enhances single-channel open probability (P_o); and confers isoform-dependent distinctive inactivation signatures (Buraei and Yang 2010). The $\alpha_2\delta$ subunits are extracellular and tethered to the plasma membrane by a glycosylphosphatidylinositol (GPI) anchor (Dolphin 2013; Wu et al. 2016). Functionally, the $\alpha_2\delta$ subunits also stabilize pore-forming α_1 subunits at the surface, but only when $Ca_V\beta$ is also co-expressed (Cassidy et al. 2014). The various VGCC isoforms have distinctive tissue and cell distributions and physiological roles (Zamponi et al. 2015). In this review, we focus on genetically-encoded inhibitors for HVGCCs.

2 HVGCC Blockers as Research Tools and Therapeutics

The availability of selective VGCC blockers has been instrumental in advancing knowledge about the physiological roles of distinct VGCC isoforms, and in some cases, also provides important therapeutics. L-type Ca^{2+} channels (LTCCs; $Ca_V1.1$–$Ca_V1.4$) are blocked by three classes of drugs – dihydropyridines (e.g., nisoldipine), phenylalkylamines (e.g., verapamil), and benzothiazepines (e.g., diltiazem), which inhibit distinct LTCC isoforms and splice variants with differing potencies, and display state- and use-dependent block (Hockerman et al. 1997; Zamponi et al. 2015; Zhao et al. 2019). All three LTCC blocker classes are used to treat hypertension and angina, made possible by preferential block of smooth muscle over cardiac LTCCs due to the comparatively depolarized resting membrane potential of smooth muscle, and expression of distinct $Ca_V1.2$ splice variants. Verapamil and diltiazem are used for supraventricular tachycardias, based on their use-dependent block of $Ca_V1.2$ channels in the atrioventricular node.

$Ca_V2.2$ channels are important targets for the treatment of pain due to their role in synaptic transmission in somatosensory and central neurons. $Ca_V2.2$ channels are effectively blocked by toxins from venomous marine cone snails, ω-conotoxin GVIA, and ω-conotoxin MVIIA, which physically occlude the pore (Gao et al. 2021; Olivera et al. 1994). Ziconotide (Prialt) is a synthetic ω-conotoxin MVIIA used as a second-line treatment, for severe chronic pain, administered by intrathecal injection (Schmidtko et al. 2010). $Ca_V2.1$ channels are inhibited by the funnel web spider toxin, ω-agatoxin IVA, and $Ca_V2.3$ is blocked by the tarantula spider toxin SNX-482. $Ca_V2.3$ inhibitors may have applications for treatment of seizures, pain, and Parkinson's disease (Benkert et al. 2019; Zamponi et al. 2015).

Gabapentin and pregabalin are gabapentinoid drugs used to treat chronic pain and epilepsy and exert their therapeutic effects by binding $\alpha_2\delta$-1 and $\alpha_2\delta$-2 subunits (Field et al. 2006; Gee et al. 1996). Small molecules that disrupt the HVGCC α_1/$Ca_V\beta$ interaction have been developed and shown to inhibit $Ca_V2.2$ channel trafficking and display analgesia in rodent models of pain (Chen et al. 2018; Khanna et al. 2019).

3 Gaps in the Landscape of HVGCC Blockers

Within individual cells Ca^{2+} influx through VGCCs is often associated with several divergent functional outputs. For example, in a single neuron, Ca^{2+} influx through VGCCs can regulate cellular excitability, control neurotransmitter release, tune Ca^{2+}-dependent enzymes, and signal to the nucleus to direct transcription (Augustine et al. 2003; Dolphin and Lee 2020). The rich diversity of signaling functions of Ca^{2+} influx through VGCCs is in part due to local mechanisms that not only shape the nature of the signal, but also decode it to contribute to a global response. From this perspective, there are gaps in the availability of VGCC inhibitors that can more specifically eliminate Ca^{2+} influx through specific VGCC complexes spanning the nano-, micro-, and macro-scales. Here are a few specific examples. At the molecular level, currently available L-type Ca^{2+} channel blockers do not discriminate perfectly among the four $Ca_V1.1$ – $Ca_V1.4$ channel isoforms (Fig. 1a). There has been an interest, for example, in developing inhibitors with specificity for $Ca_V1.3$ over $Ca_V1.2$ as a means to more selectively achieve central objectives (treating Parkinson's disease) without confounding effects on the heart and blood vessels (Guzman et al. 2010; Kang et al. 2012; Liss and Striessnig 2019). Nevertheless, this has been challenging to achieve with small molecules (Filippini et al. 2023).

At the level of VGCC complexes, the functional logic and roles of $Ca_V\beta$ (and $\alpha_2\delta$) molecular diversity are incompletely understood (Fig. 1b). Distinct $Ca_V\beta$ isoforms have some overlapping functions such as the capacity to promote forward trafficking of HVGCC α_1 subunits, increase single-channel P_o, and produce hyperpolarizing shifts in the voltage dependence of activation (Buraei and Yang 2010). Nevertheless, data from knockout mice and model organism experiments support distinctive physiological roles for different $Ca_V\beta$ isoforms (Burgess et al. 1997; Gregg et al. 1996; Murakami et al. 2002; Weissgerber et al. 2006). A clear example of both overlapping and distinctive functions of $Ca_V\beta$ isoforms is demonstrated in β_1-null zebrafish (*relaxed*) which are paralyzed and die early after birth due to loss of skeletal excitation-contraction (EC) coupling (Schredelseker et al. 2005). Swapping in $Ca_V\beta_{2a}$ for $Ca_V\beta_{1a}$ restored whole-cell $Ca_V1.1$ current but did not recover tetrad formation or restore EC coupling (Schredelseker et al. 2009). In cells such as neurons that express multiple $Ca_V\beta$ isoforms (in contrast to skeletal muscle) the functional logic of $Ca_V\beta$ molecular diversity from gene knockout experiments is less clear-cut due to the phenomenon of subunit reshuffling and the partially overlapping $Ca_V\beta$ functions (McEnery et al. 1998). The ability to inhibit Ca_V channel complexes on the basis of the identity of associated $Ca_V\beta$ isoforms would provide

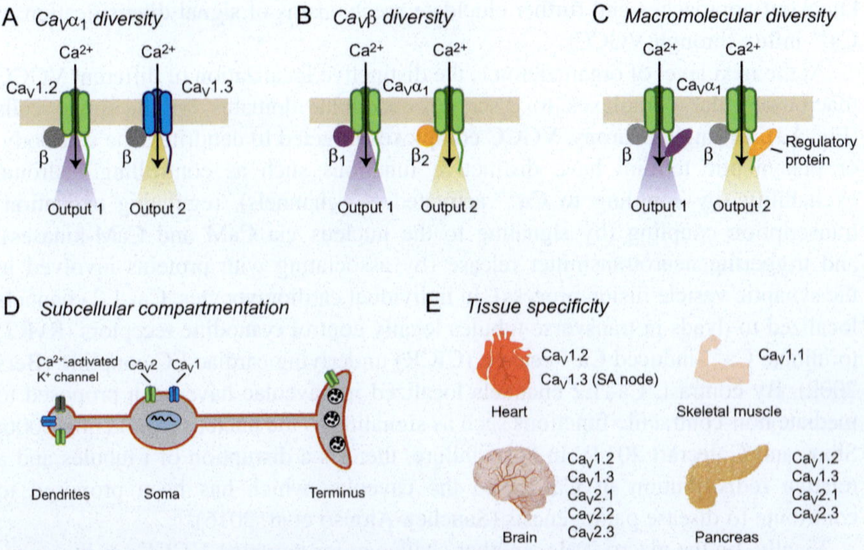

Fig. 1 Gaps in the current landscape of available high-voltage-gated Ca^{2+} channel blockers. (**a**) There is a deficiency of blockers that can distinguish between some HVGCC α_1-subunit isoforms, such as $Ca_V1.2$ and $Ca_V1.3$. (**b**) The different $Ca_V\beta$ subunit isoforms (β_1-β_4) have overlapping as well as unique functions. However, molecules that can block HVGCCs based on the identity of their associated $Ca_V\beta$ isoform are lacking. (**c**) HVGCCs associate with distinct regulatory and effector molecules to form discrete macromolecular signaling complexes in cells. The capacity to selectively inhibit such distinct HVGCC macromolecular signaling complexes is virtually non-existent but important for deepened insights into HVGCC signaling mechanisms and functions. (**d**) In single neurons, HVGCCs are localized to separate subcellular compartments where they engage in distinctive functions. Blockers that can selectively inhibit HVGCCs with subcellular precision are unavailable. (**e**) Different tissues and organs have unique as well as overlapping distribution of HVGCC isoforms. The presence of particular isoforms in more than one tissue can complicate targeting them therapeutically. Genetically-encoded Ca^{2+} channel inhibitors may uniquely address this limitation because they can be locally expressed in a particular cell type, tissue, or organ

complementary information to conventional gene knockout/knockdown and α_1 inhibition approaches to yield a more complete picture of the functional consequences of VGCC molecular diversity (Morgenstern et al. 2022).

Beyond immediate subunits, VGCCs can be associated with other proteins that either modulate their activity or sense and decode the local Ca^{2+} to initiate a downstream response (Fig. 1c). An example of this is the emerging view that in cardiac myocytes the physiologically important increase in L-type Ca^{2+} current in response to β-adrenergic agonists is due to relief from inhibition of a sub-population of $Ca_V1.2$ channels that are constitutively held in a low-P_o gating mode by the small G-protein Rad (Liu et al. 2020; Papa et al. 2021). Similar to as stated above, a capacity to inhibit VGCC macromolecular complexes on the basis of associated regulatory or signaling proteins would provide complementary information to gene knockout approaches and further elucidate mechanisms of signal diversification of Ca^{2+} influx through VGCCs.

At the next level of organization is the distinctive localization of different VGCC macromolecular complexes to discrete subcellular domains within single cells (Fig. 1d). In single neurons, VGCC complexes targeted to dendrites, the cell body, or presynaptic termini have distinctive functions such as controlling neuronal excitability (by coupling to Ca^{2+} activated K^+ channels), regulating excitation-transcription coupling (by signaling to the nucleus via CaM and CaM-kinases), and triggering neurotransmitter release (by associating with proteins involved in the synaptic vesicle fusion process). In individual cardiomyocytes, $Ca_V1.2$ channels localized to dyads in transverse tubules locally control ryanodine receptors (RyR2) to initiate Ca^{2+}-induced Ca^{2+} release (CICR) underlying cardiac EC coupling (Bers 2002). By contrast, $Ca_V1.2$ channels localized in caveolae have been proposed to mediate non-contractile functions such as signaling to the nucleus (Molkentin 2006; Shaw and Colecraft 2013). In heart failure, there is a disruption of t-tubules and a relative redistribution of $Ca_V1.2$ to the caveolae which has been proposed to contribute to disease pathogenesis (Sanchez-Alonso et al. 2016).

Finally, on the macro-scale, another challenge for targeting VGCCs as therapeutic targets is that individual channel isoforms are typically expressed in more than one cell type, tissues, or organs in the body (Zamponi et al. 2015) (Fig. 1e). Thus, it is difficult to achieve tissue- or organ-specific inhibition of a particular VGCC isoform leading to a propensity for on-target but off-tissue side effects due to depressing currents in unintended tissues or organs.

The foregoing discussion illustrates significant gaps in the availability of versatile and selective inhibitors of HVGCCs across various spatial scales – from individual α_1 isoforms; to complexes with distinct α_1 and auxiliary subunit isoforms; to macromolecular complexes involving non-core subunit regulatory or effector proteins; to divergent subcellular localizations; to cell type-specific inhibitors; and tissue or organ-specific regulation. Addressing these gaps with small molecules is difficult and likely not possible in many instances. By contrast, it may be possible to address many of these limitations with designer genetically-encoded HVGCC inhibitors (GECCIs) which have the key advantage that their expression may be precisely directed in desired cell types and even to selected subcellular sites (Colecraft 2020; Xu and Colecraft 2009).

4 Characteristics of Some Natural GECCIs

It is instructive when thinking of designer GECCIs to draw inspiration from the myriad natural proteins that inhibit distinct HVGCCs and thereby powerfully regulate physiology and can be (or have been) exploited for therapy. We consider here some examples of such natural GECCIs in terms of their functional roles and molecular mechanisms that provide guiding principles for designing novel GECCIs with unique properties for tailored applications.

G-protein inhibition of Ca_V2 channels: $Ca_V2.1$–$Ca_V2.3$ channels display a voltage-dependent type of inhibition characterized by a slowing of channel activation kinetics (kinetic slowing) and right-shift in voltage dependence of activation that is mediated by Gβγ released upon agonist binding to some G-protein coupled receptors (GPCRs). This form of modulation enables regulation of synaptic plasticity and is a prominent contributor to the mechanisms of opiate-induced analgesia. The modulation involves direct binding of Gβγ to Ca_V2 family α_1 subunits and shifting their gating from a willing to reluctant mode signature that can be directly visualized in single-channel recordings (Boland and Bean 1993; Colecraft et al. 2001; Herlitze et al. 1996; Ikeda 1996; Patil et al. 1996). Experiments using chimeras between $Ca_V2.2$ and the Gβγ-insensitive $Ca_V1.2$ revealed $Ca_V2.2$ α_{1B} N-terminus and I-II loop to be crucial to the regulation, with the N-terminus acting as a Gβ-γ-gated inhibitory module (Agler et al. 2005).

CaM regulation of Ca_V channels: Calmodulin (CaM) is constitutively bound to several Ca_V1/Ca_V2 channels and enables feedback regulation of the channels by permeating Ca^{2+} ions (Ben-Johny and Yue 2014; Peterson et al. 1999; Zuhlke et al. 1999). In $Ca_V1.2$ and $Ca_V1.3$, incoming Ca^{2+} binding to CaM induces a negative-feedback Ca^{2+}-dependent inactivation (CDI) which limits further Ca^{2+} influx. In ventricular cardiomyocytes, eliminating CDI by over-expressing a mutant CaM (CaM_{1234}) that does not bind Ca^{2+} causes markedly prolonged cardiac action potentials that would be pathologic (Alseikhan et al. 2002). Indeed, patients with CaM mutations in which $Ca_V1.2$ CDI is compromised display a marked QT prolongation and recurrent cardiac arrest (Crotti et al. 2013; Limpitikul et al. 2014). $Ca_V2.2$ and $Ca_V2.3$ channels also undergo CaM-dependent CDI, while $Ca_V2.1$ displays both CDI and Ca^{2+}-dependent facilitation (CDF) both of which also use resident CaM as the Ca^{2+} sensor (DeMaria et al. 2001; Liang et al. 2003). In $Ca_V1.3$ channels, apoCaM itself binding to the channel markedly increases single-channel P_o, and Ca^{2+} binding to CaM simply relieves this initial constitutive enhancement of current (Adams et al. 2014).

RGK protein inhibition of HVGCCs: RGK (Rad/Rem/Rem2/Gem/Kir) proteins are members of the superfamily of Ras-like G-proteins that act as GTP-regulated switches to control numerous cell processes. The role of RGK proteins as regulators of HVGCCs was first revealed when Gem was identified as a $Ca_V\beta$-binding protein in a yeast two-hybrid assay, and Gem over-expression strongly inhibited heterologously expressed Ca_V1 and Ca_V2 channels in a constitutive manner (Beguin et al. 2001). Subsequently, over-expressing the other members of the family – Rad, Rem, and Rem2 – were also found to constitutively inhibit both heterologously expressed and endogenous Ca_V1/Ca_V2 channels in cardiomyocytes, neurons, and pancreatic β cells (Chen et al. 2005; Finlin et al. 2003, 2005; Xu et al. 2010; Yang and Colecraft 2013; Yang et al. 2010). In atrial and ventricular cardiomyocytes, a sub-population of $Ca_V1.2$ channels are held in a low-P_o gating mode by constitutively bound Rad. β-adrenergic agonists activate protein kinase A which phosphorylates Rad to disengage from the $Ca_V1.2$ channel complex and relieve the constitutive inhibition (Liu et al. 2020; Papa et al. 2022). This explains the long sought-after and elusive mechanism of PKA-mediated enhancement of $Ca_V1.2$ in heart that is crucial for the physiological fight-or-flight response. Adenoviral-mediated focal delivery and expression of Gem in the atrioventricular (AV) node in a porcine model of atrial fibrillation resulted in a decreased atrio-ventricular (AV) node conduction and reduced heart rate, suggesting potential therapeutic utility of RGK proteins as GECCIs (Murata et al. 2004).

5 Engineering New Functionalities into Natural GECCIs

One approach to designing GECCIs with unique functions is to engineer new capabilities into proteins that are natural inhibitors of VGCCs. To this end, RGK proteins provide particularly fruitful examples. RGKs potently and non-selectively block all HVGCCs in a constitutive manner. We wondered whether it was possible to engineer them with two new features – 1) selectivity for particular HVGCC isoforms, and 2) acute control of their inhibitory function with a small molecule. Achieving this was enabled by an in-depth appreciation of nuances in the structure-function mechanisms by which distinct RGK proteins inhibit HVGCCs.

Engineering isoform selectivity into RGK proteins: To determine whether binding of RGK proteins to $Ca_V\beta$ was necessary for their ability to inhibit HVGCCs, we expressed selected Ca_V1/Ca_V2 isoforms with a mutant $Ca_V\beta$, $\beta_{2a,TM}$ that selectively lost interaction with RGK proteins, but retained the ability to bind α_1-subunits and form functional channels (Puckerin et al. 2018; Yang et al. 2012). Gem and Rem2 were found to lose the ability to inhibit Ca_V1/Ca_V2 channels reconstituted with $\beta_{2a,TM}$, indicating they suppressed currents using a β-binding-dependent mechanism. Rem similarly lost the capacity to inhibit $Ca_V1.3$, $Ca_V2.1$, and $Ca_V2.2$ channels reconstituted with $\beta_{2a,TM}$ indicating a purely β-binding-dependent mechanism of inhibition. By contrast, Rem retained a capacity to partially suppress $Ca_V1.2 + \beta_{2a,TM}$ channels, revealing the coexistence of both β-binding-dependent and β-binding-independent mechanisms of inhibition. Rem C-terminus directly binds to $Ca_V1.2$ α_{1C} subunit N-terminus and this interaction mediates the β-binding-independent inhibition mechanism (Yang et al. 2012). A mutant Rem that cannot bind $Ca_V\beta$, Rem[R200A/L227A], selectively inhibited reconstituted Cav1.2 channels, while having no effect on other Ca_V1/Ca_V2 channels (Puckerin et al. 2018). The absolute selectivity of Rem[R200A/L227A] for $Ca_V1.2$ over $Ca_V1.3$ compares favorably with small-molecule L-type Ca^{2+} channel blockers which do not display such a binary difference in efficacy, though they do engage the two channels with variations in potency. Rad was also found to inhibit $Ca_V1.2$ and $Ca_V2.2$ (but not $Ca_V1.3$ or $Ca_V2.1$) using both β-binding-dependent and β-binding-independent mechanisms. Accordingly, Rad[R208A/L235A], which cannot bind $Ca_V\beta$, was a selective inhibitor of $Ca_V1.2$ and $Ca_V2.2$ (but not $Ca_V1.3$ and $Ca_V2.1$) (Puckerin et al. 2018). Beyond reconstituted channels, Rem[R200A/L227A] suppressed endogenous $Ca_V1.2$ in adult guinea pig ventricular myocytes and mouse somatosensory dorsal root ganglion (DRG) neurons, while Rad[R208A/L235A] inhibited $Ca_V1.2$ in cardiomyocytes and $Ca_V1.2/Ca_V2.2$ in DRG neurons (Puckerin et al. 2018).

Engineering acutely inducible RGK-based GECCIs: In order to create an inducible RGK-based inhibitor that could be acutely regulated by an orthogonal signal such as a small molecule or light, a key finding was that RGK proteins target autonomously to the plasma membrane using a polybasic region in their C-termini. Removing this polybasic C-terminus eliminates both the capacity of RGK proteins to localize to the plasma membrane and to inhibit HVGCCs (Chen et al. 2005; Correll et al. 2007; Yang et al. 2010). We exploited this feature to create a genetically-encoded molecule for inducibly inactivating Ca_V channels (GEMIICC) by fusing C1 domain from protein kinase Cγ (PKCγ) to the N-terminus of truncated Rem lacking the last 32 residues of the C-terminus ($Rem_{1\text{-}265}$). $C1_{PKC\gamma}$-$Rem_{1\text{-}265}$ when expressed in cells is cytosolic but rapidly translocated to the plasma membrane in response to the small-molecule phorbol ester, phorbol 12,13-dibutyrate (PdBu), and concomitantly inhibited reconstituted $Ca_V1.2$ and $Ca_V2.2$ channels (Yang et al. 2007). The efficacy and selectivity of GEMIICCs could be tuned by varying the placement of the $C1_{PKC\gamma}$ module relative to $Rem_{1\text{-}265}$ (Yang et al. 2013).

A similar principle was used to generate a truncated Rem variant (optoRGK) whose association with the plasma membrane could be controlled by light, permitting optical control of HVGCC inhibition (Ma et al. 2018).

Channel inactivation by membrane-tethering an associated protein (ChIMP): In $Ca_V2.2$ channels reconstituted with $\beta_{2a,TM}$, $C1_{PKC\gamma}$-$Rem_{1\text{-}265}$ was ineffective in suppressing current when induced to associate with the plasma membrane by PdBu (Yang et al. 2012). This indicated that beyond the requirement for targeting to the plasma membrane, Rem-based GEMIICCs also need to bind $Ca_V\beta$ in order to inhibit HVGCCs. This dual requirement prompted a hypothesis that membrane binding of Rem (and other RGKs) inhibited channels by tugging on the I-II loop via the associated $Ca_V\beta$. We tested this hypothesis by attaching $C1_{PKC\gamma}$ directly to the C-terminus of $Ca_V\beta_3$ which enabled it to be inducibly associated with the membrane with PdBu. Channels reconstituted with β_3-$C1_{PKC\gamma}$ were inducibly inhibited by PdBu, with the strength and kinetics of inhibition tunable by adjusting the length of the $Ca_V\beta_3$ disordered C-terminus – shortening the disordered C-terminus increased the speed and strength of inhibition, with distinctive efficacies for $Ca_V1.2$, $Ca_V2.1$, and $Ca_V2.2$ channels (Yang et al. 2013).

We found that this approach termed ChIMP (channel inactivation by membrane-tethering an associated protein) could be generalized to other cytosolic proteins associated with HVGCC intracellular domains. For example, fusing $C1_{PKC\gamma}$ to 14-3-3, catalytically dead truncated Ca^{2+}-calmodulin-dependent protein kinase II ($CaMKII_{1\text{-}274,K42M}$), and CaM generated chimeric proteins that displayed PdBu-induced inhibition of HVGCCs, with distinctive efficacies for different channel isoforms (Yang et al. 2013). Further, fusing an 18-residue palmitoylated membrane-targeting peptide (mem) from neuromodulin to 14-3-3 generated a constitutive inhibitor of endogenous Ca_V1/Ca_V2 channels in murine DRG neurons and nerve-growth-factor (NGF)-differentiated rat pheochromocytoma (PC12) cells (Yang et al. 2013).

6 De Novo Genetically-Encoded Inhibitors

Although engineered RGK proteins have yielded efficacious GECCIs with unique properties, these may have significant limitations in their practical deployment because of the likelihood of confounding off-target effects. For example, beyond interacting with and modulating HVGCCs, RGK proteins, 14-3-3, CaM, and CaMKII are all known to regulate several other proteins. Thus, the identification of novel proteins that can bind HVGCC subunits and are not constitutively encoded within the human genome is an important aspect of designing more selective GECCIs with therapeutic potential.

Morgenstern et al. (2019) immunized llamas with purified β_1 and β_3 subunits which produce single variable domain (V_{HH}) antibodies in response to antigens (Morgenstern et al. 2019). The antigen recognition part of the antibody, known as a nanobody, forms a small protein scaffold of 12–15 kDa with three complementarity determining regions of varying lengths that participate in binding antigen. By contrast to conventional antibodies that are large (~150 kDa) and have disulfide-linked heavy and light-variable chains that disassemble in the reducing intracellular environment, nanobodies are stable and can specifically bind their target protein in the cytosol. A number of nanobody binders were identified using phage display and ELISA, including nb.F3, which bound all four $Ca_V\beta$ isoforms with nM affinity and 1:1 stoichiometry (Fig. 2a, b). When co-expressed with HVGCCs, nb.F3 had no impact on either channel targeting to the plasma membrane or on whole-cell currents, indicating it is functionally inert. Fusing the catalytic HECT domain of the E3 ubiquitin ligase, NEDD4L, to nb.F3 generated a construct named Ca_V-aβlator that strongly inhibits HVGCCs reconstituted with any $Ca_V\beta$ isoform by eliminating them from the cell surface in a ubiquitination-dependent manner (Fig. 2c, d). When deployed in DRG neurons that express multiple Cav1/Cav2 channels, Ca_V-aβlator virtually eliminated HVGCCs without affecting LVGCCs. Furthermore, expression of Ca_V-aβlator in adult guinea pig ventricular cardiomyocytes also eliminated endogenous $Ca_V1.2$ channels. Confocal imaging of $Ca_V1.2$ in Ca_V-aβlator-expressing cardiomyocytes revealed that though overall $Ca_V1.2$ fluorescence was undiminished the endogenous channels were redistributed away from dyadic junctions as indicated by a loss of colocalization with ryanodine receptors (RyR2). Instead, $Ca_V1.2$ fluorescence signal coincided with anti-Rab7 immunofluorescence, indicating a net relocation of the channel to Rab7-positive late endosomes. Ca_V-aβlator also strongly inhibited HVGCC currents in murine pancreatic beta-cells that express $Ca_V1.2$, $Ca_V1.3$, and $Ca_V2.1$, further demonstrating its use as a potent genetically-encoded inhibitor of HVGCCs across diverse cellular contexts (Morgenstern et al. 2019).

To explore the use of designer GECCIs as potential therapeutics, we investigated the capacity of Ca_V-aβlator to produce analgesia in a mouse model of chronic pain (Sun et al. 2022). Adult mice were injected subcutaneously in the left hind paw with AAV9 encoding either Ca_V-aβlator or tdTomato (control), and 4 weeks later, the spared nerve injury (SNI) model of neuropathic pain was initiated by ligating two branches of the sciatic nerve, the tibial and common peroneal nerves, while sparing

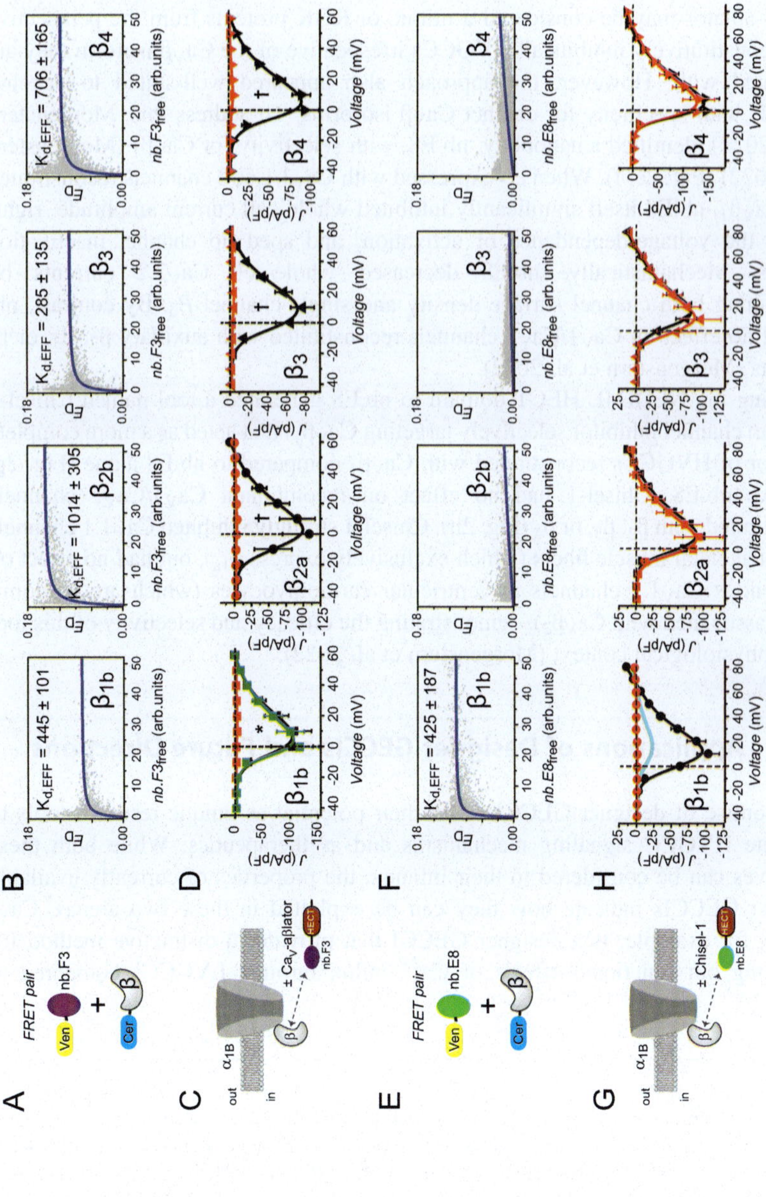

Fig. 2 Properties of nanobody-based HVGCC inhibitors – Ca_V-aβlator and Chisel-1. (**a**) Cartoon of Venus-nb.F3 + Cerulean-$Ca_V\beta$ FRET pair. (**b**) Flow cytometry FRET (flow-FRET) scatterplots with FRET efficiency (E_D) plotted as a function of free nb.F3 concentration. Blue lines fit a 1:1 binding model and

the sural nerve. Following SNI, control mice displayed a long-lasting mechanical allodynia and hyperalgesia to thermal and cold stimuli. By contrast, mice injected with Ca_V-aβlator AAV9 showed significantly less pain to mechanical, heat, and cold stimuli over the duration of behavioral assessments, with no apparent adverse effects (Sun et al. 2022) (Fig. 3).

Ca_V-aβlator may be considered a mimic of RGK proteins from the perspective that it constitutively inhibits all HVGCCs irrespective of the $Ca_V\beta$ isoform they are associated with. However, the approach also appeared well-suited to develop GECCIs with specificity for distinct $Ca_V\beta$ isoforms. To address this, Morgenstern et al. (2022) identified a nanobody, nb.E8, with selectivity for $Ca_V\beta_1$ (Morgenstern et al. 2022) (Fig. 2e, f). When co-expressed with Ca_V1/Ca_V2 channels reconstituted with $Ca_V\beta_1$, nb.E8 itself significantly inhibited whole-cell current amplitude, right-shifted the voltage-dependence of activation, and sped up channel inactivation kinetics. Mechanistically, nb.E8 decreased whole-cell $Ca_V2.2$ currents by diminishing both channel surface density and single-channel P_o. By contrast, nb. E8 had no effect on Ca_V1/Ca_V2 channels reconstituted with auxiliary β_2, β_3, or β_4 subunits (Morgenstern et al. 2022).

Fusing the NEDD4L HECT domain to nb.E8 generated a tool named Chisel-1 (calcium channel inhibitor selectively targeting $Ca_V\beta_1$) that acted as a more complete inhibitor of HVGCCs reconstituted with $Ca_V\beta_1$ compared to nb.E8 alone (Fig. 2g, h). Like nb.E8, Chisel-1 had no effect on recombinant Ca_V1/Ca_V2 channels reconstituted with β_2, β_3, or β_4 (Fig. 2h). Chisel-1 strongly inhibited $Ca_V1.1$ channels in adult skeletal muscle fibers (which exclusively express β_{1a}), but had no effect on endogenous $Ca_V1.2$ channels in ventricular cardiomyocytes (which are predominantly associated with $Ca_V\beta_2$), demonstrating the efficacy and selectivity of this tool in the physiological context (Morgenstern et al. 2022).

7 Applications of Designer GECCIs and Future Directions

The promise of designer GECCIs is in their potential as unique research tools to elucidate HVGCC signaling mechanisms and as therapeutics. While both these objectives can be considered in their infancy, the properties of currently available designer GECCIs indicate how they can be exploited in these two arenas. Ca_V-aβlator, for example, is a designer GECCI that provides a distinctive method for separating out functional effects of Ca^{2+} influx through LVGCCs compared to

Fig. 2 (continued) show that nb.F3 binds all four $Ca_V\beta$ isoforms. $K_{d,EFF}$, effective dissociation constant. (**c**) Cartoon of $\alpha_{1B} + \beta_x \pm Ca_V$-aβlator. (**d**) Current density vs voltage (*J-V*) relationships for cells expressing $\alpha_{1B} + \beta_x + \alpha_2\delta$-1 in the absence (black symbols) or presence (red symbols) of Ca_V-aβlator. (**e**) Cartoon of Venus-nb.E8 + Cerulean-$Ca_V\beta$. (**f**) Flow-FRET scatterplots and binding curves show nb.E8 exclusively binds $Ca_V\beta_1$. (**g**) Cartoon of $\alpha_{1B} + \beta_x \pm$ Chisel-1. (**h**) Current density vs voltage (*J-V*) relationships for cells expressing $\alpha_{1B} + \beta_x + \alpha_2\delta$-1 in the absence (black symbols) or presence (red symbols) of Chisel-1

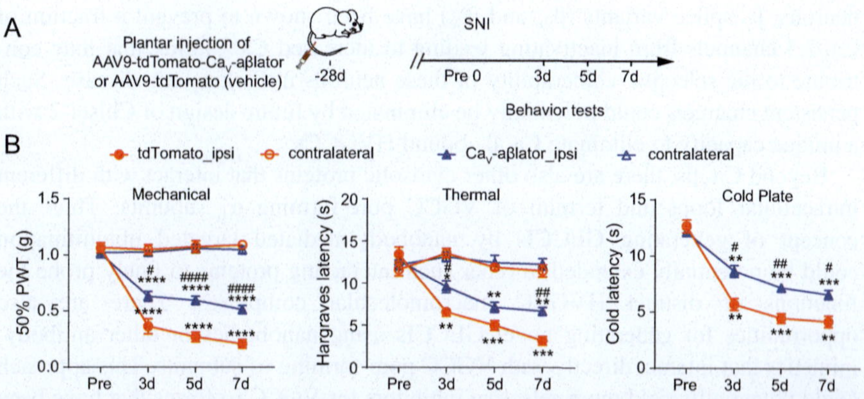

Fig. 3 Ca_V-aβlator produces analgesia in a spared nerve injury (SNI) model of chronic pain. (**a**) Cartoon of experimental design. Mice were subcutaneously injected in the hind paw with AAV9 encoding either tdTomato-p2A-Ca_V-aβlator or tdTomato alone (control) 4 weeks prior to SNI. (**b**) Behavior tests for nocifensive responses indicate SNI triggered development of mechanical, thermal, and cold hypersensitivity in the ipsilateral (ipsi) but not contralateral (contra) side of the mice. The hypersensitivity was significantly reduced in Ca_V-aβlator injected mice ($n = 6$–9 mice per group)

HVGCCs in cells that express multiple $Ca_V1/Ca_V2/Ca_V3$ channels. This capability is difficult to achieve cleanly with small molecules which often have imperfect selectivity. For example, the L-type Ca^{2+} channel blocker nifedipine also significantly inhibits LVGCCs.

The mechanisms of action of Ca_V-aβlator and Chisel-1 involve removal of entire Ca_V1/Ca_V2 complexes based on the associated $Ca_V\beta$ subunit (Morgenstern et al. 2019, 2022). This is a unique functionality that is qualitatively different from conventional approaches such as knockout or knockdown of $Ca_V\beta$ subunits. In the latter case, α_1 subunits are left intact and available to interact with other $Ca_V\beta$ isoforms that may be expressed in the cells, a process referred to as $Ca_V\beta$ reshuffling. An example of the discriminative power of Ca_V-aβlator is in adult cardiomyocytes in which it has been shown that association with a $Ca_V\beta$ is not necessary for forming functional $Ca_V1.2$ channels in the sarcolemma (Yang et al. 2019). Thus, it was possible that in adult ventricular cardiomyocytes there were separate populations of $Ca_V1.2$ with $Ca_V\beta$ bound or lacking $Ca_V\beta$, respectively. Ca_V-aβlator expressed in adult cardiomyocytes eliminates L-type Ca^{2+} current indicating that virtually all $Ca_V1.2$ channels contributing to the whole-cell current are associated with a $Ca_V\beta$ subunit.

Chisel-1 expressed in hippocampal neurons strongly suppressed KCl-depolarization-evoked nuclear phosphoCREB staining, revealing a prominent role of $Ca_V\beta_1$-bound HVGCCs in excitation-transcription coupling. Future design of Chisels-2/3/4 would create new selective GECCIs that enable new insights into the functional roles of $Ca_V\beta$ molecular diversity. In dopaminergic substantia nigra neurons, β_2 splice variants (β_{2a} and β_{2e}) have been shown to prevent a fraction of $Ca_V2.3$ channels from inactivating leading to increased Ca^{2+} influx that may contribute to the selective vulnerability of these neurons in Parkinson's disease. Such persistent channels could potentially be eliminated by future design of Chisel-2 with a unique capacity to eliminate $Ca_V\beta_2$-bound HVGCCs.

Beyond $Ca_V\beta$s, there are also other cytosolic proteins that interact with different intracellular loops and termini of VGCC pore-forming α_1 subunits. Thus, the concept of generating GECCIs by nanobody-mediated targeted ubiquitination could be potentially extended to other such interacting proteins to finely probe the functions of distinct HVGCC macromolecular complexes. There are also opportunities for generating novel GECCIs using nanobodies or other antibody-mimetics that interact directly with VGCC pore-forming α_1 subunits. This approach could potentially yield more selective inhibitors for VGCC isoforms that have been recalcitrant to specific small-molecule blockers, e.g., $Ca_V1.3$ vs $Ca_V1.2$ selective blockers.

Acknowledgments We thank members of the Colecraft who have contributed to the work on genetically-encoded high-voltage-gated Ca^{2+} channel inhibitors. This work was supported by grants RO1HL142111 and RO1-HL121253 from the NIH (to H.M.C.).

References

Adams PJ, Ben-Johny M, Dick IE, Inoue T, Yue DT (2014) Apocalmodulin itself promotes ion channel opening and Ca(2+) regulation. Cell 159:608–622

Agler HL, Evans J, Tay LH, Anderson MJ, Colecraft HM, Yue DT (2005) G protein-gated inhibitory module of N-type (ca(v)2.2) ca2+ channels. Neuron 46:891–904

Alseikhan BA, DeMaria CD, Colecraft HM, Yue DT (2002) Engineered calmodulins reveal the unexpected eminence of Ca2+ channel inactivation in controlling heart excitation. Proc Natl Acad Sci U S A 99:17185–17190

Augustine GJ, Santamaria F, Tanaka K (2003) Local calcium signaling in neurons. Neuron 40:331–346

Beguin P, Nagashima K, Gonoi T, Shibasaki T, Takahashi K, Kashima Y, Ozaki N, Geering K, Iwanaga T, Seino S (2001) Regulation of Ca2+ channel expression at the cell surface by the small G-protein kir/Gem. Nature 411:701–706

Ben-Johny M, Yue DT (2014) Calmodulin regulation (calmodulation) of voltage-gated calcium channels. J Gen Physiol 143:679–692

Benkert J, Hess S, Roy S, Beccano-Kelly D, Wiederspohn N, Duda J, Simons C, Patil K, Gaifullina A, Mannal N et al (2019) Cav2.3 channels contribute to dopaminergic neuron loss in a model of Parkinson's disease. Nat Commun 10:5094

Bers DM (2002) Cardiac excitation-contraction coupling. Nature 415:198–205

Boland LM, Bean BP (1993) Modulation of N-type calcium channels in bullfrog sympathetic neurons by luteinizing hormone-releasing hormone: kinetics and voltage dependence. J Neurosci 13:516–533

Buraei Z, Yang J (2010) The {beta} subunit of voltage-gated Ca2+ channels. Physiol Rev 90:1461–1506

Burgess DL, Jones JM, Meisler MH, Noebels JL (1997) Mutation of the Ca2+ channel beta subunit gene Cchb4 is associated with ataxia and seizures in the lethargic (lh) mouse. Cell 88:385–392

Cassidy JS, Ferron L, Kadurin I, Pratt WS, Dolphin AC (2014) Functional exofacially tagged N-type calcium channels elucidate the interaction with auxiliary alpha2delta-1 subunits. Proc Natl Acad Sci U S A 111:8979–8984

Catterall WA, Perez-Reyes E, Snutch TP, Striessnig J (2005) International Union of Pharmacology. XLVIII. Nomenclature and structure-function relationships of voltage-gated calcium channels. Pharmacol Rev 57:411–425

Chen H, Puhl HL 3rd, Niu SL, Mitchell DC, Ikeda SR (2005) Expression of Rem2, an RGK family small GTPase, reduces N-type calcium current without affecting channel surface density. J Neurosci 25:9762–9772

Chen X, Liu D, Zhou D, Si Y, Xu D, Stamatkin CW, Ghozayel MK, Ripsch MS, Obukhov AG, White FA et al (2018) Small-molecule CaValpha1CaVbeta antagonist suppresses neuronal voltage-gated calcium-channel trafficking. Proc Natl Acad Sci U S A 115:E10566–E10575

Colecraft HM (2020) Designer genetically encoded voltage-dependent calcium channel inhibitors inspired by RGK GTPases. J Physiol 598:1683–1693

Colecraft HM, Brody DL, Yue DT (2001) G-protein inhibition of N- and P/Q-type calcium channels: distinctive elementary mechanisms and their functional impact. J Neurosci 21: 1137–1147

Correll RN, Pang C, Finlin BS, Dailey AM, Satin J, Andres DA (2007) Plasma membrane targeting is essential for Rem-mediated Ca2+ channel inhibition. J Biol Chem 282:28431–28440

Crotti L, Johnson CN, Graf E, De Ferrari GM, Cuneo BF, Ovadia M, Papagiannis J, Feldkamp MD, Rathi SG, Kunic JD et al (2013) Calmodulin mutations associated with recurrent cardiac arrest in infants. Circulation 127:1009–1017

DeMaria CD, Soong TW, Alseikhan BA, Alvania RS, Yue DT (2001) Calmodulin bifurcates the local Ca2+ signal that modulates P/Q-type Ca2+ channels. Nature 411:484–489

Dolphin AC (2013) The alpha2delta subunits of voltage-gated calcium channels. Biochim Biophys Acta 1828:1541–1549

Dolphin AC, Lee A (2020) Presynaptic calcium channels: specialized control of synaptic neurotransmitter release. Nat Rev Neurosci 21:213–229

Field MJ, Cox PJ, Stott E, Melrose H, Offord J, Su TZ, Bramwell S, Corradini L, England S, Winks J et al (2006) Identification of the alpha2-delta-1 subunit of voltage-dependent calcium channels as a molecular target for pain mediating the analgesic actions of pregabalin. Proc Natl Acad Sci U S A 103:17537–17542

Filippini L, Ortner NJ, Kaserer T, Striessnig J (2023) Ca(v) 1.3-selective inhibitors of voltage-gated L-type Ca(2+) channels: fact or (still) fiction? Br J Pharmacol 180:1289–1303

Finlin BS, Crump SM, Satin J, Andres DA (2003) Regulation of voltage-gated calcium channel activity by the Rem and Rad GTPases. Proc Natl Acad Sci U S A 100:14469–14474

Finlin BS, Mosley AL, Crump SM, Correll RN, Ozcan S, Satin J, Andres DA (2005) Regulation of L-type Ca2+ channel activity and insulin secretion by the Rem2 GTPase. J Biol Chem 280: 41864–41871

Gao S, Yao X, Yan N (2021) Structure of human Cav2.2 channel blocked by the painkiller ziconotide. Nature 596:143–147

Gee NS, Brown JP, Dissanayake VU, Offord J, Thurlow R, Woodruff GN (1996) The novel anticonvulsant drug, gabapentin (Neurontin), binds to the alpha2delta subunit of a calcium channel. J Biol Chem 271:5768–5776

Gregg RG, Messing A, Strube C, Beurg M, Moss R, Behan M, Sukhareva M, Haynes S, Powell JA, Coronado R et al (1996) Absence of the beta subunit (cchb1) of the skeletal muscle dihydropyridine receptor alters expression of the alpha 1 subunit and eliminates excitation-contraction coupling. Proc Natl Acad Sci U S A 93:13961–13966

Guzman JN, Sanchez-Padilla J, Wokosin D, Kondapalli J, Ilijic E, Schumacker PT, Surmeier DJ (2010) Oxidant stress evoked by pacemaking in dopaminergic neurons is attenuated by DJ-1. Nature 468:696–700

Herlitze S, Garcia DE, Mackie K, Hille B, Scheuer T, Catterall WA (1996) Modulation of Ca2+ channels by G-protein beta gamma subunits. Nature 380:258–262

Hockerman GH, Johnson BD, Abbott MR, Scheuer T, Catterall WA (1997) Molecular determinants of high affinity phenylalkylamine block of L-type calcium channels in transmembrane segment IIIS6 and the pore region of the alpha1 subunit. J Biol Chem 272:18759–18765

Ikeda SR (1996) Voltage-dependent modulation of N-type calcium channels by G-protein beta gamma subunits. Nature 380:255–258

Kang S, Cooper G, Dunne SF, Dusel B, Luan CH, Surmeier DJ, Silverman RB (2012) CaV1.3-selective L-type calcium channel antagonists as potential new therapeutics for Parkinson's disease. Nat Commun 3:1146

Khanna R, Yu J, Yang X, Moutal A, Chefdeville A, Gokhale V, Shuja Z, Chew LA, Bellampalli SS, Luo S et al (2019) Targeting the CaValpha-CaVbeta interaction yields an antagonist of the N-type CaV2.2 channel with broad antinociceptive efficacy. Pain

Liang H, DeMaria CD, Erickson MG, Mori MX, Alseikhan BA, Yue DT (2003) Unified mechanisms of Ca2+ regulation across the Ca2+ channel family. Neuron 39:951–960

Limpitikul WB, Dick IE, Joshi-Mukherjee R, Overgaard MT, George AL Jr, Yue DT (2014) Calmodulin mutations associated with long QT syndrome prevent inactivation of cardiac L-type Ca(2+) currents and promote proarrhythmic behavior in ventricular myocytes. J Mol Cell Cardiol 74:115–124

Liss B, Striessnig J (2019) The potential of L-type calcium channels as a drug target for neuroprotective therapy in Parkinson's disease. Annu Rev Pharmacol Toxicol 59:263–289

Liu G, Papa A, Katchman AN, Zakharov SI, Roybal D, Hennessey JA, Kushner J, Yang L, Chen BX, Kushnir A et al (2020) Mechanism of adrenergic CaV1.2 stimulation revealed by proximity proteomics. Nature 577:695–700

Ma G, Liu J, Ke Y, Liu X, Li M, Wang F, Han G, Huang Y, Wang Y, Zhou Y (2018) Optogenetic control of voltage-gated calcium channels. Angew Chem Int Ed Engl 57:7019–7022

McEnery MW, Copeland TD, Vance CL (1998) Altered expression and assembly of N-type calcium channel alpha1B and beta subunits in epileptic lethargic (lh/lh) mouse. J Biol Chem 273:21435–21438

Molkentin JD (2006) Dichotomy of Ca2+ in the heart: contraction versus intracellular signaling. J Clin Invest 116:623–626

Morgenstern TJ, Park J, Fan QR, Colecraft HM (2019) A potent voltage-gated calcium channel inhibitor engineered from a nanobody targeted to auxiliary CaVbeta subunits. Elife 8

Morgenstern TJ, Nirwan N, Hernandez-Ochoa EO, Bibollet H, Choudhury P, Laloudakis JD, Johny MB, Bannister RA, Schneider MF, Minor DL, Colecraft HM (2022) Selective posttranslational inhibition of CaVβ1-associated voltage-dependent calcium channels with a functionalized nanobody. Nat Commun

Murakami M, Fleischmann B, De Felipe C, Freichel M, Trost C, Ludwig A, Wissenbach U, Schwegler H, Hofmann F, Hescheler J et al (2002) Pain perception in mice lacking the beta3 subunit of voltage-activated calcium channels. J Biol Chem 277:40342–40351

Murata M, Cingolani E, McDonald AD, Donahue JK, Marban E (2004) Creation of a genetic calcium channel blocker by targeted gem gene transfer in the heart. Circ Res 95:398–405

Olivera BM, Miljanich GP, Ramachandran J, Adams ME (1994) Calcium channel diversity and neurotransmitter release: the omega-conotoxins and omega-agatoxins. Annu Rev Biochem 63: 823–867

Papa A, Kushner J, Hennessey JA, Katchman AN, Zakharov SI, Chen BX, Yang L, Lu R, Leong S, Diaz J et al (2021) Adrenergic CaV1.2 activation via Rad phosphorylation converges at alpha1C I-II loop. Circ Res 128:76–88

Papa A, Zakharov SI, Katchman AN, Kushner JS, Chen BX, Yang L, Liu G, Jimenez AS, Eisert RJ, Bradshaw GA et al (2022) Rad regulation of Ca(V)1.2 channels controls cardiac fight-or-flight response. Nat Cardiovasc Res 1:1022–1038

Patil PG, de Leon M, Reed RR, Dubel S, Snutch TP, Yue DT (1996) Elementary events underlying voltage-dependent G-protein inhibition of N-type calcium channels. Biophys J 71:2509–2521

Peterson BZ, DeMaria CD, Adelman JP, Yue DT (1999) Calmodulin is the Ca2+ sensor for Ca2+-dependent inactivation of L-type calcium channels. Neuron 22:549–558

Puckerin AA, Chang DD, Shuja Z, Choudhury P, Scholz J, Colecraft HM (2018) Engineering selectivity into RGK GTPase inhibition of voltage-dependent calcium channels. Proc Natl Acad Sci U S A 115:12051–12056

Sanchez-Alonso JL, Bhargava A, O'Hara T, Glukhov AV, Schobesberger S, Bhogal N, Sikkel MB, Mansfield C, Korchev YE, Lyon AR et al (2016) Microdomain-specific modulation of L-type calcium channels leads to triggered ventricular arrhythmia in heart failure. Circ Res 119:944–955

Schmidtko A, Lotsch J, Freynhagen R, Geisslinger G (2010) Ziconotide for treatment of severe chronic pain. Lancet 375:1569–1577

Schredelseker J, Di Biase V, Obermair GJ, Felder ET, Flucher BE, Franzini-Armstrong C, Grabner M (2005) The beta 1a subunit is essential for the assembly of dihydropyridine-receptor arrays in skeletal muscle. Proc Natl Acad Sci U S A 102:17219–17224

Schredelseker J, Dayal A, Schwerte T, Franzini-Armstrong C, Grabner M (2009) Proper restoration of excitation-contraction coupling in the dihydropyridine receptor beta1-null zebrafish relaxed is an exclusive function of the beta1a subunit. J Biol Chem 284:1242–1251

Shaw RM, Colecraft HM (2013) L-type calcium channel targeting and local signalling in cardiac myocytes. Cardiovasc Res 98:177–186

Sun L, Tong CK, Morgenstern TJ, Zhou H, Yang G, Colecraft HM (2022) Targeted ubiquitination of sensory neuron calcium channels reduces the development of neuropathic pain. Proc Natl Acad Sci U S A 119:e2118129119

Weissgerber P, Held B, Bloch W, Kaestner L, Chien KR, Fleischmann BK, Lipp P, Flockerzi V, Freichel M (2006) Reduced cardiac L-type Ca2+ current in Ca(V)beta2−/− embryos impairs cardiac development and contraction with secondary defects in vascular maturation. Circ Res 99:749–757

Wu J, Yan Z, Li Z, Qian X, Lu S, Dong M, Zhou Q, Yan N (2016) Structure of the voltage-gated calcium channel Ca(v)1.1 at 3.6 A resolution. Nature 537:191–196

Xu X, Colecraft HM (2009) Engineering proteins for custom inhibition of Ca(V) channels. Physiology (Bethesda) 24:210–218

Xu X, Marx SO, Colecraft HM (2010) Molecular mechanisms, and selective pharmacological rescue, of Rem-inhibited CaV1.2 channels in heart. Circ Res 107:620–630

Yang T, Colecraft HM (2013) Regulation of voltage-dependent calcium channels by RGK proteins. Biochim Biophys Acta 1828:1644–1654

Yang T, Suhail Y, Dalton S, Kernan T, Colecraft HM (2007) Genetically encoded molecules for inducibly inactivating CaV channels. Nat Chem Biol 3:795–804

Yang T, Xu X, Kernan T, Wu V, Colecraft HM (2010) Rem, a member of the RGK GTPases, inhibits recombinant CaV1.2 channels using multiple mechanisms that require distinct conformations of the GTPase. J Physiol 588:1665–1681

Yang T, Puckerin A, Colecraft HM (2012) Distinct RGK GTPases differentially use alpha(1)- and auxiliary beta-binding-dependent mechanisms to inhibit Ca(V)1.2/Ca(V)2.2 channels. PloS One 7:e37079

Yang T, He LL, Chen M, Fang K, Colecraft HM (2013) Bio-inspired voltage-dependent calcium channel blockers. Nat Commun 4:2540

Yang L, Katchman A, Kushner J, Kushnir A, Zakharov SI, Chen BX, Shuja Z, Subramanyam P, Liu G, Papa A et al (2019) Cardiac CaV1.2 channels require beta subunits for beta-adrenergic-mediated modulation but not trafficking. J Clin Invest 129:647–658

Zamponi GW, Striessnig J, Koschak A, Dolphin AC (2015) The physiology, pathology, and pharmacology of voltage-gated calcium channels and their future therapeutic potential. Pharmacol Rev 67:821–870

Zhao Y, Huang G, Wu J, Wu Q, Gao S, Yan Z, Lei J, Yan N (2019) Molecular basis for ligand modulation of a mammalian voltage-gated Ca(2+) channel. Cell 177:1495–1506.e1412

Zuhlke RD, Pitt GS, Deisseroth K, Tsien RW, Reuter H (1999) Calmodulin supports both inactivation and facilitation of L-type calcium channels. Nature 399:159–162

Part III

Disease Insight from Channelopathies: Consequences for Pharmacology and Therapeutic Strategies

CACNA1C-Related Channelopathies

Kevin G. Herold, John W. Hussey, and Ivy E. Dick

Contents

Abstract

The *CACNA1C* gene encodes the pore-forming subunit of the $Ca_V1.2$ L-type Ca^{2+} channel, a critical component of membrane physiology in multiple tissues, including the heart, brain, and immune system. As such, mutations altering the function of these channels have the potential to impact a wide array of cellular functions. The first mutations identified within *CACNA1C* were shown to cause a severe, multisystem disorder known as Timothy syndrome (TS), which is characterized by neurodevelopmental deficits, long-QT syndrome, life-threatening cardiac arrhythmias, craniofacial abnormalities, and immune deficits. Since this initial description, the number and variety of disease-associated

Kevin G. Herold and John W. Hussey contributed equally to this work.

K. G. Herold · J. W. Hussey · I. E. Dick (✉)
Department of Physiology, University of Maryland School of Medicine, Baltimore, MD, USA
e-mail: ied@som.umaryland.edu

J. Striessnig (ed.), *Voltage-gated Ca^{2+} Channels: Pharmacology, Modulation and their Role in Human Disease*, Handbook of Experimental Pharmacology 279,
https://doi.org/10.1007/164_2022_624

mutations identified in *CACNA1C* have grown tremendously, expanding the range of phenotypes observed in affected patients. *CACNA1C* channelopathies are now known to encompass multisystem phenotypes as described in TS, as well as more selective phenotypes where patients may exhibit predominantly cardiac or neurological symptoms. Here, we review the impact of genetic mutations on $Ca_V1.2$ function and the resultant physiological consequences.

Keywords

CaV1.2 · CACNA1C · Channelopathy · L-type calcium channel · Timothy syndrome

1 Introduction

L-type calcium channels (LTCCs) are expressed throughout the body, where they play critical roles in the physiological function of numerous systems, including the heart, brain, smooth muscle, and the immune system (Berger and Bartsch 2014; Harvey and Hell 2013). Activation of these channels controls the entry of Ca^{2+} into the cell, which both shapes the morphology of the action potential and initiates numerous downstream signaling processes. Of the four L-type channel subtypes, $Ca_V1.2$ exhibits the widest expression across multiple tissues including the heart, brain, smooth muscle, endocrine and immune systems (Zamponi et al. 2015). Thus, disruption of these channels has the potential to impact a myriad of cellular functions across multiple organ systems. Advancements in genomic technologies have resulted in strong associations between $Ca_V1.2$ variants and clinical phenotypes and have identified an increasing number and variety of genetic mutations affecting the functionality of these channels.

2 $Ca_V1.2$ in Cellular Physiology

LTCCs play a major role in excitation-contraction coupling in all muscle tissues. $Ca_V1.2$ is renowned for its role in the heart, where Ca^{2+} entry through the channel activates the ryanodine receptor (RyR2), thereby triggering the release of calcium from the sarcoplasmic reticulum (SR) in a process known as Ca^{2+}-induced Ca^{2+} release (CICR) (Bers 1991; Lederer et al. 1990). In addition, $Ca_V1.2$ plays a major role in shaping the cardiac action potential (AP) (Nerbonne and Kass 2005). Acting largely during the plateau phase, $Ca_V1.2$ channels are a major determinant of the duration of the cardiac AP (Harvey and Hell 2013; Fabiato 1983). In smooth muscle, the $Ca_V1.2$ channel triggers contraction of the muscle, making it a primary element in blood-pressure control.

In the brain, $Ca_V1.2$ comprises nearly 90% of all LTCCs (Sinnegger-Brauns et al. 2009). They are highly expressed within the hippocampus, cerebral cortex, and cerebellum and are often detected in postsynaptic dendrites. $Ca_V1.2$ channels play a role in excitation-transcription coupling (Ma et al. 2012) as well as neuronal plasticity and have been shown to play a critical role in long-term potentiation and memory (Moosmang et al. 2005; Striessnig et al. 2014). In addition, mouse models have described a role for $Ca_V1.2$ in anxiety (Lee et al. 2012).

LTCCs, including $Ca_V1.2$, have also been found to play a role in the immune system, although their role is less well understood. $Ca_V1.2$ has been shown to be highly expressed in T lymphocytes and has been shown to play a role in B cells and dendritic cells (Matza and Flavell 2009; Davenport et al. 2015). While the precise role of $Ca_V1.2$ in these cells remains ambiguous, genetic mutations which disrupt the function of the channel result in a clear immune phenotype, demonstrating the importance of the channel in immune function (Splawski et al. 2004). Moreover, $Ca_V1.2$ is expressed in chondrocytes and osteoblasts, and has been shown to be a requirement for normal mandibular development. Interestingly, this role for $Ca_V1.2$ in non-excitable cell types represents a relatively new avenue of research, which was preceded by identification of $Ca_V1.2$ mutations which caused marked phenotypes in non-excitable tissues (Splawski et al. 2004; Pitt et al. 2021). As such, the function of these voltage-gated channels in non-excitable cells remains under investigation and promises new mechanisms for $Ca_V1.2$ function (Pitt et al. 2021).

Concurrent with this idea of $Ca_V1.2$ function in non-excitable tissues, it has been increasingly recognized that the channel is capable of multiple cellular functions beyond directly controlling Ca^{2+} entry. In neurons, it has been shown that a voltage-dependent conformational change is required to initiate excitation-transcription coupling and is required for normal synaptic function (Li et al. 2016). Moreover, a cryptic promoter has been identified in the *CACNA1C* gene which results in expression of a C-terminal portion of the channel, which acts as a transcription regulator (Gomez-Ospina et al. 2006, 2013). Thus, our understanding of the role of $Ca_V1.2$ in normal and patho-physiology continues to expand.

3 $Ca_V1.2$ Regulation

In order to function across a broad distribution of different cell types, $Ca_V1.2$ channels can be tuned by multiple processes, including feedback regulation, alternative splicing, and subunit composition. To precisely control Ca^{2+} influx in various tissues and in response to a diverse array of stimuli, $Ca_V1.2$ channels utilize two major forms of feedback regulation: voltage-dependent inactivation (VDI) and Ca^{2+}-dependent inactivation (CDI). VDI is thought to proceed through a hinged-lid mechanism (Stotz et al. 2000), where the I-II linker docks with residues within the distal S6 regions, thus occluding the channel. CDI, on the other hand, represents as allosteric process, where channels open with a reduced open probability in response to a rise in cytosolic Ca^{2+}. The Ca^{2+} sensor for $Ca_V1.2$ CDI is calmodulin (CaM), a modulatory protein which is bound to the C-tail of the channel in its Ca^{2+} free state (apoCaM). Upon channel opening, Ca^{2+} binds to the resident CaM, resulting in a conformational rearrangement leading to reduced opening and CDI.

$Ca_V1.2$ channels can further be modified by alternative splicing. Sequencing has revealed 90 distinct *CACNA1C* transcript variants within the brain, with variable splicing profiles across different regions and tissues (Clark et al. 2020). Biophysical evaluation of many of the major splice variants has revealed substantial effects on channel gating and regulation, illustrating unique channel properties tailored to tissue expression or developmental state (Tang et al. 2004; Liao et al. 2005, 2007, 2015).

Finally, subunit composition can impact channel function. Like other voltage-gated calcium channels, $Ca_V1.2$ is generally expressed as a multi-subunit protein (Catterall et al. 2005). The *CACNA1C* gene encodes the pore-forming alpha subunit, which typically associates with an auxiliary β subunit and an $\alpha_2\delta$ subunit. These subunits are known to modulate multiple channel properties, including trafficking, gating, and second messenger regulation (Buraei and Yang 2010). These channels are also associated with a γ subunit in select tissues, which is capable of modulating channel function (Wei et al. 1991; Klugbauer et al. 2002). Disruption of any of these auxiliary proteins can result in significant disease (Ablinger et al. 2020; Dolphin 2018; Burashnikov et al. 2010; Breitenkamp et al. 2014; C. Cross-Disorder Group of the Psychiatric Genomics 2013). Here, we focus on the role of the pore-forming alpha subunit in the pathogenesis of $Ca_V1.2$ channelopathies.

4 Association with Neuropsychiatric Disorders

Multiple genome wide association studies (GWAS) have found that *CACNA1C* is highly associated with several psychiatric disorders (Bhat et al. 2012). *CACNA1C* is commonly identified in GWAS of schizophrenia (SCZ) and bipolar disorder (BP) (Craddock and Sklar 2013; Green et al. 2010, 2013; G. C. B. D. W. G. Psychiatric 2011). Further, polymorphisms in *CACNA1C* have been associated with additional psychiatric diseases, such as major depression (Casamassima et al. 2010; Sullivan et al. 2009) and autism spectrum disorder (ASD) (Li et al. 2015; Sayad et al. 2019). Cross-association of *CACNA1C* with multiple psychiatric diseases has identified commonalities across BP, major depression, attention deficit hyperactivity disorder, SCZ, and autism spectrum disorder (ASD) (Berger and Bartsch 2014; Pouget et al. 2019), with *CACNA1C* identified as having a high association with BP, SCZ, and major depression. Moreover, pathway analysis identified numerous Ca^{2+} channel activity genes within association for five major psychiatric diseases (Pouget et al. 2019).

The most common neuropsychiatric-associated single nucleotide polymorphism (SNP) within *CACNA1C* has been identified as rs1006737 within intron 3 of the channel. This risk allele has been associated with BP (Ferreira et al. 2008; Keers et al. 2009; Sklar et al. 2008), SCZ (Zhu et al. 2019), and major depression (Green et al. 2010), making it one of the most consistent associations in neuropsychiatric GWAS (Yoshimizu et al. 2015). This risk allele has been evaluated for impact on brain structure, where healthy individuals harboring the allele demonstrated an increase in multiple traits including depression, anxiety, startle reactivity, and decreased verbal fluency (Erk et al. 2014; Roussos et al. 2011; Krug et al. 2010). Moreover, rs1006737 has been associated with structural changes in the amygdala and prefrontal cortical regions (Koch et al. 2019; Lancaster et al. 2016; Wolf et al. 2014), as well as changes in brain connectivity and gray matter structure and function (Janiri et al. 2021). Interestingly, evidence exists for both gain (Bigos et al. 2010; Gershon et al. 2014) and loss (Dao et al. 2010; Terrillion et al. 2017) of function effects on *CACNA1C* in psychiatric disorders. Recent analysis of post-mortem brain from schizophrenia patients identified a decrease in gene expression of $Ca_V1.2$ (Schmitt et al. 2022). Thus, the pathogenic mechanism for $Ca_V1.2$ in neuropsychiatric disorders remains to be elucidated. Interestingly, the rs1006737 risk allele is also associated with changes in other systems where $Ca_V1.2$ is known to play a critical role, including blood pressure (Zhang et al. 2021), demonstrating a potential wider impact of the allele outside the brain.

5 Timothy Syndrome

Timothy syndrome (TS) was initially identified in the early 1990s as a severe form of long-QT syndrome, comorbid with syndactyly (Napolitano et al. 1993; Reichenbach et al. 1992; Marks et al. 1995). This rare clinical disorder would eventually be the first documented $Ca_V1.2$ channelopathy and has been designated as long-QT type 8 (LQT8). TS is a debilitating disease characterized by pronounced cardiac deficits, syndactyly, immune deficiencies, craniofacial abnormalities, and neurological symptoms (Splawski et al. 2004, 2005). Of particular clinical concern are the cardiac symptoms, which include early afterdepolarizations, AV block, *torsades de pointes*, ventricular tachycardia, ventricular fibrillation, and long-QT syndrome (LQTS) (Splawski et al. 2005; Jacobs et al. 2006). The prolongation of the QT interval (615–690 ms) makes LQT8 amongst the most severe type of LQTS yet described. This severe prolongation is largely responsible for the highly lethal nature of TS, which has an average age of mortality of approximately two and half years.

The genetic mechanisms underlying TS are informed by splice variations of the *CACNA1C* gene. Mature $Ca_V1.2$ channels can take the form of a multitude of *CACNA1C* spice isoforms, and of particular relevance to TS are the mutually exclusive exons 8 and 8A. Some ambiguity has arisen in the literature regarding the identity of these exons, as different groups have used different terminology (Splawski et al. 2004; Liao et al. 2005; Soldatov 1992). For the sake of clarity, the convention of using "exon 8" to refer to the upstream exon, and "exon 8a" to refer to the downstream one, will be used. The first causative mutation for TS was identified as G406R within exon 8a of $Ca_V1.2$, which places the mutation near the intracellular end of the domain I S6 transmembrane domain (Splawski et al. 2004) (Fig. 1). In cardiac tissue, the ratio of exon 8 to exon 8a is approximately 4:1, and exon 8a is also expressed throughout other tissues that contain $Ca_V1.2$, a fact that contributes to the multisystem nature of TS.

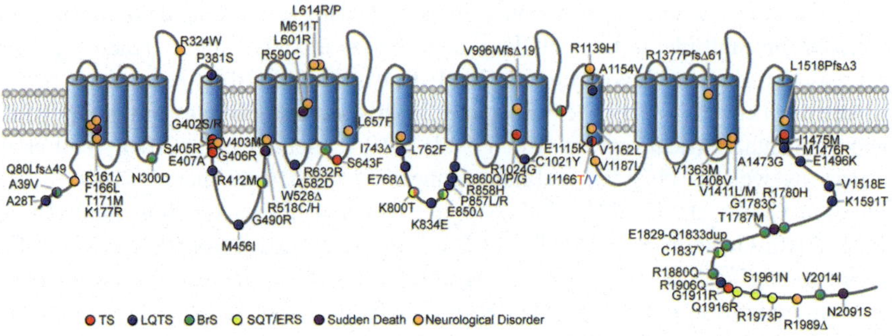

Fig. 1 *CACNA1C* mutations. Cartoon depicting the membrane topology of the pore-forming subunit of $Ca_V1.2$ indicating the locus of *CACNA1C* mutations across the channel. Phenotypes associated with each mutation are colored accordingly

Biophysical studies revealed that channels containing the G406R mutation show a loss of VDI (Fig. 2a), decreased CDI (Fig. 2b), and a hyperpolarizing shift in the voltage dependence of channel activation (Fig. 2c) (Splawski et al. 2004, 2005; Barrett and Tsien 2008; Raybaud et al. 2006; Dick et al. 2016), all relatively clear components of a gain-of-function (GOF) phenotype. This channel behavior helps explain the effect of G406R on cardiac tissues, as the plateau phase of the cardiac action potential is enhanced by the excess entry of Ca^{2+} during depolarization. This, in turn, results in the characteristic prolongation of the QT interval observed in TS patients. Further, a mouse overexpressing rabbit $Ca_V1.2$ channels harboring the mutation indicated that even proportionally small amounts of the mutated channel were sufficient to cause large alterations to intracellular calcium concentrations, with increased SR Ca^{2+} leak and SR Ca^{2+} load, higher diastolic calcium, and increased Ca^{2+} spark activity in ventricular myocytes (Drum et al. 2014). The vast majority of patients harboring a G406R mutation in exon 8a (sometimes referred to as TS type 1 or TS1) suffer from the full range of characteristic symptoms of TS, but this is not universally true (Sepp et al. 2017).

Shortly following the initial characterization of the G406R mutation in exon 8a of $Ca_V1.2$, two more mutations in *CACNA1C* were observed in patients suffering similar symptoms (Splawski et al. 2005). In 2005, a patient was observed to harbor a G406R mutation not in exon 8a, but in the more highly expressed alternative exon 8 (Splawski et al. 2005). Concurrently, a second mutation, G402S, was also identified within exon 8. Both mutations produced similar cardiac deficits, with QT intervals reaching as high as 730 ms; however, they lacked the syndactyly characteristic of the original TS patients. This form of LQT8 therefore was designated as TS type 2 (TS2). TS2 patients harboring the G406R mutations generally present with a more severe cardiac pathology, with an earlier age of lethality as compared to TS1. This is likely due to the greater abundance of exon 8 in cardiac as compared to exon 8a.

In the case of the patient harboring the G402S mutation in exon 8, a similar but somewhat less severe LQT phenotype was observed, and similar neurological

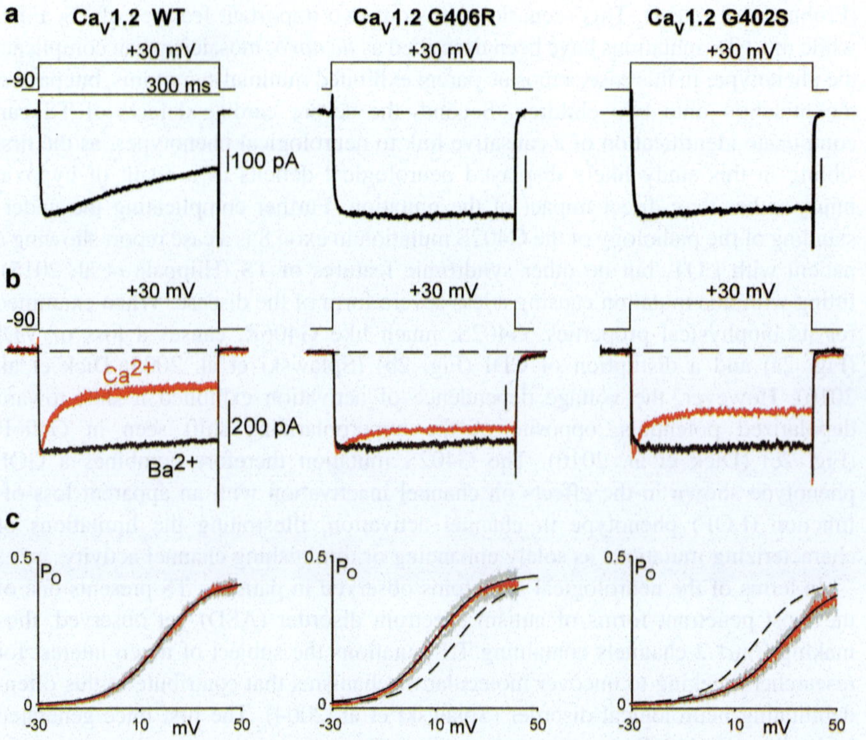

Fig. 2 TS associated *CACNA1C* mutations alter the gating of Ca$_V$1.2 channels. (**a**) Exemplar Ba^{2+} currents measured from Ca$_V$1.2 channels display typical VDI, seen as the decay in current over the time course of a 300 ms depolarization. Channels are expressed with the β_{1B} auxiliary subunit, which is permissive of robust VDI. Both the G406R and G402S mutations nearly eliminate VDI in these channels. (**b**) Exemplar Ba^{2+} (black) and Ca^{2+} (red) currents measured from Ca$_V$1.2 channels enabling visualization of CDI, seen as the stronger decay in the Ca^{2+} current as compared to the Ba^{2+} current. Ca$_V$1.2 is expressed with the β_{2A} auxiliary subunit which decreases the amount of VDI, thus enabling measurement of CDI in relative isolation. Both the G406R and G402S mutations significantly blunt CDI. (**c**) Open-probability curves measured through single channel recordings of Ca$_V$1.2 opening as a function of voltage. Data is displayed in red, with SEM in gray, and a Boltzmann fit to the data shown as the solid black curve. The G406R mutation left shifts the activation of the channel as compared to WT, which is reproduced as the black dashed line (middle panel). The G402S mutation causes a right shift in channel activation (right panel) as compared to the WT data reproduced as the dashed black line. Reproduced with permission from Dick et al. (2016), *Nat. Commun*)

symptoms were described (Splawski et al. 2005). However, the patient suffered from a cardiac arrest, which could somewhat complicate the understanding of the underlying effects of the mutation on neuronal function and development. Subsequent patients harboring the G402S mutation in exon 8 have been shown to lack the neurological phenotypes typically associated with TS. In one case, a child harboring the G402S mutation exhibited neurological impairments, however the patient's affected sibling was neurologically healthy despite harboring the same mutation

(Frohler et al. 2014). This scenario points out two important features of TS. First, while most TS mutations have been described as *de novo*, mosaicism can complicate the phenotype. In this case, a mosaic parent exhibited minimal symptoms, but passed the mutation onto two children. Second, the severe cardiac defects of TS can complicate identification of a causative link to neurological phenotypes, as the first sibling in this study likely displayed neurological deficits as a result of hypoxic injury rather than direct impact of the mutation. Further complicating the understanding of the pathology of the G402S mutation in exon 8 is a case report showing a patient with LQT, but no other syndromic features of TS (Hiippala et al. 2015), fitting with this mutation causing a less severe form of the disease. When examined for its biophysical properties, G402S, much like G406R, causes a loss of VDI (Fig. 2a) and a disruption of CDI (Fig. 2b) (Splawski et al. 2005; Dick et al. 2016). However, the voltage dependence of activation exhibited a shift toward depolarized potentials, opposite of the hyperpolarizing shift seen in G406R (Fig. 2c) (Dick et al. 2016). The G402S mutation therefore combines a GOF phenotype shown in the effects on channel inactivation with an apparent loss-of-function (LOF) phenotype in channel activation, illustrating the limitations of characterizing mutations as solely enhancing or diminishing channel activity.

In terms of the neurological symptoms observed in patients, TS presents one of the most penetrant forms of autism spectrum disorder (ASD) yet observed, thus making $Ca_V1.2$ channels containing TS mutations the subject of much interest for researchers seeking to uncover molecular mechanisms that contribute to this often-debilitating neurological disorder (Splawski et al. 2004). The first mice generated harboring the G406R mutation in exon 8a were engineered to contain a neomycin cassette to ensure survival to maturity (Bader et al. 2011). Behavioral evaluation revealed traits analogous to those seen to be at the core of ASD, including social and learning impairments, repetitive behavior patterns, and altered vocalizations (Bader et al. 2011). Brain tissue from this TS mouse model demonstrated increased development and myelination of cortical oligodendrocyte progenitor cells (OPCs) from these mice (Cheli et al. 2018). Notably, altered myelination patterns and increased white matter volume have been noted in human cases of ASD (Courchesne et al. 2001; Carmody and Lewis 2010). Additionally, cortical neurons isolated from these mice showed activity-dependent dendritic retraction, a result which was replicated in neurons differentiated from induced pluripotent stem cell (iPSCs) derived from a TS patient (Pasca et al. 2011). The G406R mutation was also shown to result in abnormal coupling of AKAP150 to the channel (Cheng et al. 2011), and AKAP150's association with the channel is thought to play a role in mediating the ability of $Ca_V1.2$ channels to couple cellular excitation to gene transcription (Murphy et al. 2014).

Following the initial description of TS1 and TS2 mutations, an ever-growing number of mutations have been identified in $Ca_V1.2$, often within constitutively expressed exons. While a number of these mutations have been shown to alter channel properties and are observed in patients suffering from symptoms observed in TS, many of these mutations do not recapitulate the full multisystem TS disease phenotype. Mutations such as A1473G (Gillis et al. 2012), I1166T (Boczek et al.

2015a), I1166V (Wemhoner et al. 2015), and E407A (Colson et al. 2019) all occupy positions in or adjacent to S6 domains (Fig. 1), much like the originally described G406R and G402S mutations, but notably, I1166V and A407A appear to have primarily or exclusively cardiac symptoms (see below). Other mutations, such as S643F, appear to cause both cardiac and neurological symptoms but without syndactyly (Ozawa et al. 2018). Moreover, R1024G appears to cause neurological symptoms and syndactyly, but not LQTS (Kosaki et al. 2018). The expanding array of disease-associated mutations observed in $Ca_V1.2$ and the heterogeneity in clinical outcomes presents a challenge to researchers attempting to fully characterize the scope and pattern of these channelopathic impairments.

The multisystem nature of TS symptoms suggests that $Ca_V1.2$ functionality extends beyond its well-described role in excitable tissues, and indeed, RNA-seq data show that the channel is widely expressed throughout the body (Lachmann et al. 2018). Prior to the implementation of molecular techniques that could identify the genetic identity of particular calcium channels, calcium currents and calcium dynamics were shown to influence the activity of T lymphocytes (Alcover et al. 1986). Additionally, functional $Ca_V1.2$ channels were identified in osteoblasts (Chesnoy-Marchais and Fritsch 1988; Meszaros et al. 1996), and exposure of these cells to the LTCC agonist BAY K 8644 stimulated the secretion of the bone matrix protein, osteocalcin, with evidence suggesting a role of the channels in bone remodeling (Guggino et al. 1989). One of the symptoms of TS is mandibular malformation, and mice expressing $Ca_V1.2$ channels harboring TS mutations show enlarged mandibles as compared to those expressing wildtype channels (Ramachandran et al. 2013). Evidence from mice and zebra fish indicates that $Ca_V1.2$ acts on mandibular development through calcineurin signaling. Thus, the phenotype observed in TS patients has increased our understanding of the role of $Ca_V1.2$ in physiology, demonstrating a clear importance for the channel in non-excitable tissues (Pitt et al. 2021).

6 Cardiac-Selective *CACNA1C* Mutations

As the variety of known mutations in *CACNA1C* has continued to expand, so too does the spectrum of phenotypes associated with these mutations. While TS was first described as a multisystem disorder, it was not long before *CACNA1C* mutations were identified as causative of cardiac-selective phenotypes (Burashnikov et al. 2010; Boczek et al. 2015b). Mutations in *CACNA1C* (Fig. 1) have been identified among patients with Brugada syndrome (BrS), idiopathic ventricular fibrillation (IVF), short-QT syndrome (SQT), and early repolarization syndrome (ERS) (Burashnikov et al. 2010; Antzelevitch et al. 2007; Fukuyama et al. 2013). Unlike the full TS phenotype, these patients do not exhibit extra-cardiac symptoms. Moreover, these syndromes appear to result from a loss of function of the $Ca_V1.2$ channel. Lack of Ca^{2+} entry during the cardiac AP readily explains the decreased AP duration identified in ERS. In addition, when BrS is the result of a *CACNA1C* mutation, it often includes overlapping features of SQT (Burashnikov et al. 2010; Antzelevitch et al. 2007; Fukuyama et al. 2013), consistent with a loss of $Ca_V1.2$ current. Thus, a loss of $Ca_V1.2$ current can impact the morphology of the cardiac AP, resulting in arrhythmia via similar mechanisms as the gain-of-function mutations, but with opposing direction. While these loss-of-function mutations can manifest as several overlapping syndromes, the mechanism of pathogenesis does not appear to significantly affect other tissues where $Ca_V1.2$ is highly expressed.

The cardiac-selectivity of *CACNA1C* mutations, however, is not specific only to loss-of-function mutations. Even for gain-of-function mutations in *CACNA1C*, the multisystem effects described for canonical TS are not always present. Mutations in *CACNA1C* have increasingly been described in patients exhibiting cardiac-selective phenotype, sometimes called cardiac-only TS (Boczek et al. 2015b) or atypical TS (Frohler et al. 2014; Bauer et al. 2021). Initially these patients were described to exhibit severe LQTS (LQT8), in complex with additional cardiac deficits including hypertrophic cardiomyopathy (HCM) or congenital heart defects (Boczek et al. 2013, 2015b). Later, atrial fibrillation and sick sinus syndrome have also been described as occurring in cardiac-only TS patients, further expanding the spectrum of cardiac disorders in the absence of the full multisystem phenotype (Gakenheimer-Smith et al. 2021). In fact, since the first description of TS, more mutations in *CACNA1C* have been identified with cardiac-selective LQTS as compared to the full multisystem TS phenotype (Fig. 1). The pathogenesis of these LQT8 mutations is in line with other forms of LQTS, where a genetic mutation altering the function of an ion channel results in an abnormal prolongation of the action potential due to an excess entry of a depolarizing ion. However, like in the full multisystem manifestation, the QT prolongation of LQT8 patients often exceeds that of other LQTS phenotypes, making these patients highly susceptible to cardiac arrhythmia and sudden death.

CACNA1C mutations have been increasingly associated with sudden cardiac death (SCD) (Antzelevitch et al. 2007; Halvorsen et al. 2021; Hennessey et al. 2014). This finding is often explained by the severe LQTS phenotype produced by gain-of-function mutations (Splawski et al. 2004, 2005; Boczek et al. 2015b; Dufendach et al. 2018), or as a result of BrS/SQT resulting from loss-of-function mutations (Burashnikov et al. 2010; Antzelevitch et al. 2007). Multiple postmortem studies of sudden cardiac death have identified mutations in the *CACNA1C* gene (Narula et al. 2015; Sutphin et al. 2016). These studies have identified *CACNA1C* as associated with sudden cardiac death both in young and adult patients (Halvorsen et al. 2021; Narula et al. 2015), making it a strong risk factor for SCD. Thus, while *CACNA1C* mutations remain rare, their potential for severe pathological consequences is very high.

7 Neurological Phenotypes

The role of the $Ca_V1.2$ channel in the heart has long been known to be critical. It is therefore not surprising that the majority of *CACNA1C* mutations described thus far have been associated with significant cardiac effects, sometimes with concurrent neurological features. However, as more pathogenic mutations in *CACNA1C* are being identified, this cardiac-inclusive phenotype has begun to change, and mutations have been identified as causative of primarily neurological phenotypes. While it is surprising that significant changes in $Ca_V1.2$ channel gating may spare cardiac function in some cases, this finding is consistent with the numerous studies which have identified a strong association of *CACNA1C* with schizophrenia, major depressive disorder, bipolar disorder, and autism (Green et al. 2010), generally in the absence of a described cardiac phenotype.

CACNA1C variants (V1363M and splice site variant c.3717+1_3717+2insA) have been reported in individuals with epilepsy in the form of neonatal onset epileptic encephalopathy and late-onset epilepsy (Fig. 1) (Bozarth et al. 2018). The individual with the V1363M had a normal ECG, but displayed hypotonia, syndactyly, and dysmorphic facial features. The patients with the splice variant in the described study exhibited learning disabilities, congenital cardiac anomalies, and dysmorphic facial features in addition to epilepsy, also without overt cardiac symptoms. Furthermore, a mutation in a *CACNA1C* intron (rs779393130) has been described as being associated with autosomal dominant cerebellar ataxia (Chen et al. 2019). In 2018, the mutation R1024G (Fig. 1) was described as associated with developmental and speech delay but no QT prolongation (Kosaki et al. 2018). This mutation results in the neutralization of the one of the gating charges in the IIIS4 voltage sensor. Though it has not been functionally characterized, it has been suggested that it may destabilize the VSD leading to a gating pore current (Striessnig 2021). Thus, the pathogenic mechanism leading to a neuro-specific effect may be distinct from those underlying LQT8.

This class of *CACNA1C* mutations with predominantly neurological phenotypes is continuing to grow, with identification of patients with both neurodevelopmental phenotypes as well as epilepsy and/or ataxia. A recent study describes a number of novel *CACNA1C* variants in which individuals display a predominantly neurological phenotype which includes developmental delay, intellectual disability, ASD, hypotonia, epilepsy, and ataxia (Rodan et al. 2021). Interestingly, the phenotypic differences among this cohort of patients could be broadly separated into those stemming from non-truncating vs. truncating mutations (Fig. 1). The non-truncating mutations in the study all occurred de novo and included both missense variants and in-frame deletions. All individuals with such mutations displayed severe global developmental delay/intellectual disability and impaired motor function. The majority of these patients also had a spectrum of epileptic phenotypes; from medically refractory infantile epileptic encephalopathy to later onset focal or generalized seizures. Furthermore, most exhibited ataxia, orthopedic abnormalities, and hypotonia. Only a single individual showed mild QT prolongation.

Four missense mutations identified in this study (L614P, L614R, L657F, L1408V) were functionally characterized using whole-cell patch-clamp electrophysiology. The L1408V mutation is located within the IVS5 region, near the cytosolic side of the channel (Fig. 1), and causes a decrease in current density and no change in activation kinetics. L657F is located in IIS5 (Fig. 1) and causes an increase in current density and a hyperpolarizing shift in the voltage dependence of activation. In addition, L657F was associated with higher protein expression levels as seen via western blot, which could explain the increase in current density. Residue L614 is located in the IIS4 VSD (Fig. 1) and mutations L614R and L614P did not result in any detectable differences in the patch-clamp experiments, though L614P appeared to show a non-statistically significant increase in current.

Patients within this study identified as harboring truncation mutations exhibited distinct features. While the majority of these mutations were also de novo, a small fraction were paternally inherited. In addition, these patients were reported to have more selective developmental abnormalities involving expressive language. Except for a single individual, none of these patients showed aberrant early motor development, and one-third were described as having poor balance or coordination. ASD was also found to be more common in this subset of patients, and only a single individual had a history of seizures. None of the individuals who were tested with an ECG showed any cardiac abnormalities.

Finally, the mutation K800T appears to contradict the concurrence of LQTS and ASD described in TS (Endres et al. 2020). This mutation was associated with ASD and severe psychiatric symptoms including depressive and hypomanic episodes in a single patient. However, unlike in typical TS, a short-QT phenotype accompanied the neurological symptoms. Thus, as more mutations are identified within *CACNA1C*, the spectrum of phenotypes continues to expand.

8 Beyond Loss of Function/Gain of Function

Numerous studies have evaluated the impact of a variety of $Ca_V1.2$ mutations on channel function, identifying changes in current amplitude, activation, and inactivation (Splawski et al. 2004, 2005; Barrett and Tsien 2008; Dick et al. 2016; Boczek et al. 2015a, b; Wemhoner et al. 2015; Sutphin et al. 2016; Bamgboye et al. 2022a). In the heart, the impact of these gating changes correlates well with disease (Splawski et al. 2004, 2005; Dick et al. 2016). For loss-of-function mutations, most studies identified an overall decrease in current amplitude (Burashnikov et al. 2010; Antzelevitch et al. 2007; Sutphin et al. 2016; Blancard et al. 2018; Beziau et al. 2014), which can be attributed to either a loss of channels at the membrane surface or a decrease in current through the channel. This decrease in Ca^{2+} influx through $Ca_V1.2$ is sufficient to explain the shortening of the AP and resulting arrhythmia within these patients. For gain-of-function mutations, it has been shown that mutations can increase $Ca_V1.2$ current either through a shift in the voltage dependence of activation of the channel, a decrease in VDI or a decrease in CDI (Dick et al. 2016). Notably, the canonical G406R TS mutation exhibited all three of these gating changes (Fig. 2), resulting in a marked excess entry of Ca^{2+} through the channel during the cardiac AP, despite the fact that the mutant channel is expected to only account for ~10% of $Ca_V1.2$ channels in a typical human ventricle due to alternative splicing (Splawski et al. 2004). Importantly, a loss of either form of inactivation or a depolarizing shift in channel activation is predicted to be sufficient to explain the AP prolongation caused by these LQT8 mutations (Morotti et al. 2012; Faber et al. 2007). Even the G402S mutation, which exhibits a gain-of-function effect on CDI and VDI but a loss-of-function effect on channel activation (Fig. 2), is predicted to result in an overall increase in Ca^{2+} entry during the cardiac AP, explaining the LQTS phenotype of the mutation (Dick et al. 2016). However, while the cardiac features of *CACNA1C* mutations are well explained by the gating changes in the channel, the same cannot be said for the neurological phenotypes exhibited by these patients.

TS represents one of the most penetrant forms of ASD (Marcantoni et al. 2020), and various mutations in $Ca_V1.2$ have been strongly associated with neurodevelopmental delay and epilepsy. Further, GWAS have identified a strong association with the *CACNA1C* gene, yet associating these neurological phenotypes with gain- or loss-of-function effects has not been successful. One explanation includes the likelihood that neurological deficits can occur in response to either a decrease or increase in $Ca_V1.2$ current (Rodan et al. 2021). However, if we consider just the strong neurodevelopmental phenotype associated with the gain-of-function TS mutations, it appears that the ASD identified in these patients does, indeed, associate with mutations shown to cause excess Ca^{2+} entry (Bamgboye et al. 2022a). Unfortunately, this observation does not explain why many $Ca_V1.2$ mutations which have been shown to produce similar gain-of-function effects on the channel result in a cardiac, but not neuronal, phenotypes. It seems likely that consideration of the distinct mechanisms of channel disruption is required, and simple gain-of-function descriptions are not sufficient to explain the unique impact of $Ca_V1.2$ mutations on neuronal function. This lack of clear correlation between excess Ca^{2+} entry through $Ca_V1.2$ and ASD has led to the consideration that some neurological deficits may result from mechanisms that are independent from Ca^{2+} influx through the channel (Li et al. 2016; Krey et al. 2013). However, the strong impact of these mutations on channel gating and the known importance of $Ca_V1.2$ channel function in the normal function of neurons make it difficult to dismiss the likely impact of significant $Ca_V1.2$ gating changes in the brain.

The canonical G406R TS mutation caused a hyperpolarizing shift in the voltage dependence of $Ca_V1.2$ activation and deficits in both VDI and CDI (Fig. 2) and is causative of ASD in all patients. The G402S mutation caused a similar deficient in both CDI and VDI; however, the effect of this mutation on channel activation was in the opposite (depolarizing) direction (Fig. 2). Interestingly, the neurodevelopmental phenotype first described for these patients is not consistent across the patient population and evidence suggests that it may be secondary to hypoxic injury (Frohler et al. 2014). It is therefore tantalizing to speculate that this difference in activation shift direction may underlie the lack of a neurological phenotype in these patients. In fact, the left shift in channel activation has been hypothesized as being a requirement for the $Ca_V1.2$ related ASD phenotype (Marcantoni et al. 2020; Calorio et al. 2019). Detailed study of the biophysical properties of multiple $Ca_V1.2$ mutations has demonstrated a correlation between a marked left shift in channel activation and neurodevelopmental disease, and it has been shown that the left shift in channel activation correlates with an altered response of the channels to a neuronal

AP stimulus (Bamgboye et al. 2022a). Finally, this hypothesis is further strengthened by similar observations of the related $Ca_V1.3$ LTCC – where ASD occurs in patients harboring mutations which also produce a left shift in channel activation (Calorio et al. 2019). Thus, it appears that the neurodevelopmental deficits of TS patients best track with a significant left shift in channel activation (Bamgboye et al. 2022a). While this result does not preclude additional mechanisms capable of modulating neuronal function, it is interesting that it is consistent with the reported importance of a voltage-dependent conformational change in the pathogenesis of TS (Li et al. 2016).

9 Functional Channel Hotspots and the Emergence of S6-Opathies

While mutations in *CACNA1C* occur throughout the length of the channel, select regions of known channel function have been noted to correlate with patient phenotypes, hinting at hotspots for pathogenesis. One of these is located on the cytosolic II-III loop of the channel, containing six mutations (Fig. 1) clustered near a known STAC (SH3 and cysteine rich domain) binding domain of the channel (Mellor et al. 2019). Each of these mutations correlated with a significant increase in the QT interval of the patient. Interestingly, while STAC proteins have been shown to modulate $Ca_V1.2$ channel gating (Niu et al. 2018; Polster et al. 2018), it is not clear that this mechanism can fully explain the cardiac phenotype seen in these patients due to the locus of a functionally relevant STAC binding site on the C-tail of the channel and unclear expression of STAC protein in the heart. Nonetheless, the overlapping patient phenotypes of mutations located within a known binding region point toward a common pathogenic mechanism of these mutations.

Yet, another locus that appears to be emerging as a hotspot for severe disease is the distal S6 regions of the channel. Interestingly, out of the 11 mutations currently reported as producing a full, multisystem TS phenotype including both severe LQTS and neurological symptoms, seven fall on a distal S6 region (Fig. 1, red). In addition, multiple non-syndromic LQTS and neurodevelopmental associated mutations fall within the same regions (Fig. 1, blue and orange). Interestingly, this pattern has emerged not only for mutations within $Ca_V1.2$ but also for channelopathic mutations within other Ca^{2+} channels including $Ca_V3.1$, $Ca_V3.2$, $Ca_V1.3$, $Ca_V2.1$, and $Ca_V2.3$ (Lory et al. 2020; Shelley et al. 2013; Scholl et al. 2013), and within other ion channels including NALCN (Kschonsak et al. 2020) and HCN4 (Biel et al. 2016). Thus, S6-opathies may represent a growing class of ion channelopathies (Bamgboye et al. 2022a; Lory et al. 2020). As the distal S6 region is known to be a critical element controlling channel activation, it is not surprising that many of these $Ca_V1.2$ S6 mutations have demonstrated shifts in channel activation (Fig. 2) (Dick et al. 2016; Boczek et al. 2015a; Hennessey et al. 2014; Bamgboye et al. 2022a; Marcantoni et al. 2020). Moreover, the distal S6 region contains residues critical for the docking of the I-II linker region in a "hinged-lid" mechanism underlying VDI (Stotz et al. 2000; Stotz and Zamponi 2001; Tadross et al. 2010), consistent with the impact of many $Ca_V1.2$ distal S6 mutations on VDI. Additionally, deficits in CDI have been shown to result from modulation of residues within this same S6 locus, often through a mechanism secondary to changes in channel activation (Dick et al. 2016; Bamgboye et al. 2022a; Tadross et al. 2010). Thus, it is not unexpected that mutations within this region are likely to result in significant and physiologically impactful changes to channel gating, making these mutations highly deleterious. Finally, it must be considered that many disease-causing mutations identified within the S5; S4-5 linker may affect interactions with the S6 region, providing an expanded impact for altered S6 function. As the number of known mutations continues to grow, no doubt that additional regions of pathogenic significance are likely to emerge.

10 Therapeutic Implications

Treatment for TS and LQT8 has remained challenging. As the cardiac symptoms are generally the most life-threatening component of *CACNA1C* channelopathies, they constitute the main focus for treatment for these patients. However, treatments that have been effective for other forms of LQTS have had limited benefit in the context of LQT8. Beta blockers have been used with little effect, and the sodium channel blocker mexiletine has shown partial efficacy in treating the QT prolongation of TS (Gao et al. 2013; Tunca Sahin and Ergul 2018; Hermida et al. 2021). Given that the LQTS phenotype due to *CACNA1C* mutations is the result of excess Ca^{2+} entry through the channel, L-type Ca^{2+} channel blockers (CCBs) would seem well positioned for the treatment of LQT8. Multiple CCBs are available and commonly used to treat cardiac arrhythmia, hypertension, and angina pectoris. However, their efficacy in TS has remained limited, resulting in continued reliance on internal cardiac defibrillators in many patients (Jacobs et al. 2006; Hiippala et al. 2015). Verapamil in particular has been tried in the treatment of TS, however, while initial reports indicated that the drug was partially effective in reducing ventricular fibrillation, it did not reduce the QT interval (Jacobs et al. 2006) and over time resulted in increased atrial fibrillation and reduced atrial contraction (Shah et al. 2010). As a result, ranolazine, a multi-channel blocker, was added to the verapamil regimen, resulting in a decrease in atrial and ventricular fibrillation (Shah et al. 2012). Thus, it appears that targeting the $Ca_V1.2$ channel itself has not provided adequate results for the treatment of TS.

It appears that this lack of efficacy likely stems from the mechanism of CCB block of $Ca_V1.2$ channels. Three classes of CCBs have been described, phenylalkylamines, benzothiazepines, and dihydropyridines (DHPs). Each of these drugs is known to be state-dependent blockers, such that channel opening during repeated depolarizations enhances drug binding (Lee and Tsien 1983; Hering et al. 1997). Such use-dependent block by phenylalkylamines and benzothiazepines is thought to result from open and inactivated state block, where opening of the channel provides access of the drug to the receptor site (Catterall et al. 2020). Block by DHPs is proposed to result from preferential binding to the inactivated state of the channel. However, many TS mutations reduce channel inactivation, resulting in a loss of use-dependent block and an overall decrease in efficacy of the CCBs. Verapamil has

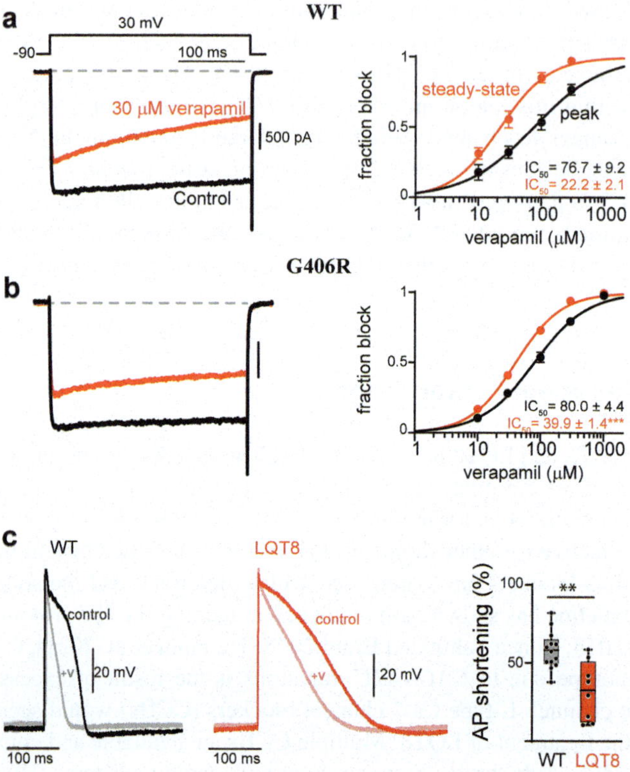

Fig. 3 *CACNA1C* mutations reduce the efficacy of verapamil. (**a**) Verapamil concentration-response curves measured for WT $Ca_V1.2$ in HEK 293 cells. *Left*: Exemplar Ba^{2+} current before (black) and after (red) the addition of 30 μM verapamil. Accumulation of block can be seen during the 300 ms depolarization in the presence of verapamil (red). *Right:* concentration-response curves can be generated by measuring the effect of the drug on the peak current amplitude immediately following depolarization (black), or at the end of the voltage step after 300 ms depolarization (red). The increased block measured from the steady-state value (red) is indicative of use-dependence of the drug. (**b**) The G406R mutation significantly reduces the use-dependent block of verapamil, seen as an increase in steady-state IC_{50}. (**c**) *Left*: Exemplar AP recordings of WT iPSC-derived cardiomyocytes before (black) and after (gray) the addition of 10 μM verapamil. *Middle*: Exemplar AP recording from an iPSC-derived cardiomyocyte from a patient harboring the G406R mutation (red, LQT8) shows significant prolongation of the action potential as compared to WT (left, black). Addition of 10 μM verapamil (pink) reduces the AP duration to a lesser extent as compared to WT iPSC-derived cardiomyocytes. *Right:* Quantification of the percent of AP shortening in WT vs. LQT8 (G406R) containing iPSC-derived cardiomyocytes demonstrates a significant deficit in verapamil's ability to shorten the AP in the context of patient-derived cells harboring the G406R mutation (LQT8). Reproduced with permission from (Bamgboye et al. 2022b)

been shown to exhibit a decrease in use-dependent block in the context of multiple pathogenic $Ca_V1.2$ mutations (Bamgboye et al. 2022b). This can be readily seen as a decrease in the activity-dependent shift of the concentration-response curve measured in HEK cells expressing WT versus G406R $Ca_V1.2$ channels (Fig. 3a, b) and has been shown to persist in the context of iPSC-derived cardiomyocytes (Bamgboye et al. 2022b). Importantly, this result translated to a decrease in AP shortening in iPSC-derived cardiomyocytes derived from a TS (LQT8) patient harboring the G406R mutation (Fig. 3c). Likewise, DHPs have been shown to exhibit decreased efficacy in the context of pathogenic $Ca_V1.2$ mutations in HEK cells (Splawski et al. 2004; Bamgboye et al. 2022b; Sheng et al. 2012). Thus, the loss of inactivation due to the deleterious mutation causes a major challenge for treatment of these channelopathies, limiting the usefulness of currently available CCBs.

Due to this lack of efficacy, research continues to explore alternative mechanisms for the treatment of these patients. One approach is the use of roscovitine, a cyclin-dependent kinase inhibitor and atypical LTCC blocker that has been reported to reverse the phenotype associated with TS in heterologous systems (Yarotskyy and Elmslie 2007; Yarotskyy et al. 2009, 2010). Roscovitine has been shown to enhance VDI of LTCCs (Yarotskyy and Elmslie 2007; Yarotskyy et al. 2009, 2010; Yazawa et al. 2011; Angelini et al. 2021) and suppress early afterdepolarizations, which can lead to ventricular arrhythmias (Angelini et al. 2021). It has also been reported that roscovitine can rescue the TS phenotype in both iPSC-derived cardiomyocytes and neurons differentiated from a TS patient (Pasca et al. 2011; Yazawa et al. 2011; Song et al. 2015, 2017). In iPSC-derived CMs, roscovitine was able to reduce the irregular timing and amplitude of Ca^{2+} transients and significantly increased VDI (Yazawa et al. 2011; Song et al. 2015). Efforts continue to evaluate novel mechanisms for the treatment of TS, including the use of genetic approaches and identification of novel small molecule therapies.

References

Ablinger C, Geisler SM, Stanika RI, Klein CT, Obermair GJ (2020) Neuronal alpha2delta proteins and brain disorders. Pflugers Arch 472:845–863

Alcover A, Weiss MJ, Daley JF, Reinherz EL (1986) The T11 glycoprotein is functionally linked to a calcium channel in precursor and mature T-lineage cells. Proc Natl Acad Sci U S A 83:2614–2618

Angelini M et al (2021) Suppression of ventricular arrhythmias by targeting late L-type Ca2+ current. J Gen Physiol 153

Antzelevitch C et al (2007) Loss-of-function mutations in the cardiac calcium channel underlie a new clinical entity characterized by ST-segment elevation, short QT intervals, and sudden cardiac death. Circulation 115:442–449

Bader PL et al (2011) Mouse model of Timothy syndrome recapitulates triad of autistic traits. Proc Natl Acad Sci U S A 108:15432–15437

Bamgboye MA et al (2022a) CaV1.2 channelopathic mutations evoke diverse pathophysiological mechanisms. J Gen Physiol 154

Bamgboye MA et al (2022b) Impaired CaV1.2 inactivation reduces the efficacy of calcium channel blockers in the treatment of LQT8. J Mol Cell Cardiol 173:92–100

Barrett CF, Tsien RW (2008) The Timothy syndrome mutation differentially affects voltage- and calcium-dependent inactivation of CaV1.2 L-type calcium channels. Proc Natl Acad Sci U S A 105:2157–2162

Bauer R, Timothy KW, Golden A (2021) Update on the molecular genetics of Timothy syndrome. Front Pediatr 9:668546

Berger SM, Bartsch D (2014) The role of L-type voltage-gated calcium channels Cav1.2 and Cav1.3 in normal and pathological brain function. Cell Tissue Res 357:463–476

Bers DM (1991) Ca regulation in cardiac muscle. Med Sci Sports Exerc 23:1157–1162

Beziau DM et al (2014) Complex Brugada syndrome inheritance in a family harbouring compound SCN5A and CACNA1C mutations. Basic Res Cardiol 109:446

Bhat S et al (2012) CACNA1C (Cav1.2) in the pathophysiology of psychiatric disease. Prog Neurobiol 99:1–14

Biel S et al (2016) Mutation in S6 domain of HCN4 channel in patient with suspected Brugada syndrome modifies channel function. Pflugers Arch 468:1663–1671

Bigos KL et al (2010) Genetic variation in CACNA1C affects brain circuitries related to mental illness. Arch Gen Psychiatry 67:939–945

Blancard M et al (2018) An African loss-of-function CACNA1C variant p.T1787M associated with a risk of ventricular fibrillation. Sci Rep 8:14619

Boczek NJ et al (2013) Exome sequencing and systems biology converge to identify novel mutations in the L-type calcium channel, CACNA1C, linked to autosomal dominant long QT syndrome. Circ Cardiovasc Genet 6:279–289

Boczek NJ et al (2015a) Novel Timothy syndrome mutation leading to increase in CACNA1C window current. Heart Rhythm 12:211–219

Boczek NJ et al (2015b) Identification and functional characterization of a novel CACNA1C-mediated cardiac disorder characterized by prolonged QT intervals with hypertrophic cardiomyopathy, congenital heart defects, and sudden cardiac death. Circ Arrhythm Electrophysiol 8: 1122–1132

Bozarth X et al (2018) Expanding clinical phenotype in CACNA1C related disorders: from neonatal onset severe epileptic encephalopathy to late-onset epilepsy. Am J Med Genet A 176:2733–2739

Breitenkamp AF et al (2014) Rare mutations of CACNB2 found in autism spectrum disease-affected families alter calcium channel function. PLoS One 9:e95579

Buraei Z, Yang J (2010) The ss subunit of voltage-gated Ca2+ channels. Physiol Rev 90:1461–1506

Burashnikov E et al (2010) Mutations in the cardiac L-type calcium channel associated with inherited J-wave syndromes and sudden cardiac death. Heart Rhythm 7:1872–1882

C. Cross-Disorder Group of the Psychiatric Genomics (2013) Identification of risk loci with shared effects on five major psychiatric disorders: a genome-wide analysis. Lancet 381:1371–1379

Calorio C et al (2019) Impaired chromaffin cell excitability and exocytosis in autistic Timothy syndrome TS2-neo mouse rescued by L-type calcium channel blockers. J Physiol 597:1705–1733

Carmody DP, Lewis M (2010) Regional white matter development in children with autism spectrum disorders. Dev Psychobiol 52:755–763

Casamassima F et al (2010) L-type calcium channels and psychiatric disorders: a brief review. Am J Med Genet B Neuropsychiatr Genet 153B:1373–1390

Catterall WA, Perez-Reyes E, Snutch TP, Striessnig J (2005) International Union of Pharmacology. XLVIII. Nomenclature and structure-function relationships of voltage-gated calcium channels. Pharmacol Rev 57:411–425

Catterall WA, Lenaeus MJ, Gamal El-Din TM (2020) Structure and pharmacology of voltage-gated sodium and calcium channels. Annu Rev Pharmacol Toxicol 60:133–154

Cheli VT et al (2018) Enhanced oligodendrocyte maturation and myelination in a mouse model of Timothy syndrome. Glia 66:2324–2339

Chen J, Sun Y, Liu X, Li J (2019) Identification of a novel mutation in the CACNA1C gene in a Chinese family with autosomal dominant cerebellar ataxia. BMC Neurol 19:157

Cheng EP et al (2011) Restoration of normal L-type Ca2+ channel function during Timothy syndrome by ablation of an anchoring protein. Circ Res 109:255–261

Chesnoy-Marchais D, Fritsch J (1988) Voltage-gated sodium and calcium currents in rat osteoblasts. J Physiol 398:291–311

Clark MB et al (2020) Long-read sequencing reveals the complex splicing profile of the psychiatric risk gene CACNA1C in human brain. Mol Psychiatry 25:37–47

Colson C et al (2019) Unusual clinical description of adult with Timothy syndrome, carrier of a new heterozygote mutation of CACNA1C. Eur J Med Genet 62:103648

Courchesne E et al (2001) Unusual brain growth patterns in early life in patients with autistic disorder: an MRI study. Neurology 57:245–254

Craddock N, Sklar P (2013) Genetics of bipolar disorder. Lancet 381:1654–1662

Dao DT et al (2010) Mood disorder susceptibility gene CACNA1C modifies mood-related behaviors in mice and interacts with sex to influence behavior in mice and diagnosis in humans. Biol Psychiatry 68:801–810

Davenport B, Li Y, Heizer JW, Schmitz C, Perraud AL (2015) Signature channels of excitability no more: L-type channels in immune cells. Front Immunol 6:375

Dick IE, Joshi-Mukherjee R, Yang W, Yue DT (2016) Arrhythmogenesis in Timothy syndrome is associated with defects in Ca(2+)-dependent inactivation. Nat Commun 7:10370

Dolphin AC (2018) Voltage-gated calcium channel alpha 2delta subunits: an assessment of proposed novel roles. F1000Res 7

Drum BM, Dixon RE, Yuan C, Cheng EP, Santana LF (2014) Cellular mechanisms of ventricular arrhythmias in a mouse model of Timothy syndrome (long QT syndrome 8). J Mol Cell Cardiol 66:63–71

Dufendach KA et al (2018) Clinical outcomes and modes of death in Timothy syndrome: a multicenter international study of a rare disorder. JACC Clin Electrophysiol 4:459–466

Endres D et al (2020) New Cav1.2 channelopathy with high-functioning autism, affective disorder, severe dental enamel defects, a short QT interval, and a novel CACNA1C loss-of-function mutation. Int J Mol Sci 21

Erk S et al (2014) Hippocampal and frontolimbic function as intermediate phenotype for psychosis: evidence from healthy relatives and a common risk variant in CACNA1C. Biol Psychiatry 76: 466–475

Faber GM, Silva J, Livshitz L, Rudy Y (2007) Kinetic properties of the cardiac L-type Ca2+ channel and its role in myocyte electrophysiology: a theoretical investigation. Biophys J 92: 1522–1543

Fabiato A (1983) Calcium-induced release of calcium from the cardiac sarcoplasmic reticulum. Am J Phys 245:C1–C14

Ferreira MA et al (2008) Collaborative genome-wide association analysis supports a role for ANK3 and CACNA1C in bipolar disorder. Nat Genet 40:1056–1058

Frohler S et al (2014) Exome sequencing helped the fine diagnosis of two siblings afflicted with atypical Timothy syndrome (TS2). BMC Med Genet 15:48

Fukuyama M et al (2013) L-type calcium channel mutations in Japanese patients with inherited arrhythmias. Circ J 77:1799–1806

G. C. B. D. W. G. Psychiatric (2011) Large-scale genome-wide association analysis of bipolar disorder identifies a new susceptibility locus near ODZ4. Nat Genet 43:977–983

Gakenheimer-Smith L et al (2021) Expanding the phenotype of CACNA1C mutation disorders. Mol Genet Genomic Med 9:e1673

Gao Y et al (2013) Inhibition of late sodium current by mexiletine: a novel pharmotherapeutical approach in timothy syndrome. Circ Arrhythm Electrophysiol 6:614–622

Gershon ES et al (2014) A rare mutation of CACNA1C in a patient with bipolar disorder, and decreased gene expression associated with a bipolar-associated common SNP of CACNA1C in brain. Mol Psychiatry 19:890–894

Gillis J et al (2012) Long QT, syndactyly, joint contractures, stroke and novel CACNA1C mutation: expanding the spectrum of Timothy syndrome. Am J Med Genet A 158A:182–187

Gomez-Ospina N, Tsuruta F, Barreto-Chang O, Hu L, Dolmetsch R (2006) The C terminus of the L-type voltage-gated calcium channel Ca(V)1.2 encodes a transcription factor. Cell 127:591–606

Gomez-Ospina N et al (2013) A promoter in the coding region of the calcium channel gene CACNA1C generates the transcription factor CCAT. PLoS One 8:e60526

Green EK et al (2010) The bipolar disorder risk allele at CACNA1C also confers risk of recurrent major depression and of schizophrenia. Mol Psychiatry 15:1016–1022

Green EK et al (2013) Replication of bipolar disorder susceptibility alleles and identification of two novel genome-wide significant associations in a new bipolar disorder case-control sample. Mol Psychiatry 18:1302–1307

Guggino SE, Lajeunesse D, Wagner JA, Snyder SH (1989) Bone remodeling signaled by a dihydropyridine- and phenylalkylamine-sensitive calcium channel. Proc Natl Acad Sci U S A 86:2957–2960

Halvorsen M et al (2021) De novo mutations in childhood cases of sudden unexplained death that disrupt intracellular Ca(2+) regulation. Proc Natl Acad Sci U S A 118

Harvey RD, Hell JW (2013) CaV1.2 signaling complexes in the heart. J Mol Cell Cardiol 58:143–152

Hennessey JA et al (2014) A CACNA1C variant associated with reduced voltage-dependent inactivation, increased CaV1.2 channel window current, and arrhythmogenesis. PLoS One 9: e106982

Hering S et al (1997) Molecular mechanism of use-dependent calcium channel block by phenylalkylamines: role of inactivation. Proc Natl Acad Sci U S A 94:13323–13328

Hermida A et al (2021) Long-term follow-up of a patient with type 2 Timothy syndrome and the partial efficacy of mexiletine. Gene 777:145465

Hiippala A, Tallila J, Myllykangas S, Koskenvuo JW, Alastalo TP (2015) Expanding the phenotype of Timothy syndrome type 2: an adolescent with ventricular fibrillation but normal development. Am J Med Genet A 167A:629–634

Jacobs A, Knight BP, McDonald KT, Burke MC (2006) Verapamil decreases ventricular tachyarrhythmias in a patient with Timothy syndrome (LQT8). Heart Rhythm 3:967–970

Janiri D et al (2021) Genetic neuroimaging of bipolar disorder: a systematic 2017-2020 update. Psychiatr Genet 31:50–64

Keers R, Farmer AE, Aitchison KJ (2009) Extracting a needle from a haystack: reanalysis of whole genome data reveals a readily translatable finding. Psychol Med 39:1231–1235

Klugbauer N, Welling A, Specht V, Seisenberger C, Hofmann F (2002) L-type Ca2+ channels of the embryonic mouse heart. Eur J Pharmacol 447:279–284

Koch K et al (2019) CACNA1C risk variant affects microstructural connectivity of the amygdala. Neuroimage Clin 22:101774

Kosaki R, Ono H, Terashima H, Kosaki K (2018) Timothy syndrome-like condition with syndactyly but without prolongation of the QT interval. Am J Med Genet A 176:1657–1661

Krey JF et al (2013) Timothy syndrome is associated with activity-dependent dendritic retraction in rodent and human neurons. Nat Neurosci 16:201–209

Krug A et al (2010) Effect of CACNA1C rs1006737 on neural correlates of verbal fluency in healthy individuals. NeuroImage 49:1831–1836

Kschonsak M et al (2020) Structure of the human sodium leak channel NALCN. Nature 587:313–318

Lachmann A et al (2018) Massive mining of publicly available RNA-seq data from human and mouse. Nat Commun 9:1366

Lancaster TM, Foley S, Tansey KE, Linden DE, Caseras X (2016) CACNA1C risk variant is associated with increased amygdala volume. Eur Arch Psychiatry Clin Neurosci 266:269–275

Lederer WJ et al (1990) Excitation-contraction coupling in heart cells. Roles of the sodium-calcium exchange, the calcium current, and the sarcoplasmic reticulum. Ann N Y Acad Sci 588:190–206

Lee KS, Tsien RW (1983) Mechanism of calcium channel blockade by verapamil, D600, diltiazem and nitrendipine in single dialysed heart cells. Nature 302:790–794

Lee AS et al (2012) Forebrain elimination of cacna1c mediates anxiety-like behavior in mice. Mol Psychiatry 17:1054–1055

Li J et al (2015) Schizophrenia related variants in CACNA1C also confer risk of autism. PLoS One 10:e0133247

Li B, Tadross MR, Tsien RW (2016) Sequential ionic and conformational signaling by calcium channels drives neuronal gene expression. Science 351:863–867

Liao P, Yong TF, Liang MC, Yue DT, Soong TW (2005) Splicing for alternative structures of Cav1.2 Ca2+ channels in cardiac and smooth muscles. Cardiovasc Res 68:197–203

Liao P et al (2007) A smooth muscle Cav1.2 calcium channel splice variant underlies hyperpolarized window current and enhanced state-dependent inhibition by nifedipine. J Biol Chem 282:35133–35142

Liao P et al (2015) Alternative splicing generates a novel truncated Cav1.2 channel in neonatal rat heart. J Biol Chem 290:9262–9272

Lory P, Nicole S, Monteil A (2020) Neuronal Cav3 channelopathies: recent progress and perspectives. Pflugers Arch 472:831–844

Ma H, Cohen S, Li B, Tsien RW (2012) Exploring the dominant role of Cav1 channels in signalling to the nucleus. Biosci Rep 33:97–101

Marcantoni A, Calorio C, Hidisoglu E, Chiantia G, Carbone E (2020) Cav1.2 channelopathies causing autism: new hallmarks on Timothy syndrome. Pflugers Arch 472:775–789

Marks ML, Whisler SL, Clericuzio C, Keating M (1995) A new form of long QT syndrome associated with syndactyly. J Am Coll Cardiol 25:59–64

Matza D, Flavell RA (2009) Roles of Ca(v) channels and AHNAK1 in T cells: the beauty and the beast. Immunol Rev 231:257–264

Mellor GJ et al (2019) Type 8 long QT syndrome: pathogenic variants in CACNA1C-encoded Cav1.2 cluster in STAC protein binding site. Europace 21:1725–1732

Meszaros JG, Karin NJ, Akanbi K, Farach-Carson MC (1996) Down-regulation of L-type Ca2+ channel transcript levels by 1,25-dihyroxyvitamin D3. Osteoblastic cells express L-type alpha1C Ca2+ channel isoforms. J Biol Chem 271:32981–32985

Moosmang S et al (2005) Role of hippocampal Cav1.2 Ca2+ channels in NMDA receptor-independent synaptic plasticity and spatial memory. J Neurosci 25:9883–9892

Morotti S, Grandi E, Summa A, Ginsburg KS, Bers DM (2012) Theoretical study of L-type Ca(2+) current inactivation kinetics during action potential repolarization and early afterdepolarizations. J Physiol 590:4465–4481

Murphy JG et al (2014) AKAP-anchored PKA maintains neuronal L-type calcium channel activity and NFAT transcriptional signaling. Cell Rep 7:1577–1588

Napolitano C, Splawski I, Timothy KW, Bloise R, Priori SG (1993) In: Pagon RA et al (eds) GeneReviews(R), Seattle

Narula N, Tester DJ, Paulmichl A, Maleszewski JJ, Ackerman MJ (2015) Post-mortem whole exome sequencing with gene-specific analysis for autopsy-negative sudden unexplained death in the young: a case series. Pediatr Cardiol 36:768–778

Nerbonne JM, Kass RS (2005) Molecular physiology of cardiac repolarization. Physiol Rev 85: 1205–1253

Niu J et al (2018) Allosteric regulators selectively prevent Ca(2+)-feedback of CaV and NaV channels. elife 7

Ozawa J et al (2018) A novel CACNA1C mutation identified in a patient with Timothy syndrome without syndactyly exerts both marked loss- and gain-of-function effects. HeartRhythm Case Rep 4:273–277

Pasca SP et al (2011) Using iPSC-derived neurons to uncover cellular phenotypes associated with Timothy syndrome. Nat Med 17:1657–1662

Pitt GS, Matsui M, Cao C (2021) Voltage-gated calcium channels in nonexcitable tissues. Annu Rev Physiol 83:183–203

Polster A et al (2018) Stac proteins suppress Ca(2+)-dependent inactivation of neuronal l-type Ca (2+) channels. J Neurosci 38:9215–9227

Pouget JG et al (2019) Cross-disorder analysis of schizophrenia and 19 immune-mediated diseases identifies shared genetic risk. Hum Mol Genet 28:3498–3513

Ramachandran KV et al (2013) Calcium influx through L-type CaV1.2 Ca2+ channels regulates mandibular development. J Clin Invest 123:1638–1646

Raybaud A et al (2006) The role of the GX9GX3G motif in the gating of high voltage-activated Ca2+ channels. J Biol Chem 281:39424–39436

Reichenbach H, Meister EM, Theile H (1992) The heart-hand syndrome. A new variant of disorders of heart conduction and syndactylia including osseous changes in hands and feet. Kinderarztl Prax 60:54–56

Rodan LH et al (2021) Phenotypic expansion of CACNA1C-associated disorders to include isolated neurological manifestations. Genet Med 23:1922–1932

Roussos P, Giakoumaki SG, Georgakopoulos A, Robakis NK, Bitsios P (2011) The CACNA1C and ANK3 risk alleles impact on affective personality traits and startle reactivity but not on cognition or gating in healthy males. Bipolar Disord 13:250–259

Sayad A et al (2019) Association study of sequence variants in voltage-gated Ca2+ channel subunit alpha-1C and autism spectrum disorders. Rep Biochem Mol Biol 8:56–62

Schmitt A et al (2022) Post-mortem gene expression of calcium channels Cav1.2 and Cav1.3 in schizophrenia. Eur Arch Psychiatry Clin Neurosci 272:1135–1137

Scholl UI et al (2013) Somatic and germline CACNA1D calcium channel mutations in aldosterone-producing adenomas and primary aldosteronism. Nat Genet 45:1050–1054

Sepp R et al (2017) Timothy syndrome 1 genotype without syndactyly and major extracardiac manifestations. Am J Med Genet A 173:784–789

Shah DP, Baez-Escudero JL, Weisberg IL, Beshai JF, Burke MC (2010) Ranolazine safely decreases ventricular and atrial fibrillation in timothy syndrome (LQT8). Pacing Clin Electrophysiol

Shah DP, Baez-Escudero JL, Weisberg IL, Beshai JF, Burke MC (2012) Ranolazine safely decreases ventricular and atrial fibrillation in Timothy syndrome (LQT8). Pacing Clin Electrophysiol 35:e62–e64

Shelley C, Whitt JP, Montgomery JR, Meredith AL (2013) Phosphorylation of a constitutive serine inhibits BK channel variants containing the alternate exon "SRKR". J Gen Physiol 142:585–598

Sheng X et al (2012) Two mechanistically distinct effects of dihydropyridine nifedipine on CaV1.2 L-type Ca(2)(+) channels revealed by Timothy syndrome mutation. Eur J Pharmacol 685:15–23

Sinnegger-Brauns MJ et al (2009) Expression and 1,4-dihydropyridine-binding properties of brain L-type calcium channel isoforms. Mol Pharmacol 75:407–414

Sklar P et al (2008) Whole-genome association study of bipolar disorder. Mol Psychiatry 13:558–569

Soldatov NM (1992) Molecular diversity of L-type Ca2+ channel transcripts in human fibroblasts. Proc Natl Acad Sci U S A 89:4628–4632

Song L et al (2015) Dual optical recordings for action potentials and calcium handling in induced pluripotent stem cell models of cardiac arrhythmias using genetically encoded fluorescent indicators. Stem Cells Transl Med 4:468–475

Song L, Park SE, Isseroff Y, Morikawa K, Yazawa M (2017) Inhibition of CDK5 alleviates the cardiac phenotypes in timothy syndrome. Stem Cell Rep 9:50–57

Splawski I et al (2004) Ca(V)1.2 calcium channel dysfunction causes a multisystem disorder including arrhythmia and autism. Cell 119:19–31

Splawski I et al (2005) Severe arrhythmia disorder caused by cardiac L-type calcium channel mutations. Proc Natl Acad Sci U S A 102:8089–8096. Discussion 8086–8088

Stotz SC, Zamponi GW (2001) Identification of inactivation determinants in the domain IIS6 region of high voltage-activated calcium channels. J Biol Chem 276:33001–33010

Stotz SC, Hamid J, Spaetgens RL, Jarvis SE, Zamponi GW (2000) Fast inactivation of voltage-dependent calcium channels. A hinged-lid mechanism? J Biol Chem 275:24575–24582

Striessnig J (2021) Voltage-gated Ca(2+)-channel alpha1-subunit de novo missense mutations: gain or loss of function – implications for potential therapies. Front Synaptic Neurosci 13:634760

Striessnig J, Pinggera A, Kaur G, Bock G, Tuluc P (2014) L-type Ca(2+) channels in heart and brain. Wiley Interdiscip Rev Membr Transp Signal 3:15–38

Sullivan PF et al (2009) Genome-wide association for major depressive disorder: a possible role for the presynaptic protein piccolo. Mol Psychiatry 14:359–375

Sutphin BS et al (2016) Molecular and functional characterization of rare CACNA1C variants in sudden unexplained death in the young. Congenit Heart Dis 11:683–692

Tadross MR, Ben Johny M, Yue DT (2010) Molecular endpoints of Ca2+/calmodulin- and voltage-dependent inactivation of Ca(v)1.3 channels. J Gen Physiol 135:197–215

Tang ZZ et al (2004) Transcript scanning reveals novel and extensive splice variations in human l-type voltage-gated calcium channel, Cav1.2 alpha1 subunit. J Biol Chem 279:44335–44343

Terrillion CE et al (2017) Reduced levels of Cacna1c attenuate mesolimbic dopamine system function. Genes Brain Behav 16:495–505

Tunca Sahin G, Ergul Y (2018) A case report: is mexiletine usage effective in the shortening of QTC interval and improving the T-wave alternans in Timothy syndrome? Ann Noninvasive Electrocardiol 23:e12522

Wei XY et al (1991) Heterologous regulation of the cardiac Ca^{2+} channel alpha 1 subunit by skeletal muscle beta and gamma subunits. Implications for the structure of cardiac L-type Ca^{2+} channels. JBiolChem 266:21943–21947

Wemhoner K et al (2015) Gain-of-function mutations in the calcium channel CACNA1C (Cav1.2) cause non-syndromic long-QT but not Timothy syndrome. J Mol Cell Cardiol 80:186–195

Wolf C et al (2014) CACNA1C genotype explains interindividual differences in amygdala volume among patients with schizophrenia. Eur Arch Psychiatry Clin Neurosci 264:93–102

Yarotskyy V, Elmslie KS (2007) Roscovitine, a cyclin-dependent kinase inhibitor, affects several gating mechanisms to inhibit cardiac L-type (Ca(V)1.2) calcium channels. Br J Pharmacol 152: 386–395

Yarotskyy V, Gao G, Peterson BZ, Elmslie KS (2009) The Timothy syndrome mutation of cardiac CaV1.2 (L-type) channels: multiple altered gating mechanisms and pharmacological restoration of inactivation. J Physiol 587:551–565

Yarotskyy V et al (2010) Roscovitine binds to novel L-channel (CaV1.2) sites that separately affect activation and inactivation. J Biol Chem 285:43–53

Yazawa M et al (2011) Using induced pluripotent stem cells to investigate cardiac phenotypes in Timothy syndrome. Nature 471:230–234

Yoshimizu T et al (2015) Functional implications of a psychiatric risk variant within CACNA1C in induced human neurons. Mol Psychiatry 20:162–169

Zamponi GW, Striessnig J, Koschak A, Dolphin AC (2015) The physiology, pathology, and pharmacology of voltage-gated calcium channels and their future therapeutic potential. Pharmacol Rev 67:821–870

Zhang H et al (2021) CACNA1C rs1006737 SNP increases the risk of essential hypertension in both Chinese Han and ethnic Russian people of Northeast Asia. Medicine (Baltimore) 100: e24825

Zhu D et al (2019) CACNA1C (rs1006737) may be a susceptibility gene for schizophrenia: an updated meta-analysis. Brain Behav 9:e01292

CACNA1D-Related Channelopathies: From Hypertension to Autism

Nadine J. Ortner

Contents

Abstract

Tightly controlled Ca^{2+} influx through voltage-gated Ca^{2+} channels (Cavs) is indispensable for proper physiological function. Thus, it is not surprising that Cav loss and/or gain of function have been implicated in human pathology. Deficiency of Cav1.3 L-type Ca^{2+} channels (LTCCs) causes deafness and bradycardia, whereas several genetic variants of *CACNA1D*, the gene encoding the pore-forming α1 subunit of Cav1.3, have been linked to various disease phenotypes, such as hypertension, congenital hypoglycemia, or autism. These variants include not only common polymorphisms associated with an increased disease risk, but also rare de novo missense variants conferring high risk. This review provides a concise summary of disease-associated *CACNA1D* variants, whereas the main focus lies on de novo germline variants found in individuals with a neurodevelopmental disorder of variable severity. Electrophysiological recordings revealed activity-enhancing gating changes induced by these de

N. J. Ortner (✉)
Department of Pharmacology and Toxicology, Institute of Pharmacy, Center for Molecular Biosciences Innsbruck, University of Innsbruck, Innsbruck, Austria
e-mail: Nadine.ortner@uibk.ac.at

J. Striessnig (ed.), *Voltage-gated Ca^{2+} Channels: Pharmacology, Modulation and their Role in Human Disease*, Handbook of Experimental Pharmacology 279,
https://doi.org/10.1007/164_2022_626

novo variants, and tools to predict their pathogenicity and to study the resulting pathophysiological consequences will be discussed. Despite the low number of affected patients, potential phenotype–genotype correlations and factors that could impact the severity of symptoms will be covered. Since increased channel activity is assumed as the disease-underlying mechanism, pharmacological inhibition could be a treatment option. In the absence of Cav1.3-selective blockers, dihydropyridine LTCC inhibitors clinically approved for the treatment of hypertension may be used for personalized off-label trials. Findings from in vitro studies and treatment attempts in some of the patients seem promising as outlined. Taken together, due to advances in diagnostic sequencing techniques the number of reported *CACNA1D* variants in human diseases is constantly rising. Evidence from in silico, in vitro, and in vivo disease models can help to predict the pathogenic potential of such variants and to guide diagnosis and treatment in the clinical practice when confronted with patients harboring *CACNA1D* variants.

Keywords

CACNA1D · Cav1.3 Ca^{2+} channels · Genetic variants · Neurodevelopmental disorders

1 Introduction

Voltage-gated Ca^{2+} channels (Cavs) conduct Ca^{2+} influx across the plasma membrane in a tightly controlled manner, which is crucial for proper Ca^{2+} signaling. Cavs can be found in almost all electrically excitable cells, where they are involved in multiple physiological processes, such as muscle contraction, endocrine and neurotransmitter secretion, regulation of cellular excitability, or excitation-transcription coupling (Catterall 2011; Zamponi et al. 2015). To be able to fulfill such a wide spectrum of cellular processes, three main families of 10 different pore-forming Cav α1 subunits exist, each encoded by individual genes: Cav1 (L-type; Cav1.1–Cav1.4), Cav2 (P/Q-type Cav2.1; N-type Cav2.2; R-type Cav2.3), and Cav3 (T-type; Cav3.1–Cav3.3). Their functional heterogeneity is further increased by the association of the α1 subunit with auxiliary subunits (β, α2δ, γ) and other interaction partners, alternative splicing, posttranslational modifications, expression patterns, and subcellular localization (reviewed in Zamponi et al. 2015). The importance of the strict control of Ca^{2+} influx is evident from the fact that mutations in all Cav genes (as well as of auxiliary subunits), causing loss or gain of channel function, have been linked to human diseases (Striessnig 2021; Zamponi et al. 2015). This review focuses on Cav1.3 L-type Ca^{2+} channels (LTCCs; α1 encoded by the *CACNA1D* gene) that are expressed throughout the body, including the cochlear, heart, endocrine tissues and the nervous system. Pathology associated with a homozygous loss of Cav1.3 function is well-studied, causing congenital deafness and bradycardia in mice and humans (sinoatrial node dysfunction and deafness syndrome, SANDD; OMIM # 614896) (Baig et al. 2011; Platzer et al. 2000). Next generation sequencing-based genetic screening has advanced the discovery of genetic causes for rare and more common diseases. Genetic *CACNA1D* variants could thereby be linked to pathologies such as hypertension, disturbed glucose metabolism, and brain disorders, thus expanding the pathophysiological spectrum associated with *CACNA1D*. These genetic variants will be summarized in the first section of this review and then special focus will be laid on high-risk de novo missense *CACNA1D* variants found in the germline of individuals with a neurodevelopmental syndrome. Although a definite proof for their disease-causing nature is still missing, several lines of evidence point to pathogenicity as outlined in this review. Also, information from other Cavs and somatic de novo *CACNA1D* mutations found in aldosterone-producing adrenal lesions can further help to assess the pathogenic potential of newly identified variants. Electrophysiological studies with some of these variants confirmed functional changes of channel gating predicting enhanced Cav1.3 activity as the disease-underlying mechanism. These gating changes and model systems to study the underlying molecular mechanisms as well as cellular consequences will be discussed. The limited number of patients identified so far hinders a meaningful phenotype–genotype correlation. Individuals with de novo germline variants share common clinical features, but the severity and manifestation of associated symptoms can vary and potential mechanisms that could underlie this variability will be outlined. Lastly, the important question if available LTCC targeting drugs could be repurposed for the off-label treatment of affected patients will be assessed. Taken together, this review aims at giving an overview on i) the current evidence supporting the pathogenic role of *CACNA1D* variants for human diseases, ii) tools for the assessment of pathogenicity and study of mutation-induced functional consequences, and iii) potential pharmacological interventions.

2 *CACNA1D* (Cav1.3) Genetic Variants Associated with Human Disease

Genetic studies have linked single nucleotide variants in intronic as well as coding regions of the *CACNA1D* gene to various disease phenotypes including deafness, bradycardia, hypertension, diabetes, or autism spectrum disorder (ASD) (Tables 1 and 2). Some of them rather represent common small effect variants that are associated with an increased disease risk, whereas for others a clear pathogenicity can be assumed. A valuable tool in this regard is the gnomAD database, which is a collection of 125,748 exome sequences and 15,708 whole-genome sequences from individuals without severe pediatric disease, i.e. apparently healthy controls (https://gnomad.broadinstitute.org/). The presence of a variant in gnomAD (given in Tables 1 and 2 for the here discussed variants) or other control populations does not automatically exclude pathogenicity, but can be indicative of a lower pathogenic effect. Conversely, the absence in healthy controls can support the notion of disease causality. Moreover, the location and nature of the mutation itself, as well as functional data – if available – should be considered (see also next section). The following section gives an overview and basic assessment of genetic disease-

Table 1 Genetic variants of *CACNA1D* associated with disease risk

Associated phenotype	Reference	Genetic variant	GnomAD allele count (frequency); # of HOMs	Channel domain	Comment
SANDD syndrome	(Baig et al. 2011; Liaqat et al. 2019)	G403dup	2× (0.000008)	I – S6	Loss of function; in alternative exon 8b; conserved in Cav1 + Cav2; channel still expressed, but non-functional
	(Abu Rayyan et al. 2020; Liaqat et al. 2019)	A376V	4× (0.00002)	I – S5-S6 linker (pore loop)	Loss of function; in alternative exon 8b; conserved in all Cavs; 3D modeling showed an additional hydrophobic interaction in the mutant
Blood pressure	(Lu et al. 2015; Lule et al. 2019; Yang et al. 2019)	Intronic SNP rs9810888	16,438 (0.5244); 4,307× HOM		Found in Chinese adults and children, Ugandan early adolescents
Hypertension	(Zhu et al. 2015)	Intronic SNP rs10049492	11,554 (0.3689); 2,314× HOM		Suggestive association for hypertension-related traits
	(Curtis 2021; Sun et al. 2022)	F1899del (F1943del)	673× (0.0024); 3× HOM	C-terminus	Increased risk of hypertension
	(Bao et al. 2020)	D307G	–	I – S5-S6 linker (pore loop)	Family history of hypertension; also associated with pre-eclampsia; rat model shows elevated blood pressure (Wang et al. 2022b)
		V877I (V897I)	10× (0.00004)	III – S0	Only exacerbated hypertensive phenotype of D307G rats (Wang et al. 2022b)
		R910H (R930H)	123× (0.0004); 1× HOM	III S1-S2 loop	6× R910C (R930C) in gnomAD
		N1234S (N1254S)	150× (0.0005); 1× HOM	IV – S2	3× N1234I (N1254I) in gnomAD

		* R1457Q (R1492Q)	1× (0.000004)	C-terminus	Family history of hypertension; also associated with pre-eclampsia; only exacerbated hypertensive phenotype of D307G rats (Wang et al. 2022b); start of C-terminus, conserved in Cav1 + Cav2
		H1725R (1760R)	32× (0.0001)	C-terminus	
		R1885H (R1929H)	2× (0.000008)	C-terminus	Shows some conservation (Wu et al. 2016); 1× R1885C (R1929C) in gnomAD
	(Takahashi et al. 2021)	Intronic SNP rs3774427	3,208 (0.1023); 436× HOM		Suggestive association with resistant hypertension in Japanese population
	(Kamide et al. 2009)	Intronic SNP rs312481	25,447 (0.8111); 10,325× HOM		Japanese patients with essential hypertension; associated with antihypertensive effects of DHPs; GG carriers most sensitive to DHP effects for reducing blood pressure
		Intronic SNP rs3774426	9,880 (0.3155); 1,740× HOM		Japanese patients with essential hypertension; associated with antihypertensive effects of DHPs; CC carriers
Insulin secretion	(Reinbothe et al. 2013)	Intronic SNP rs312480	25,341 (0.8440); 10,832× HOM		5'UTR of exon 1; C allele associated with reduced Cav1.3 mRNA expression and insulin secretion
Type 2 diabetes	(Reinbothe et al. 2013)	Intronic SNP rs312486	25,012 (0.7976); 9,963× HOM		G allele associated with increased risk of type 2 diabetes

(continued)

Table 1 (continued)

Associated phenotype	Reference	Genetic variant	GnomAD allele count (frequency); # of HOMs	Channel domain	Comment
		Intronic SNP rs9841978	11,407 (0.3640); 2,123× HOM		G allele associated with increased risk of type 2 diabetes
Cocaine dependence	(Martinez-Rivera et al. 2017)	Intronic SNP rs4687735	30,506 (0.9712); 14,818× HOM		Strongest association in European-Americans; trend association in African-Americans
		Intronic SNP rs2926559	30,169 (0.9709); 14,651× HOM		Strong association in European-Americans; trend association in African-Americans
		Intronic SNP rs898407	30,191 (0.9696); 14,643× HOM		Strong association in European-Americans; trend association in African-Americans
Epilepsy	(Tumiene et al. 2018)	# R498L	–	I – II linker	In alternative exon 11; no functional changes detected in vitro (Hofer et al. 2021); 1× R498del, 8× R498W and 37× R498Q in gnomAD
	(Klassen et al. 2011)	# P1318S	1,508× (0.0053); 11× HOM	IV – S3-S4 linker	In alternative exon 32
		I1759N (I1794N)	8× (0.00003)	C-terminus	1× I1759M (I1794N) in gnomAD
		A519P (A539P)	–	I – II linker	3× A519T (A539T) in gnomAD; near a regulatory polybasic motif (Kaur et al. 2015)
		Q1598X (Q1633X)	–	C-terminus	STOP after I1597 of the calmodulin-binding "IQ" motif; either non-function or, if functionally expressed, reduced Ca^{2+} dependent inactivation

		E1293A (E1328A)	–	IV – S3-S4 linker	Present in control subjects
		E1658K (E1693K)	186× (0.0007)	C-terminus	Present in control subjects
Schizophrenia	(Wang et al. 2022a)	G330R	–	I – S5-S6 linker (pore loop)	Conserved in Cav1.1, Cav1.2, Cav1.3, Cav2.1, Cav2.3
		M716V (M736V)	10× (0.00004)	II – S5-S6 linker (pore loop)	Conserved in Cav1
Schizophrenia/ASD	(Wang et al. 2022a)	V728I (V748I)	115× (0.0004)	II – S6	Reported 1× in an APA (Wang et al. 2015); conserved in Cav1
		E1660del (E1695del)	39× (0.0001)	C-terminus	Conserved in Cav1
ASD	(Wang et al. 2022a)	C1436X (C1471X)	–	IV – S6	STOP in IV-S6; conserved in Cav1 and Cav2; likely resulting in a non-functional channel
	(De Rubeis et al. 2014)	A59V	–	N-terminus	NSCaTE motif, Calmodulin-binding; involved in Ca^{2+}-dependent inactivation (Dick et al. 2008)
		S1953L (S1997L)	10× (0.00004)	C-terminus	1× S1953P (S1997P) in gnomAD
		R1997H (R2041H)	6× (0.00002)	C-terminus	7× R1997C (R2041C) in gnomAD

(continued)

Table 1 (continued)

Associated phenotype	Reference	Genetic variant	GnomAD allele count (frequency); # of HOMs	Channel domain	Comment
Moderate hearing loss, intellectual disability, seizures, autistic behaviors, and hyperactivity	(Garza-Lopez et al. 2018)	Q547H (Q567H)	–	II – S1-S2 linker	Patient homozygous for variant (both parents heterozygous); induced channel gating changes (type-3); conserved in all Cavs

Genomic *CACNA1D* variants associated with human pathologies (de novo missense variants are listed in Table 2). The residues are numbered according to the genome reference sequence EU363339 and the corresponding residue in the genome reference sequence NM_000720 used in the gnomAD database is given in parenthesis. Allele frequency = allele count divided by allele number as reported in gnomAD. #, R498L and P1318 are located in additional exons 11 and 32, respectively, which are not present in EU363339, therefore, numbering of NM_000720 is used. *, in the original manuscript (Bao et al. 2020) inconsistent information for this variant was given, but the correct information is NM_000720 c.G4475A p.R1492Q (personal communication). HOM, homozygous individuals. Channel domains/homologous repeats I-IV; "SX" refers to transmembrane segment X (1–6). SANDD syndrome, sinoatrial node and deafness syndrome (OMIM # 614896); DHPs, dihydropyridines; APA, aldosterone-producing adenoma; ASD, autism spectrum disorder

Table 2 Summary of somatic (APCCs/APAs) and germline de novo missense variants of *CACNA1D* with predicted pathogenicity

Position	Variant	Reference	Gating change	Molecular modeling	Times reported	Reported in gnomAD	Pathogenicity
V123	V123A	(Wu et al. 2021)			1		Likely pathogenic
E124	E124K	(Omata et al. 2018)			1		Likely pathogenic
R240	R240C	(Wu et al. 2021)			1		Likely pathogenic
L248	L248F	(Omata et al. 2018)			1		Pathogenic
	L248P	(Wu et al. 2021)			1		Pathogenic
V259	V259A	(Semenova et al. 2018)			Germline (1×)		Pathogenic
	Seizures, hypotonia, primary aldosteronism, intellectual impairment/disability, developmental delay, facial dysmorphism, microcephaly						
	V259D	(Azizan et al. 2013; Fernandes-Rosa et al. 2014; Gagnon et al. 2018; Nanba et al. 2018)	Type-2 (Azizan et al. 2013)		5		Pathogenic
	V259G	(Nanba et al. 2018; Omata et al. 2018)			3		Pathogenic
L271	L271H	(De Mingo Alemany et al. 2020)			Germline (1×)		Pathogenic
	Hypotonia, primary aldosteronism, hypoglycemic hyperinsulinism, developmental delay, maternal pre-eclampsia with HELLP syndrome, facial dysmorphism						
L272	L272R	(Omata et al. 2018)			1		Likely pathogenic
V309	V309A	(Nanba et al. 2019)			1	V309I (84×; 1× HOM)	Likely pathogenic
G323	G323R	(Omata et al. 2018)			1		Likely pathogenic
V401	V401L	(Pinggera et al. 2017)	Type-2 (Pinggera et al. 2017)		Germline (1×), exon 8a		Pathogenic

(continued)

Table 2 (continued)

Position	Variant	Reference	Gating change	Molecular modeling	Times reported	Reported in gnomAD	Pathogenicity
	ASD, seizures, limb spasticity, hypotonia, intellectual impairment/disability, autoaggression/self-injury						
	V401L	(Akerstrom et al. 2015; Nanba et al. 2019; Omata et al. 2018)	Type-2 (Pinggera et al. 2017)		4 (exon 8a)		Pathogenic
G403	G403D	(Flanagan et al. 2017; Scholl et al. 2013)	Type-1 (Scholl et al. 2013)		Germline (2×), exon 8b	G403dup (2×) – loss of function (Baig et al. 2011)	Pathogenic
	Patient 1, (Scholl et al. 2013): Seizures, limb spasticity, hypotonia, primary aldosteronism, intellectual impairment/disability, developmental delay, sinus bradycardia/biventricular hypertrophy, ventricular septum defect, patent foramen ovale, cortical blindness, cerebral palsy, transient hypoglycemia on day2; *patient 2*, (Flanagan et al. 2017): (ASD), seizures, limb spasticity, hypotonia, hypoglycemic hyperinsulinism, intellectual impairment/disability, developmental delay, at age 3 not ambulatory not verbal						
	G403R	(Akerstrom et al. 2015; Azizan et al. 2013; Fernandes-Rosa et al. 2014; Guo et al. 2020; Hacini et al. 2021; Kitamoto et al. 2018; Kitamoto et al. 2016; Murakami et al. 2019; Nanba et al. 2018; Nanba et al. 2019; Nanba et al. 2020; Nishimoto et al. 2015; Omata et al. 2017; Omata et al. 2018; Scholl et al. 2013; Scholl et al. 2015; Tan et al. 2017; Yamazaki et al. 2017; Zheng et al. 2015; Zhou et al. 2021)	Type-1 (Azizan et al. 2013; Scholl et al. 2013)		74 (exon 8a: 42; exon 8b: 16; not reported: 16)	G403dup (2×) – loss of function (Baig et al. 2011)	Pathogenic
G407	G407R	(Iossifov et al. 2012)	Type-1 (Pinggera et al. 2015; Pinggera et al. 2018)	(Pinggera et al. 2015)	Germline (1×), exon 8a		Pathogenic

	ASD, anxiety, depression						
S410	S410L	(Backman et al. 2019; Omata et al. 2017; Omata et al. 2018)			3		Pathogenic
E412	E412D	(Yamazaki et al. 2017)			1		Likely pathogenic
G457	G457R	(Omata et al. 2017)			1	G457R (1×), G457del (2×)	Uncertain
R510 (R530)	R510X	(Omata et al. 2017)			1	R510Q (2×)	* Likely benign
P548 (P568)	P548L	(Omata et al. 2017)			1		Likely pathogenic
M570 (M590)	M570I	(Wu et al. 2021)			1		Likely pathogenic
V584 (V604)	V584I	(De Rubeis et al. 2014)			Germline (1×)	V584I (8×), V584F (1×)	Likely benign
	ASD						
L613 (L633)	L613Q	(Nishimoto et al. 2015)			1		Likely pathogenic
R619 (R639)	R619P	(Nanba et al. 2019)		(Ortner et al. 2020)	1	R619W (1×)	Likely pathogenic
	R619W	(Nishimoto et al. 2015)		(Ortner et al. 2020)	1	R619W (1×)	Uncertain
I650 (I670)	I650N	(Zhou et al. 2021)			1	I650T (1×)	Likely pathogenic
S652 (S672)	S652L	(Deciphering Developmental Disorders 2015)	Type-2 (Hofer et al. 2020)	(Hofer et al. 2020)	Germline (2×, monozygotic twins)	S652W (3×)	Pathogenic
	Patient 1: ASD, seizures (no recurrence), intellectual impairment/disability, autoaggression/self-injury, developmental delay, undescended tests, facial dysmorphism; *patient 2*: ASD, intellectual impairment/disability, autoaggression/self-injury, facial dysmorphism						

(continued)

Table 2 (continued)

Position	Variant	Reference	Gating change	Molecular modeling	Times reported	Reported in gnomAD	Pathogenicity
	S652L	(Fernandes-Rosa et al. 2014; Nanba et al. 2019; Nanba et al. 2020; Omata et al. 2018; Wu et al. 2021; Yamazaki et al. 2017)	Type-2 (Hofer et al. 2020)	(Hofer et al. 2020)	7	S652W (3×)	Pathogenic
L653 (L673)	L653P	(Omata et al. 2018)			1		Likely pathogenic
L655 (L675)	L655P	(Fernandes-Rosa et al. 2014; Murakami et al. 2019; Murakami et al. 2017)			3	L655F (1×)	Pathogenic
S724 (S744)	S724L	(Omata et al. 2018)			1		Likely pathogenic
V728 (V748)	V728I	(Wang et al. 2015)			1	V728I (115×)	Likely benign
Y741 (Y761)	Y741C	(Fernandes-Rosa et al. 2014)			1		Likely pathogenic
F747 (F767)	F747C	(Nanba et al. 2016; Nanba et al. 2019; Omata et al. 2017)			3		Pathogenic
	F747L	(Akerstrom et al. 2015; Azizan et al. 2013; Fernandes-Rosa et al. 2014; Guo et al. 2020; Nanba et al. 2018; Nanba et al. 2019; Nishimoto et al. 2015; Omata et al. 2017; Omata et al. 2018; Tan et al. 2017; Yamazaki et al. 2017; Zhou et al. 2021)	Type-2 (Pinggera et al. 2018)		27		Pathogenic

	F747S	(Guo et al. 2020)			1		Pathogenic
	F747S	(Török et al. 2022)	Type-2 (Török et al. 2022)	(Török et al. 2022)	Germline (1×)		Pathogenic
	Unusual jitteriness triggered upon contact, hypoglycemia, hypertension (hyperaldosteronism), unable to suck, swallow and breath efficiently at birth (CPAP treatment and nasogastric feeding)						
	F747V	(Fernandes-Rosa et al. 2014; Guo et al. 2020; Murakami et al. 2019; Nanba et al. 2016; Nanba et al. 2018; Nanba et al. 2019; Nanba et al. 2020; Nishimoto et al. 2015; Omata et al. 2017; Omata et al. 2018; Scholl et al. 2015; Yamazaki et al. 2017)			28		Pathogenic
L748 (L768)	L748S	(Omata et al. 2018)			1		Likely pathogenic
A749 (A769)	A749G	(O'Roak et al. 2012)	Type-2 (Limpitikul et al. 2016; Pinggera et al. 2015; Pinggera et al. 2018)	(Pinggera et al. 2015)	Germline (1×)		Pathogenic
	ASD, intellectual impairment/disability, sleep disorder, impaired motor skills, clumsy, uncoordinated, night incontinence						
	A749T	(Ortner et al. 2020; Strauss et al. 2018)			Germline (2×)		Pathogenic
	Patient 1 (Ortner et al. 2020): ASD, hypotonia, intellectual impairment/disability, sleep disorder, autoaggression/self-injury, developmental delay, binocular vision disorder; *patient 2* (Strauss et al. 2018): no clinical details reported						
I750 (I770)	I750M	(Scholl et al. 2013)	Type-2 (Azizan et al. 2013; Scholl et al. 2013)		Germline (1×)		Pathogenic
	Seizures, limb spasticity, primary aldosteronism, intellectual impairment/disability, sleep disorder, developmental delay, ventricular hypertrophy, cerebral palsy, spastic quadriplegia, movement disorder with verbal outbursts						
	I750M	(Azizan et al. 2013; Fernandes-Rosa et al. 2014; Kitamoto et al. 2018; Kitamoto et al. 2016;	Type-2 (Azizan et al. 2013; Scholl et al. 2013)		15		Pathogenic

(continued)

Table 2 (continued)

Position	Variant	Reference	Gating change	Molecular modeling	Times reported	Reported in gnomAD	Pathogenicity
		Nanba et al. 2018; Nanba et al. 2019; Scholl et al. 2013; Scholl et al. 2015; Yamazaki et al. 2017; Zhou et al. 2021)					
	I750F	(Fernandes-Rosa et al. 2014; Nanba et al. 2019)			2		Pathogenic
V752 (V772)	V752G	(Yamazaki et al. 2017)			1		Likely pathogenic
N754 (N774)	N754S	(Guo et al. 2020)			1		Likely pathogenic
755–757 (775–777)	755-757del	(Omata et al. 2018)			1	A756T (1×)	Likely pathogenic
S969 (S989)	S969L	(Nanba et al. 2020; Omata et al. 2018)			2		Pathogenic
V979 (V999)	V979D	(Fernandes-Rosa et al. 2014)			1	V979I (1×)	Likely pathogenic
K981 (K1001)	K981N	(Fernandes-Rosa et al. 2014)			1		Likely pathogenic
R990 (R1010)	R990G	(Hacini et al. 2021; Nanba et al. 2019; Zhou et al. 2021)			4		Pathogenic
	R990H	(Azizan et al. 2013; Guo et al. 2020; Nanba et al. 2018; Nanba et al. 2020; Omata et al. 2017; Omata et al. 2018; Tan et al. 2017; Yamazaki et al. 2017)	Type-4 (Monteleone et al. 2017; Pinggera et al. 2018)	(Monteleone et al. 2017)	13		Pathogenic
R993 (R1013)	R993M	(Guo et al. 2020)			1		Pathogenic
	R993S	(Yamazaki et al. 2017)			1		Pathogenic

	R993T	(Nanba et al. 2019)			3		Pathogenic
A998 (A1018)	A998I	(Fernandes-Rosa et al. 2014; Nanba et al. 2020)			4		Pathogenic
	A998V	(Fernandes-Rosa et al. 2014; Guo et al. 2020; Nanba et al. 2018; Nanba et al. 2019; Nanba et al. 2020; Omata et al. 2017; Omata et al. 2018; Yamazaki et al. 2017)			14		Pathogenic
C1007 (C1027)	C1007R	(Nanba et al. 2019)			1		Likely pathogenic
A1011 (A1031)	A1011T	(Omata et al. 2018)			1		Likely pathogenic
I1015 (I1035)	I1015S	(Nanba et al. 2019)			1		Pathogenic
	I1015T	(Nanba et al. 2020)			1		Pathogenic
	I1015V	(Omata et al. 2018; Yamazaki et al. 2017)			2		Pathogenic
F1147 (F1167)	F1147C	(Omata et al. 2017)			1		Pathogenic
	F1147L	(Omata et al. 2017; Omata et al. 2018; Zhou et al. 2021)			3		Pathogenic
V1151 (V1171)	V1151A	(Guo et al. 2020)			1		Pathogenic
	V1151F	(Fernandes-Rosa et al. 2014; Gagnon et al. 2018; Guo et al. 2020; Murakami et al. 2019; Nanba et al. 2018; Nanba et al. 2019)			8		Pathogenic

(continued)

Table 2 (continued)

Position	Variant	Reference	Gating change	Molecular modeling	Times reported	Reported in gnomAD	Pathogenicity
I1152 (I1172)	I1152N	(Fernandes-Rosa et al. 2014)			1		Likely pathogenic
V1153 (V1173)	V1153G	(Tan et al. 2017)	Type-2 (Tan et al. 2017)		1	V1153I (3×)	Pathogenic
	V1153N	(Hacini et al. 2021)			1	V1153I (3×)	Pathogenic
R1183 (R1203)	R1183H	(Omata et al. 2018)			1		Likely pathogenic
F1248 (F1268)	F1248L	(Omata et al. 2017; Omata et al. 2018)			4		Pathogenic
D1273 (D1293)	D1273N	(Omata et al. 2018)			1		Likely pathogenic
P1336 (P1371)	P1336R	(Azizan et al. 2013; Fernandes-Rosa et al. 2014; Omata et al. 2017; Zhou et al. 2021)	Type-3 (Azizan et al. 2013)		5		Pathogenic
V1338 (V1373)	V1338D	(Hacini et al. 2021)			1		Pathogenic
	V1338M	(Akerstrom et al. 2015; Fernandes-Rosa et al. 2014; Guo et al. 2020; Nanba et al. 2018; Nanba et al. 2020; Nishimoto et al. 2015; Omata et al. 2017; Omata et al. 2018; Scholl et al. 2015; Wu et al. 2021; Zhou et al. 2021)			20		Pathogenic
I1352 (I1387)	I1352T	(Omata et al. 2017)			1		Likely pathogenic

M1354 (M1389)	M1354I	(Azizan et al. 2013; Fernandes-Rosa et al. 2014)	No gating change (Pinggera et al. 2018)		2		+ Pathogenic/ uncertain
T1376 (T1411)	T1376M	(Jiao et al. 2020)			Germline (1×)		Likely pathogenic
	ASD						
V1447 (V1482)	V1447L	(Long et al. 2019)			Germline (1×)		Likely pathogenic
	ASD, epilepsy						
P1499 (P1534)	P1499L	(Omata et al. 2017)			1		Likely pathogenic
G1527 (G1562)	G1527R	(Wu et al. 2021)			1		Likely pathogenic
T1835 (T1879)	T1835I	(Omata et al. 2018)			1		Likely pathogenic
W1836 (W1880)	W1836X	(Omata et al. 2017)			1		# Likely pathogenic
A2065 (A2109)	A2065T	(Wu et al. 2021)			1		Likely pathogenic

For the interpretation of genomic de novo variants the ACMG guidelines were applied (Richards et al. 2015), while for somatic variants the criteria proposed in Ortner et al. (2020) were used. The given residue position refers to reference sequence EU_363339 (the position in the reference sequence NM_000720, used in the gnomAD database, is given in parenthesis). The residue number differs due to a different exon composition (EU_363339 contains exon 8a, but not exons 11, 32, and 44 which are present in NM_000720). "X" indicates a STOP. Specific residue substitutions were not given in Murakami et al. (2019) and Murakami et al. (2017), but provided via personal communication. G457R ("uncertain") was found together with the F1248L mutation ("pathogenic") in one APCC sample. N754S ("likely pathogenic") was present together with F747S ("pathogenic") in one APA sample. P548L, P1499L, and W1836X (all "likely pathogenic") were found together in one APCC sample. *, R510X likely causes a truncated, not functional channel (loss of function) and was found together with the F1147C mutation (1× in APA, not in gnomAD, "pathogenic") in one sample. +, M1354I would be categorized as pathogenic, but no functional gating changes were observed. #, W1836X results in a channel lacking the distal C-terminal regulatory domain (DCRD), predicted to enhance Ca^{2+}-dependent inactivation and voltage-sensitivity due to the disruption of an automodulatory interaction with the proximal C-terminal regulatory domain (PCRD)

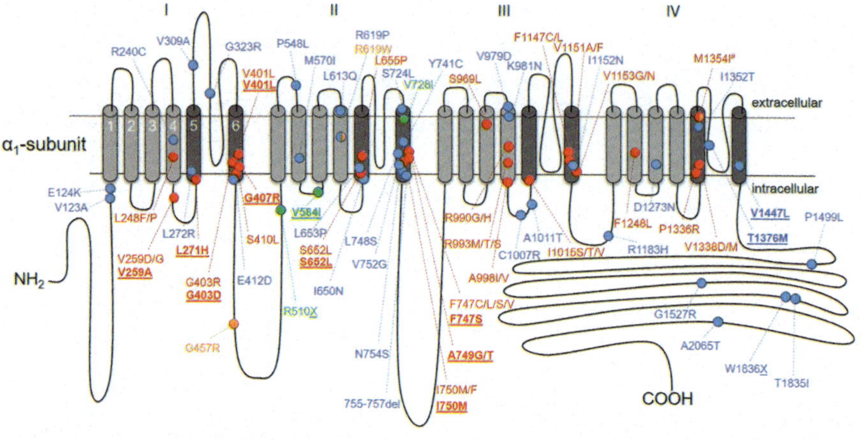

Fig. 1 Location and predicted pathogenicity of de novo *CACNA1D variants*. This graph is an updated and modified version of Suppl. Fig. 1 in Ortner et al. (2020) and shows the approximate location according to Wu et al. (2016) of de novo *CACNA1D* variants within the Cav1.3 α1 protein (encoded by the *CACNA1D* gene). 2D topology of the Cav1.3 pore-forming α1 subunit, consisting of four homologous repeats (I-IV) with 6 transmembrane segments (S) each. S1–4 build the voltage-sensing domain (VSD; shown in light gray) with the positive gating charges located in S4. S5 and S6 together with the connecting loop form the ion conducting pore (shown in dark gray), where the cytoplasmic ends of the S6 helices constitute the "activation gate." The cytoplasmic linker between S4 and S5 is important to transfer the VSD rearrangement upon depolarization to the opening of the pore. Both, somatic and germline *CACNA1D* variants are shown (germline variants indicated in bold and underlined). Their pathogenicity was predicted based on the criteria proposed in Ortner et al. (2020) (somatic) or the ACGM guidelines (Richards et al. 2015) (germline). Color code: red – pathogenic; blue–likely pathogenic; green – likely benign; orange – uncertain

associated *CACNA1D* variants, which are also summarized in Tables 1 and 2 (de novo variants are listed in Table 2).

Homozygous loss of Cav1.3 function causes congenital deafness and bradycardia (SANDD syndrome; OMIM #614896; clinically silent in the heterozygous state) (Baig et al. 2011). This phenotype is in line with the crucial role of Cav1.3 for vesicle release at ribbon synapses in cochlear inner hair cells and their pacemaker function in the sinoatrial node (SAN) (Brandt et al. 2003; Mangoni et al. 2003; Platzer et al. 2000). Two associated *CACNA1D* variants, G403dup and A376V, have been found in Pakistani and/or Palestinian families (Abu Rayyan et al. 2020; Baig et al. 2011; Liaqat et al. 2019) (Table 1). The residue numbering is according to the human *CACNA1D* reference sequence EU_363339, which is used throughout the manuscript if not mentioned otherwise. G403 is located in the transmembrane segment 6 (S6) of the first Cav1.3 homologous repeat, which is part of the conductive pore region (Fig. 1). Electrophysiological recordings revealed that G403dup containing channel constructs are still expressed at the membrane, but don't conduct Ca^{2+}, likely due to an uncoupling of the voltage-sensor movement and pore-opening (Baig et al. 2011). For A376V, present in the pore loop of domain I, functional data is missing but 3D modeling suggested a novel hydrophobic interaction in the mutant as a potential molecular mechanism (Liaqat et al. 2019). Cav1.3 knockout mice recapitulate the human phenotype and helped to uncover (patho)physiological roles of Cav1.3 (Namkung et al. 2001; Platzer et al. 2000).

Genome-wide association studies (GWAS) revealed an association of the intronic rs9810888 *CACNA1D* single nucleotide polymorphism (SNP) with blood pressure in Chinese adults (Lu et al. 2015) and children (Yang et al. 2019) as well as Ugandan adolescents (Lule et al. 2019). Furthermore, the intronic SNP rs10049492 showed a suggestive association with hypertension-related traits (Zhu et al. 2015). Using GWAS data from the UK Biobank and FinnGen, the inframe deletion F1899del, which is listed 673× in gnomAD (2× homozygous), has been linked to increased risk of hypertension (Curtis 2021; Sun et al. 2022). Besides F1899del, seven further *CACNA1D* missense variants were identified in a targeted sequencing study in hypertensive individuals from the Chinese Han population (Bao et al. 2020) (Table 1). Two of them showed a clear family history for hypertension and pre-eclampsia (D307G and R1457Q) and except for D307G, all can be found in healthy controls (gnomAD) (Table 1). With three of them, D307G, V877I, and R1457Q, rat models were generated and indeed, in D307G rats blood pressure was elevated (Wang et al. 2022b). In contrast, V877I and R1457Q (corresponding to V936I and R1516Q in rats; NM_017298.1) only exacerbated the D307G phenotype when they were present additionally, fitting to their presumed lower pathogenic effect suggested by their presence in gnomAD. A GWAS study in Japanese patients with resistant hypertension found a suggestive association of the intronic rs3774427 *CACNA1D* SNP, which has been previously also associated with blood pressure and hypertension (Takahashi et al. 2021). Moreover, two intronic SNPs significantly correlated with blood pressure lowering effects of dihydropyridine (DHP) LTCC inhibitors in Japanese patients with essential hypertension (Kamide et al. 2009). Interestingly, more than 70 de novo missense mutations of *CACNA1D* were found somatically in adrenal aldosterone-producing cell clusters (APCCs, more recently referred to as aldosterone-producing micronodules) and aldosterone-producing adenomas (APAs) (Table 2). APCCs are thought to represent precursor lesions of APAs, which are a major cause of primary aldosteronism and resistant hypertension (for a recent review, see Scholl 2022). Cav1.3 contributes to the intracellular Ca^{2+} signal that drives aldosterone synthesis and secretion in zona glomerulosa cells of the adrenal cortex (Barrett et al. 2016). The activity-enhancing gating changes of these de novo channel variants observed in vitro (discussed in more detail below) can explain the elevated aldosterone levels in affected individuals. In contrast, D307G mutant rats had similar aldosterone levels as compared to wildtypes, which suggests an alternative pathophysiological mechanism besides hyperaldosteronism (Wang et al. 2022b). D307G is located at the beginning of the domain I S5-S6 linker and shows some conservation (in Cav1.2, Cav1.4, and Cav2.1), but if it also alters channel gating remains to be investigated.

Three intronic *CACNA1D* SNPs could be linked to glucose metabolism. The rs312480 SNP was associated with reduced Cav1.3 mRNA expression and less insulin secretion in vitro and in vivo, while the two other SNPs, rs312486 and rs9841978, showed an association with type 2 diabetes (Reinbothe et al. 2013)

(Table 1). Interestingly, in pancreatic islets of individuals with type 2 diabetes, Cav1.3 mRNA levels were also reduced (Reinbothe et al. 2013). Cav1.3 is expressed in human insulin-secreting pancreatic β cells (also in α and δ cells to a minor extent) and plays a role in exocytosis (Reinbothe et al. 2013; Tuluc et al. 2021). In agreement, a study with a Cav1.3 knockout mouse model found impaired glucose tolerance and lower insulin levels (Namkung et al. 2001) (not observed in another Cav1.3 knockout mouse line (Platzer et al. 2000)). Intriguingly, both congenital hypoglycemia and/or hyperaldosteronism (hypertension) were present in some individuals with activity-enhancing de novo missense *CACNA1D* variants in their germline (V259A, L271H, G403D, I750M; for references, see Table 2) (Ortner et al. 2020). Since the first identified patients with such germline variants suffered from primary aldosteronism and neurological symptoms including seizures (Scholl et al. 2013), the condition was termed PASNA syndrome (primary aldosteronism, seizures, and neurologic abnormalities; OMIM #615474), which is part of a broader neurodevelopmental disorder with variable phenotypic severity (Ortner et al. 2020) (discussed in more detail below).

Cav1.3 channels are expressed throughout the brain, where they are located postsynaptically and shape neuronal excitability and dendritic spine morphology as well as couple cellular activity to gene transcription (D'Arco and Dolphin 2012; Stanika et al. 2016; Zamponi et al. 2015). Given these diversified functions, it is not surprising that genetic variants of *CACNA1D* have been linked to various brain pathologies. Three intronic SNPs associated with cocaine dependence were identified in a candidate gene association analysis in the "The Study of Addiction: Genetics and Environment" (SAGE) GWAS dataset (Martinez-Rivera et al. 2017) (Table 1). Studies in mouse models supported this observation as they showed an involvement of Cav1.3 channels in drug-taking behavior (Schierberl et al. 2011, 2012). The R498L variant was found in a patient with epilepsy (Tumiene et al. 2018) (NM_000720 numbering; EU363339 reference sequence does not include the affected exon 11). Due to missing genetic information of the parents, it is unclear if the variant occurred de novo, and no gating changes were detected in vitro (Hofer et al. 2021). Together with the fact that several other variants affecting R498 are reported in gnomAD (Table 1), a pathogenic role appears unlikely. Six further *CACNA1D* variants were present in patients with idiopathic epilepsy, of which four were absent in matched controls (Klassen et al. 2011) (Table 1). However, P1318 (NM_00720, since EU363339 does not contain the affected exon 32) and I1759 are not likely to confer high risk as they are reported 1,508 (11× HOM) and eight times, respectively, in healthy controls (gnomAD). A519P is positioned near a regulatory polybasic motif in the I-II linker (Kaur et al. 2015) and Q1598X results in a truncated channel lacking the calmodulin-binding site important for channel inactivation. Thus, these point mutations might result in altered channel function. Early truncation could also suppress channel expression (Page et al. 2004) or induce nonsense-mediated mRNA decay, which would not be of pathological relevance (heterozygous Cav1.3-deficiency is clinically silent). A recent study found five rare *CACNA1D* variants in a sample set of schizophrenia and ASD patients, including one deletion, one disruptive and three missense variants (Table 1) (Wang et al.

2022a). Three of them, E1660del, M716V, and V728I, are present multiple times in healthy controls in the gnomAD database (Table 1). The disruptive C1436X mutation was not found in patients or controls in a second independent dataset, thus it was speculated that it might represent an ultra-rare variant (Wang et al. 2022a). However, C1436X is located in S6 of domain IV and the premature stop likely results in a non-functional channel (clinically irrelevant in the heterozygous state). In this second dataset, G330R was present in both, schizophrenia and ASD patients, but in a third dataset was additionally found in controls. In Han Chinese, no association of *CACNA1D* with schizophrenia was found in a two-stage case–control study (Guan et al. 2015). De Rubeis et al. detected the de novo V584I variant in an ASD patient (Table 2) as well as three further ASD-associated mutants in a case–control sample (De Rubeis et al. 2014) (Table 1). V584 is located in a channel region without clear functional indication (cytoplasmic domain II – S2-S3 loop), is only conserved in Cav1.2 and Cav1.3 LTCCs and V584I listed 8 times in gnomAD controls, thus representing a likely benign or low risk variant. A59V is absent in gnomAD and affects a residue within the N-terminal NSCaTE motif, which is involved in calmodulin-binding and thus Ca^{2+}-dependent channel inactivation (Dick et al. 2008), but potential functional changes were not tested yet. S1953L and R1997H are located in the C-terminus at sites without clear functional implication and are found 10× or 6×, respectively, in the gnomAD database, casting doubt on their pathogenicity.

The inherited Q547H variant was present in the homozygous state in a patient with moderate hearing loss, intellectual disability, seizures, autistic behaviors, and hyperactivity (Garza-Lopez et al. 2018). Q547 is conserved among all Cavs and mutation to a histidine decreased channel inactivation when tested in vitro (type-3 gating; see below), making a disease-causing role likely (Garza-Lopez et al. 2018). In agreement, gating-modifying de novo missense variants of *CACNA1D* have also been found in 13 patients with a broad neurodevelopmental syndrome with or without endocrine features (for a recent review, see Ortner et al. 2020) (Table 2). The following sections of this review will mainly focus on these variants. Some of them have been already described somatically in APAs/APCCs and electrophysiological studies indicated increased activity of the mutated channels (Table 2). This gets further substantiated by the different clinical phenotype compared to homozygous loss of Cav1.3 function (deafness and bradycardia without an obvious brain phenotype; see above). The spectrum of associated symptoms in these patients include developmental delay, ASD, intellectual impairment, hypotonia, hyperactivity, autoaggression, seizures, hyperaldosteronism (hypertension), and congenital hypoglycemia (reviewed in Ortner et al. 2020). Not of all of them are present in all affected individuals, and the phenotypic severity can vary. ASD seems to be a common symptom, as it was described in most affected individuals. Additional neurological or endocrine symptoms were present in all of them, except for two patients with ASD (A749G and G407R). Potential explanations for this variability are discussed later in this review. Importantly, this syndrome comprises several symptoms for which genetic associations have been shown as outlined above (Table 1), further supporting a role of *CACNA1D* in these disease phenotypes.

3 Prediction of Pathogenicity

For the here described de novo germline *CACNA1D* variants, definite proof for a causal role in neurodevelopmental disease is still missing, but several facts point to a disease-causing nature of these variants: i) the shared clinical phenotype; ii) their absence in unaffected parents (de novo) and healthy controls listed in the gnomAD database; iii) the recurrence of some variants or variants affecting the same residue; iv) their presence at functionally important channel regions; v) typical channel gating changes observed in vitro. The American College of Medical Genetics (ACMG) has established guidelines to classify genetic variants based on their proposed pathogenicity, taking several of these factors into account (Richards et al. 2015). According to them, ten of the de novo germline variants can be considered pathogenic (Ortner et al. 2020) (Table 2). In the absence of additional supporting evidence (e.g., functional data, recurrence), the two variants found de novo in patients with ASD (T1376M (Jiao et al. 2020)) or ASD and epilepsy (V1447L (Long et al. 2019)) are assumed likely pathogenic.

From the large number of to date over 70 somatic de novo APCC/APA variants further information for the interpretation of the pathogenic potential of newly identified variants can be drawn. These were recently collected and criteria to predict their pathogenicity were proposed (Ortner et al. 2020) (an updated version of this collection is given in Table 2 and Fig. 1). Mutations of certain residues have been detected repeatedly in APCCs/APAs, marking mutational hotspots with putatively high disease relevance. For instance, four variants of residue F747 were found recurrently in APAs/APCCs (3× F747C, 27× F747L, 28× F747V, 1× F747S; Table 1) and indeed, the F747S mutation was recently identified in the germline of a patient with a neurodevelopmental disease phenotype (Török et al. 2022). These hotspots cluster around functionally important channel regions implying mutation-induced changes of channel function. As shown in Fig. 1, these regions include the voltage-sensor, in particular the positive gating charges in transmembrane segment 4 (S4), the S4-S5 linker, the cytoplasmic ends of S5 as well as the activation gate constituted by the intracellular portions of the S6 helices (Ortner et al. 2020; Striessnig 2021). Electrophysiological studies in heterologous expression systems confirmed altered biophysical properties caused by such variants (for references, see Table 2; discussed below). Different mutations of the same residue can cause comparable changes of channel gating as observed for G403R/D (Scholl et al. 2013). Thus, functional data from a variant of a certain residue (e.g., somatic V259D (Azizan et al. 2013)) could give a clue about potential functional consequences of another one (germline V259A, not functionally characterized yet (Semenova et al. 2018)). However, this must not be true, as shown for the pathogenic S652L variant (enhanced activity) and S652W, which is listed in gnomAD and confers a weak loss of channel function (Hofer et al. 2020). This shows that not only the location, but also the nature of the amino acid exchange can determine the functional impact on channel function.

The amino acid sequence of Cavs is highly conserved, especially in regions of functional relevance, such as the positive gating charges in S4, the selectivity filter or the activation gate, which implies similar molecular gating mechanisms (for an alignment, see Supplementary Fig. 1 in Wu et al. 2016). High conservation of a residue among Cavs and disease-associated mutations of the corresponding residue in other Cav α1 subunits can further support the evaluation of novel variants (Striessnig 2021). For instance, in *CACNA1C* (encoding the Cav1.2 α1 subunit) the G402R/S and G406R mutations cause Timothy syndrome, a rare multisystem disorder with heart and nervous system pathology (OMIM # 601005) (Bauer et al. 2021; Splawski et al. 2004, 2005). These residues are highly conserved within the Cav1 and Cav2 families and correspond to G403 and G407 in *CACNA1D* that were both mutated in patients with a neurodevelopmental phenotype. Additionally, mutation of G403 to arginine was found 74 times in APA/APCC samples (Table 2). For both Cav1.2 and Cav1.3 comparable changes of the biophysical properties induced by these mutations were observed (type-1; see below). Further examples are pathogenic mutations in other Cavs that affect residues corresponding to the highly conserved *CACNA1D* A749 (Cav2.1 A711T and Cav2.3 A700T in early infantile epileptic encephalopathy and developmental delay (Helbig et al. 2018; Niu et al. 2022); Cav3.1 A961T in childhood-onset cerebellar atrophy (Barresi et al. 2020; Chemin et al. 2018)). Again, similar gating changes were found (type-2; see below). Also, the F747 residue is conserved among Cav1 and Cav2 channels and mutations of the corresponding phenylalanine in Cav1.4 (F742C (Peloquin et al. 2007)) and Cav2.3 (F698S (Helbig et al. 2018)) are associated with stationary night blindness and developmental and epileptic encephalopathy, respectively. Thus, when assessing a newly identified missense variant, also supporting evidence from somatic or other Cav mutations should be taken into account. A comprehensive summary of Cav α1 channelopathies for such comparisons can be found in a recent review (Striessnig 2021).

In silico structural modeling and molecular dynamics simulations have emerged as valuable tools to elucidate the molecular mechanisms underlying the aberrant channel gating. The cryo-EM structure of Cav1.3 was published just recently (Yao et al. 2022). Before, a Cav1.3 homology model based on the cryo-EM structure of the skeletal Cav1.1 channel (Wu et al. 2015, 2016) was used to model the structural consequences of *CACNA1D* mutations (for references, see Table 2). For all investigated variants, predicted structural changes that might underlie the abnormal gating were found. For instance, Monteleone et al. predicted that the R990H variant causes a gating-pore current in the resting state but not upon channel activation, which was experimentally confirmed using electrophysiology (type-4 gating) (Monteleone et al. 2017). Similarly, such a gating mechanism has been modeled for the likely pathogenic R619P, whereas no evidence for an ion-passing pore within the VSD was observed for R619W, a variant found 1× in an APCC and in the gnomAD database ("uncertain" pathogenicity; Table 2) (Ortner et al. 2020). Homology models of the inactivated closed, activated open and resting Cav1.3 channel state, based on cryo-EM structures of voltage-gated Ca^{2+} and Na^{+} channels captured in these states, provided further insights into the channel state-dependent effects of a certain variant (Török et al. 2022). Using this approach, Török et al. (2022) showed that the F747S substitution selectively stabilized the activated open channel state, which can explain the observed gating changes such as the prolongation of tail currents or slowing of the inactivation time course. These findings indicate that molecular modeling can also help with the prediction of pathogenicity for channel variants.

4 Models to Study the Functional Consequences of *CACNA1D* Variants

One important aspect supporting a pathogenic role of a genetic variant is the detection of functional changes induced by the mutation. Electrophysiological recordings in heterologous expression systems such as HEK293 or tsA201 cells are standardly used to study Ca^{2+} influx through distinct channel complexes in a well-defined setting. Thereby, typical changes of Cav1.3 channel gating could be confirmed for the here described de novo variants (for references, see Table 2). As already evident from the divergent clinical presentation compared to Cav1.3-deficiency, they overall enhance channel function. Hallmarks are the shift of the voltage-dependence of channel activation and inactivation toward more negative membrane potentials (Fig. 2a), changes of the inactivation kinetics (in both directions; Fig. 2b) as well as slowing of channel deactivation during membrane repolarization (Fig. 2c). The presence and intensity of these gating effects differs among variants, and a categorization based on them has been proposed (Ortner et al. 2020; Pinggera et al. 2018) (Table 2). *Type-1* variants are characterized by a strong reduction of channel inactivation, which is an important mechanism to prevent Ca^{2+} overload-induced toxicity ("inactivation-deficient" variants), with or without changes of their voltage-dependence. *Type-2* gating is characterized by the left-shift of the voltage-dependent activation curve, i.e. the earlier opening of channels during depolarizations. Channel inactivation can be (un)altered, but must be preserved. *Type-3* mutants have impaired inactivation kinetics (slower or less complete), but unchanged voltage-dependent gating. *Type-4* class variants conduct gating-pore currents, which are depolarizing currents through an artificial pore formed by rearrangements in their voltage-sensing domain (VSD). The absence of (prominent) gating changes does, however, not exclude pathogenicity, since a mutation could also act via other mechanisms such as interfering with protein interactions (e.g., with auxiliary subunits) or affecting expression levels.

In these studies, standardized square pulse protocols were used, which allow direct comparison and thus categorization of such variants. However, they are non-physiological and cannot predict Ca^{2+} load during specific cellular activity. This is of importance, as these gating changes do not confer a pure gain of function, but rather induce complex changes of channel gating that can have different outcomes depending on the respective activity pattern. Action potential (AP)-clamp, i.e. the use of distinct firing patterns as command voltage, can give a better estimation. For instance, looping a typical AP of a substantia nigra (SN) dopamine (DA) neuron recorded in a mouse brain slice was used to mimic regular pacemaking in HEK293 cells expressing Cav1.2 and Cav1.3 channel complexes (Ortner et al. 2017). These experiments showed that depending on the activation threshold, Ca^{2+} was conducted during different phases of the AP. Cav1.2 activates around the AP

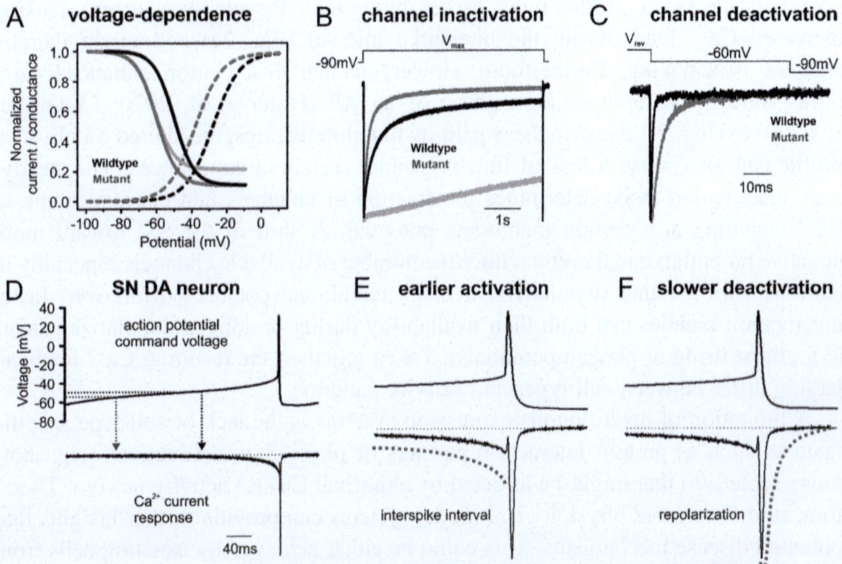

Fig. 2 Mutation-induced Cav1.3 gating changes and possible consequences during SN DA neuron firing. (**a–c**) Typical changes of channel gating of mutant Cav1.3 (gray) as observed for different Cav1.3 variants (to different degrees) upon heterologous expression (for references, see Table 2). (**a**) Most variants shift the steady-state voltage-dependence of activation (dashed lines) and inactivation (solid lines) toward more negative potentials. Also, steady-state inactivation can be less complete (light gray solid line). These shifts increase the window current, i.e. constant background Ca^{2+} influx, at negative membrane potentials (overlap of activation and inactivation curves). (**b**) Channel inactivation during prolonged depolarizations can be reduced (light gray; "type-1" inactivation-deficient mutants), unchanged or increased (dark gray). (**c**) Channel deactivation upon repolarization can be slowed down in mutants, causing prolonged tail currents. (**d–f**) Potential consequences of variant-induced gating changes during substantia nigra (SN) dopamine (DA) neuron action potential (AP) firing. (**d**) AP command voltage obtained from a mouse SN DA neuron (Ortner et al. 2017) and respective inward current during AP-clamp in Cav1.3 expressing HEK293 cells. Shifting the activation threshold from -50 mV (black horizontal dotted line) by ~5 mV toward more negative voltages (gray horizontal dotted line) can facilitate channel opening much earlier during the interspike interval of an AP cycle as indicated by the respective arrows. This can result in increased Ca^{2+} influx during the interspike interval (**e**; gray dotted line). (**f**) Slower channel deactivation kinetics can cause increased Ca^{2+} influx during the repolarization phase of an AP (gray dotted line; shown experimentally for S652L and A749G (Hofer et al. 2020; Limpitikul et al. 2016))

threshold and thus only conducted current in response to the strong depolarization during the AP spike, i.e. during the repolarization phase (Ortner et al. 2017). In contrast, Cav1.3 channels that possess a more negative voltage operation range also allowed Ca^{2+} influx at subthreshold voltages during the interspike interval, which explains how they can support pacemaking in SN DA neurons (Guzman et al. 2009; Ortner et al. 2017). Figure 2d illustrates an SN DA neuron AP (top panel) and the respective Ca^{2+} current response (lower panel) through a Cav1.3 channel that activates at ~−50 mV. Shifting the activation voltage by ~5 mV to more negative potentials allows Ca^{2+} influx much earlier during the AP cycle (gray arrow), causing increased Ca^{2+} load during the interspike interval (Fig. 2e) and might thereby enhance pacemaking. Furthermore, slower channel deactivation enhanced Ca^{2+} influx during the repolarization phase of an AP (Hofer et al. 2020; Limpitikul et al. 2016) (Fig. 2f). Next to these gain-of-function features, the altered biophysical profile can also cause a loss of function under certain circumstances. The steady-state inactivation (SSI) determines the fraction of channels that can contribute to Ca^{2+} signaling at a certain membrane potential. A shift of the SSI toward more negative potentials can thereby reduce the number of available channels especially in neurons with a rather depolarized average membrane potential. Moreover, faster inactivation kinetics can limit their availability during prolonged depolarized states (e.g., burst firing or plateau potentials). Taken together, the resulting Ca^{2+} load can largely differ between cell types and activity patterns.

A limitation of heterologous expression systems is the lack of cell-type specific features such as protein interaction partners or physiological functions (e.g., hormone secretion) that might be affected by abnormal Cav1.3 activity in vivo. Therefore, studies in more physiological model systems can provide further insights into potential disease mechanisms. This could be either achieved by isolating cells from an organism containing one of the variants, or by artificially introducing the mutant channels in a certain cell system (e.g., via transfection). When transiently transfected in the human adrenal H295R cell line, the two investigated APA *CACNA1D* variants, P1336R and V259D, caused an ~2-fold increase of the basal and angiotensin II-stimulated aldosterone production compared to the wildtype controls (Xie et al. 2016). These experiments confirmed that such Cav1.3 mutants indeed drive elevated aldosterone secretion. Neuroendocrine adrenal mouse chromaffin cells (MCCs) represent a feasible model system for the study of potential effects on neuron-like activity. MCCs fire spontaneously APs in a tonic pacemaker or bursting mode, similar to DA midbrain neurons (Carbone et al. 2019) and were already used to study effects on native firing by other mutant Cavs. Overexpression of I860M and I860N Cav3.3 T-type channels (corresponding to I750 in Cav1.3) identified in patients with neurodevelopmental phenotypes altered cellular firing toward a hyperexcitable phenotype (El Ghaleb et al. 2021). MCCs dissected from transgenic Timothy syndrome mice containing the G406R *CACNA1C* variant (TS2-neo mice) exhibited LTCC current gating changes as observed upon heterologous expression, abnormal electrical activity, and decreased catecholamine secretion (Calorio et al. 2019). This cellular system was further used to study potential beneficial effects by Ca^{2+} channel blockers. They found that overnight incubation with 0.3 μM of the dihydropyridine (DHP) nifedipine could reverse the abnormal spontaneous activity and catecholamine release in mutant MCCs (Calorio et al. 2019). Cav1.3 is expressed in MCCs and plays a crucial role for their excitability (Marcantoni et al. 2007, 2010; Vandael et al. 2012, 2015). Hints for abnormal MCC activity induced by altered Cav1.3 gating comes from a transgenic mouse line where a C-terminal automodulatory domain (CTM) in full-length Cav1.3 variants was disrupted (Cav1.3DCRD$^{-/-}$ mice (Scharinger et al. 2015)). CTM disruption causes biophysical properties as observed with C-terminally short splice variants, i.e. activation at more negative membrane potentials, higher open probability, and faster inactivation kinetics (Bock et al. 2011; Singh et al. 2008; Tan et al. 2011). The resting membrane potential of MCCs from Cav1.3DCRD$^{-/-}$ mice was more hyperpolarized, associated with a lower number of spontaneously active cells. Since *CACNA1D* variants induce similar functional changes, it is likely that they will also affect MCC firing. Murine primary hippocampal neurons have been used as a model system to study dendritic spine morphology and synapse formation (Geisler et al. 2019; Schopf et al. 2021). Overexpression of C-terminally short Cav1.3 channels, which facilitate Ca^{2+} influx (see above), was associated with elongation of dendritic spines (Stanika et al. 2015, 2016). Activity-enhancing *CACNA1D* variants may therefore also lead to an abnormal structure of dendritic spines, which has been associated with pathogenesis in ASD (Joensuu et al. 2018).

Another exciting approach are induced pluripotent stem cells (iPSCs), either genetically generated (e.g., using CRISPR/Cas) or patient-derived, that can be further differentiated into a multitude of cell types. Such cellular models were already generated with iPSCs containing Timothy syndrome *CACNA1C* (Cav1.2) mutations (cardiomyocytes, neurons, and organoids; for a recent review, see Bekdash et al. 2020). Recently, the successful generation of a patient-derived *CACNA1D* iPSC line by reprogramming peripheral mononuclear blood cells from the L271H patient was described (Tisch et al. 2022). This iPSC line will be an excellent model to further explore the pathophysiological role of gating-modifying *CACNA1D* variants in different cell types.

To finally proof that a certain variant causes a disease phenotype and to study the disease-underlying mechanisms, transgenic in vivo disease models are required. For Cav1.3 a good transferability for results from rodent models can be assumed due to the comparable phenotypes of Cav1.3-deficient mice and humans (Baig et al. 2011; Namkung et al. 2001; Platzer et al. 2000). This is further supported by the finding of elevated blood pressure in rats harboring the hypertension-associated D307G variant (Wang et al. 2022b). It was not reported if D307G rats displayed additional phenotypes associated with the symptomatic spectrum reported for *CACNA1D* variants, such as hyperinsulinism, developmental delay or autistic behaviors. To date, no further animal models containing one of the above described variants have been published yet.

5 Genotype–Phenotype Correlation

The presence and manifestation of the neurological and endocrine features differ among patients with de novo *CACNA1D* variants in their germline. Since altered Ca^{2+} influx through mutant channels is considered the pathological mechanism, one would expect that stronger gating changes correlate with a more severe phenotype. However, the currently small number of affected individuals prevents a meaningful phenotype–genotype correlation. In general, it appears likely that the shift of the voltage-dependence of activation to more negative membrane potentials is the main determinant for the severity of the disease phenotype. The two *CACNA1D* variants in the patients with ASD but without additional endocrine or neurological comorbidities (G407R and A749G) showed the weakest shifts in their half maximal activation voltage, $V_{0.5}$, by approximately −4 mV (not statistically significant) and −10 mV, respectively (Pinggera 2015; Pinggera et al. 2018). This is in contrast to ~15–23 mV negative shifts reported for the other germline variants in more severely affected individuals (Azizan et al. 2013; Hofer et al. 2020; Pinggera et al. 2017; Scholl et al. 2013). In agreement, a new germline *CACNA1D* variant with the so far most negative activation voltage range was found in a patient with a very severe brain and endocrine phenotype (F747S, Török et al. 2022). Interestingly, also for Timothy syndrome *CACNA1C* variants the shift of the activation voltage seems to play a role for the clinical presentation. Both *CACNA1C* G406R and G402S abolish channel inactivation, but the G406R variant shifts the activation voltage to more negative potentials and is associated with ASD, whereas G402S even causes a shift to more positive activation voltages and affected patients are neurologically intact (Bauer et al. 2021; Li et al. 2016; Servili et al. 2020; Splawski et al. 2005). Other aspects of the altered biophysical profile may then further aggravate (e.g., less channel inactivation) or alleviate (e.g., faster inactivation) the outcome. Three of the five *CACNA1D* variants associated with endocrine symptoms have been functionally characterized so far, and all showed slower inactivation kinetics (G403D, I750M (Azizan et al. 2013; Scholl et al. 2013); F747S (Török et al. 2022)). If this is also true for the other two, V259A and L271H, and could indeed favor endocrine dysfunction, has to be evaluated in future studies. However, several other factors that could impact the phenotype severity are more likely to account for these differences as outlined below.

The α1 subunit of Cav1.3 channels is subject to alternative splicing, which cannot only affect channel gating itself, but also interaction with protein partners or tissue expression. Thus, the presence of a variant in an alternative exon can impact its clinical outcome. The transmembrane segment S6 of domain I is encoded by the mutually exclusive alternative exons 8a and 8b. While channel gating is similar for both isoforms, tissue expression differs greatly. 8b containing constructs are predominant in cochlear inner hair cells as well as the sinoatrial node (Baig et al. 2011), in agreement with hearing loss and bradycardia caused by loss-of-function mutations present in exon 8b (G403dup, A376V). In the brain, exon 8b expression is also ~6-fold higher (Hofer et al. 2021). Besides the weak shift in the voltage-dependent activation, this could further explain the weaker phenotype, i.e. ASD without further neurological features, in the G407R patient (exon 8a) as compared to the other inactivation-deficient G403D variant (PASNA syndrome; exon 8b). Also, hypothetically, lower exon 8a levels in respective endocrine tissues, which is not known yet, could account for the lack of an endocrine phenotype in the G407R and V401L patients (both exon 8a). Exons 11 and 32 can be included additionally and stabilized voltage-dependent gating at more negative membrane potentials by ~3–4 mV when present individually or together in a C-terminally short Cav1.3 splice variant (Hofer et al. 2021). Both exons were found in mouse cochlear, adrenal chromaffin cells and whole brain preparations (Hofer et al. 2021). Exon 11 is located within the cytoplasmic loop connecting domains I and II, between the interaction site with the auxiliary β subunit and a regulatory polybasic motif (Kaur et al. 2015). Exon 32 resides in the extracellular S3-S4 connecting loop of domain IV, which is also subject to alternative splicing in other LTCCs associated with altered biophysical properties (Tang et al. 2004; Tuluc et al. 2009). Two missense variants, R498L (exon 11, (Tumiene et al. 2018)) and P1318S (exon 32, (Klassen et al. 2011)) were found in epilepsy patients, but as outlined in the first section of this review, they do not seem to confer high risk. If this is also due to their presence in only a subset of Cav1.3 constructs remains speculative. Alternative splicing within the C-terminus results in biophysically diverse splice variants (Bock et al. 2011; Singh et al. 2008; Tan et al. 2011). As mentioned above, C-terminally short isoforms lack a functional C-terminal autoregulatory domain (CTM), resulting in activation at more negative potentials, higher open probability, and enhanced Ca^{2+} dependent inactivation (Bock et al. 2011; Singh et al. 2008; Tan et al. 2011). Several *CACNA1D* variants have been studied in full-length and short Cav1.3 isoforms and similar mutation-induced effects were observed in both (Hofer et al. 2020; Pinggera et al. 2017, 2018). This means that in affected patients Cav1.3 channels operate within a broad voltage range, spanning up to ~25 mV between full-length wildtype and short mutant channels (as shown for S652L (Hofer et al. 2020)). In this context it also has to be considered that variants could affect mRNA stability and expression levels of Cav1.3 in general and of certain splice variants. This was observed for the Timothy syndrome G406R mutation present in exon 8a of *CACNA1C* (Cav1.2) that caused an abnormal persistent use of the mutated exon by preventing a developmental switch to exon 8 during neuronal differentiation (Panagiotakos et al. 2019).

Another factor that can impact the phenotype is genetic mosaicism, which means that not all cells within a person are genetically identical. Thus, a pathogenic variant that was acquired during embryonic development might only be present in distinct tissues and thereby cause a diverging phenotype. The earlier during embryogenesis the mutation occurred, the more cells will be affected, likely associated with a more severe clinical outcome. Genetic mosaicism and its impact on phenotypic variability have been recently reviewed for Timothy syndrome causing *CACNA1C* variants (Bauer et al. 2021). Moreover, the genetic background of affected individuals can further influence the clinical phenotype. The patient with the homozygous *CACNA1D* Q547H variant (Table 1) also harbored the T1480I *OTOG* mutation in the homozygous state (4× in gnomAD; no homozygotes) (Garza-Lopez et al. 2018). The symptoms of this patient fit to the described symptomatic spectrum reported for de novo *CACNA1D* germline variants, including intellectual impairment, seizures, autistic behaviors, and hyperactivity. However, this was not true for the moderate hearing loss. *OTOG* encodes the otogelin protein that is highly expressed in the inner ear (Schraders et al. 2012). $OTOG^{-/-}$ mice are deaf and homozygous mutations resulting in premature stop or affecting a highly conserved residue within a von Willebrand factor type D domain have been linked to autosomal-recessive hearing loss (Schraders et al. 2012; Simmler et al. 2000). Although T1480 does not reside within a known functional domain of the otogelin protein, T1480 also shows a high degree of conservation among different species. Since hearing loss is associated with homozygous Cav1.3-deficiency and has not been observed in other individuals with activity-enhancing *CACNA1D* variants, it appears likely that the *OTOG* mutation accounts for the moderate hearing impairment in this patient. The A749G patient also harbored a de novo mutation in a splice site of *KATNAL2* with unknown functional relevance (O'Roak et al. 2012). Another de novo splice site mutation in

KATNAL2, reported as highly disruptive (Sanders et al. 2012) as well as three de novo loss-of-function *KATNAL2* mutations were associated with autism (Neale et al. 2012). Thus, a potential role of this additional variant for the disease phenotype cannot be excluded. However, the typical gating changes observed for A749G, the high conservation of A749 among Cavs, the presence of *CACNA1D* A749T in two further patients with a neurodevelopmental phenotype as well as diseases associated with mutations of the corresponding residue in other Cavs all point to a pathogenic role of the A749G variant. The individual carrying the V401L variant also had a de novo variant in the *FASN* gene (R477H, NM_004104), which is unlikely to cause the phenotype, since it is reported as SNP (rs113931914) and is present seven times in the gnomAD database (plus 10× R477P and 1× R477C).

Besides additive pathogenic effects, different mutations could also lead to genetic compensation, thereby masking or modifying certain phenotypes. This has been shown by crossing two mutant mouse models with epilepsy associated alleles of *KCNA1* and *CACNA1A*, which have opposing functions for cellular excitability (Glasscock et al. 2007). *KCNA1* encodes the α1 subunit of voltage-gated Kv1.1 potassium channels, which are crucial for neuronal excitability by regulating action potential shape and spike repolarization (Dodson and Forsythe 2004; Tanouye et al. 1981). *CACNA1A* encodes the pore-forming α1 subunit of Cav2.1 P/Q-type Ca^{2+} channels that trigger membrane depolarization and presynaptic neurotransmitter release (Zamponi et al. 2015). *KCNA1*$^{-/-}$ mice suffer from frequent spontaneous seizures (Smart et al. 1998) and *CACNA1A* tottering mice (partial loss of function) exhibit absence-like seizures and ataxia (Fletcher et al. 1996; Imbrici et al. 2004). Double mutant mice had higher survival rates and an alleviated seizure phenotype, demonstrating genetic disease modification (Glasscock et al. 2007). Also, Cav3.1 T-type Ca^{2+} channels (*CACNA1G*) were identified as epilepsy modifiers in two transgenic mouse models of Dravet syndrome harboring genetic alterations in the voltage-gated sodium channel genes *SCN1A* and *SCN2A* (Calhoun et al. 2016, 2017). In both models (*SCN1A*$^{+/-}$ or *SCN2A*^{Q54}), decreased *CACNA1G* expression led to improved survival and reduced seizure frequency. These findings suggest that it is important to look at the whole genetic profile of affected patients and that other ion channels may serve as potential therapeutic targets.

6 Ca^{2+} Channel Blockers: Hope for a Therapy?

Functional studies with somatic (APCCs/APAs) and germline de novo *CACNA1D* variants and the absence of a comparable phenotype upon Cav1.3-deficiency point to an overall enhanced activity of mutant Cav1.3 channels as the disease-underlying mechanism. Thus, pharmacological inhibition represents a potential therapeutic option for Cav1.3 channelopathy patients with resistant hypertension or neurodevelopmental disease. No clinically applicable drugs that specifically target and inhibit Cav1.3 channels are available. One experimental compound that claimed Cav1.3-selective inhibition (compound 8; 1-(3-chlorophenethyl)-3-cyclopentylpyrimidine-2,4,6-(1H,3H,5H)-trione; (Kang et al. 2012); also termed

BPN-4689, cp-PYT or CPT in later studies) showed divergent results in follow-up studies, namely non-selective LTCC activation (Ortner et al. 2014) and non-selective weak LTCC inhibition (Huang et al. 2014)). Nevertheless, it has already been used in cellular and animal models where effects besides the proposed Cav1.3 inhibition cannot be excluded (Jiang et al. 2021; Plotkin et al. 2013; Sanchez-Padilla et al. 2014; Xie et al. 2016). Thus, further studies to identify the underlying reasons for the discrepant modes of action observed on isolated Cav1.3 currents in vitro will be required. Meanwhile, efforts to develop Cav1.3-selective inhibitory drugs are ongoing, also in light of their implications for Parkinson's disease (Liss and Striessnig 2019) or spasticity after spinal cord injury (Marcantoni et al. 2020).

At present, non-selective dihydropyridine (DHP) inhibitors of LTCCs, clinically used as antihypertensives via inhibition of vascular smooth muscle Cav1.2, are a reasonable option for off-label treatment of individuals affected by high-risk *CACNA1D* mutations. For one of the G403D patients a normalization of blood pressure was reported upon treatment with the DHP amlodipine (Scholl et al. 2013). In contrast to other DHPs, amlodipine shows poor penetration into the brain (Uchida et al. 1997), and thus might not be the best choice for treatment of neurological symptoms. In the L271H patient, oral administration of the DHP nifedipine was associated with an improvement of hypertension and muscle hypotonia (De Mingo Alemany et al. 2020). One of the individuals harboring the A749T variant was treated with immediate release isradipine, but therapy had to be stopped early (concerns for aggravation of symptoms and development of a common cold) (Ortner et al. 2020). These observations suggest that DHPs could indeed help to alleviate symptoms in these patients.

Importantly, DHPs are allosteric modulators of LTCCs and act in a voltage-dependent manner, preferentially binding to the inactivated channel state. Thus, changes of the biophysical properties of Cav1.3 channels might affect the sensitivity toward DHPs. Indeed, V401L, F747S, and S652L variants showed enhanced inhibition by the DHP isradipine in vitro (Török et al. 2022; Hofer et al. 2020; Pinggera et al. 2017). In contrast, G407R mutant channels required significantly higher isradipine concentrations compared to wildtype channels (unpublished), which could be due to the lower abundance of inactivated channels in this inactivation-deficient mutant. Regarding therapy, it has to be considered that the DHP dose that can be administered without cardiovascular side effects is limited due to the preferential inhibition of Cav1.2 channels. Cav1.3 channels have a lower drug sensitivity to DHPs compared to Cav1.2 (Koschak et al. 2001; Ortner et al. 2017; Xu and Lipscombe 2001), thus, patients with even less DHP sensitive *CACNA1D* variants might not benefit from clinically approved doses of this class of drugs. Furthermore, in light of the above described complex gating changes combining gain- and loss-of-function features, a variant could also lead to reduced Cav1.3 activity in a certain cell type or during specific cellular activity patterns. In case of symptoms arising from this cell/activity-specific loss of function, Cav1.3 inhibition would not be of therapeutic value. Another important aspect is the time point of therapy onset. Since it is a neurodevelopmental disorder, pathologic changes might have already occurred before birth. Consideration should also be given to the drug formulation. Immediate release formulations result in a fast drug onset and rapidly declining plasma peak levels. For instance, immediate release isradipine tablets cause peak concentrations of ~1 ng/ml/mg about 1.5 h post administration with an early half-life of 1.5–2 h and a terminal half-life of ~8 h (SPC Dynacirc www.drugs.com (Johnson et al. 2005; Shenfield et al. 1990; Wang et al. 2013)). In contrast, slow onset continuous release formulations result in a longer and more stable drug exposure (SPC Dynacirc CR, www.drugs.com: plateau ~7–18 h after dosing, then >50% of the peak concentration for 17–20 h; (Johnson et al. 2005; Park et al. 2009)) and are therefore the favorable choice. Moreover, they limit the risk of reflex tachycardia that can be caused by the fast onset of immediate release formulations.

7 Concluding Remarks

There is accumulating evidence that *CACNA1D* plays a role in the pathogenesis of a number of human diseases, including hypertension and brain disorders such as autism. While some (common) genetic variants confer an increased disease risk, there is compelling evidence for the causality of de novo missense *CACNA1D* variants in resistant hypertension (APAs/APCCs) and in a neurodevelopmental syndrome when present in the germline. Although several in vitro and in silico tools can help to predict the pathogenicity of genetic variants, transgenic disease models will be required to prove their disease-conferring potential. So far, the number of patients affected by germline missense variants is low, but due to the recent rise in reports, *CACNA1D* is increasingly recognized as a high-risk gene for neurodevelopmental conditions. This drives the identification of affected individuals, which will not only augment our understanding of the associated pathophysiology and phenotype–genotype correlations, but also guide diagnosis in the clinical practice. Exploratory off-label treatment of affected patients with DHPs as well as studies using in vivo disease models in future will uncover if DHPs can be used to attenuate *CACNA1D*-associated symptoms.

Acknowledgments This work was supported by the University of Innsbruck (Erika-Cremer habilitation fellowship) and by the Austrian Science Fund (FWF, grant P35087).

References

Abu Rayyan A, Kamal L, Casadei S, Brownstein Z, Zahdeh F, Shahin H, Canavati C, Dweik D, Jaraysa T, Rabie G, Carlson RJ, Gulsuner S, Lee MK, Avraham KB, Walsh T, King MC, Kanaan MN (2020) Genomic analysis of inherited hearing loss in the Palestinian population. Proc Natl Acad Sci U S A 117:20070–20076. https://doi.org/10.1073/pnas.2009628117

Akerstrom T, Willenberg HS, Cupisti K, Ip J, Backman S, Moser A, Maharjan R, Robinson B, Iwen KA, Dralle H, Volpe CD, Backdahl M, Botling J, Stalberg P, Westin G, Walz MK, Lehnert H, Sidhu S, Zedenius J, Bjorklund P, Hellman P (2015) Novel somatic mutations and distinct molecular signature in aldosterone-producing adenomas. Endocr Relat Cancer 22:735–744. https://doi.org/10.1530/ERC-15-0321

Azizan EA, Poulsen H, Tuluc P, Zhou J, Clausen MV, Lieb A, Maniero C, Garg S, Bochukova EG, Zhao W, Shaikh LH, Brighton CA, Teo AE, Davenport AP, Dekkers T, Tops B, Kusters B, Ceral J, Yeo GS, Neogi SG, McFarlane I, Rosenfeld N, Marass F, Hadfield J, Margas W, Chaggar K, Solar M, Deinum J, Dolphin AC, Farooqi IS, Striessnig J, Nissen P, Brown MJ (2013) Somatic mutations in ATP1A1 and CACNA1D underlie a common subtype of adrenal hypertension. Nat Genet 45:1055–1060. https://doi.org/10.1038/ng.2716. ng.2716 [pii]

Backman S, Akerstrom T, Maharjan R, Cupisti K, Willenberg HS, Hellman P, Bjorklund P (2019) RNA sequencing provides novel insights into the transcriptome of aldosterone producing adenomas. Sci Rep 9:6269. https://doi.org/10.1038/s41598-019-41525-2

Baig SM, Koschak A, Lieb A, Gebhart M, Dafinger C, Nurnberg G, Ali A, Ahmad I, Sinnegger-Brauns MJ, Brandt N, Engel J, Mangoni ME, Farooq M, Khan HU, Nurnberg P, Striessnig J, Bolz HJ (2011) Loss of Ca(v)1.3 (CACNA1D) function in a human channelopathy with bradycardia and congenital deafness. Nat Neurosci 14:77–84. https://doi.org/10.1038/nn.2694. nn.2694 [pii]

Bao M, Li P, Li Q, Chen H, Zhong Y, Li S, Jin L, Wang W, Chen Z, Zhong J, Geng B, Fan Y, Yang X, Cai J (2020) Genetic screening for monogenic hypertension in hypertensive individuals in a clinical setting. J Med Genet 57:571–580. https://doi.org/10.1136/jmedgenet-2019-106145

Barresi S, Dentici ML, Manzoni F, Bellacchio E, Agolini E, Pizzi S, Ciolfi A, Tarnopolsky M, Brady L, Garone G, Novelli A, Mei D, Guerrini R, Capuano A, Pantaleoni C, Tartaglia M (2020) Infantile-onset syndromic cerebellar ataxia and CACNA1G mutations. Pediatr Neurol 104:40–45. https://doi.org/10.1016/j.pediatrneurol.2019.09.005

Barrett PQ, Guagliardo NA, Klein PM, Hu C, Breault DT, Beenhakker MP (2016) Role of voltage-gated calcium channels in the regulation of aldosterone production from zona glomerulosa cells of the adrenal cortex. J Physiol 594:5851–5860. https://doi.org/10.1113/JP271896

Bauer R, Timothy KW, Golden A (2021) Update on the molecular genetics of Timothy syndrome. Front Pediatr 9:668546. https://doi.org/10.3389/fped.2021.668546

Bekdash R, Klein AD, Yazawa M (2020) Timothy syndrome iPSC modeling. Mol Cell Neurosci 107:103529. https://doi.org/10.1016/j.mcn.2020.103529

Bock G, Gebhart M, Scharinger A, Jangsangthong W, Busquet P, Poggiani C, Sartori S, Mangoni ME, Sinnegger-Brauns MJ, Herzig S, Striessnig J, Koschak A (2011) Functional properties of a newly identified C-terminal splice variant of Cav1.3 L-type Ca2+ channels. J Biol Chem 286: 42736–42748. https://doi.org/10.1074/jbc.M111.269951

Brandt A, Striessnig J, Moser T (2003) CaV1.3 channels are essential for development and presynaptic activity of cochlear inner hair cells. J Neurosci 23:10832–10840. https://doi.org/10.1523/JNEUROSCI.23-34-10832.2003

Calhoun JD, Hawkins NA, Zachwieja NJ, Kearney JA (2016) Cacna1g is a genetic modifier of epilepsy caused by mutation of voltage-gated sodium channel Scn2a. Epilepsia 57:e103–e107. https://doi.org/10.1111/epi.13390

Calhoun JD, Hawkins NA, Zachwieja NJ, Kearney JA (2017) Cacna1g is a genetic modifier of epilepsy in a mouse model of Dravet syndrome. Epilepsia 58:e111–e115. https://doi.org/10.1111/epi.13811

Calorio C, Gavello D, Guarina L, Salio C, Sassoe-Pognetto M, Riganti C, Bianchi FT, Hofer NT, Tuluc P, Obermair G, Defilippi P, Balzac F, Turco E, Bett GC, Rasmusson RL, Carbone E (2019) Impaired chromaffin cells excitability and exocytosis in autistic Timothy syndrome TS2-neo mouse rescued by L-type calcium channel blockers. J Physiol. https://doi.org/10.1113/JP277487

Carbone E, Borges R, Eiden LE, Garcia AG, Hernandez-Cruz A (2019) Chromaffin cells of the adrenal medulla: physiology, pharmacology, and disease. Compr Physiol 9:1443–1502. https://doi.org/10.1002/cphy.c190003

Catterall WA (2011) Voltage-gated calcium channels. Cold Spring Harb Perspect Biol 3:a003947. https://doi.org/10.1101/cshperspect.a003947. a003947 [pii]

Chemin J, Siquier-Pernet K, Nicouleau M, Barcia G, Ahmad A, Medina-Cano D, Hanein S, Altin N, Hubert L, Bole-Feysot C, Fourage C, Nitschke P, Thevenon J, Rio M, Blanc P,

Vidal C, Bahi-Buisson N, Desguerre I, Munnich A, Lyonnet S, Boddaert N, Fassi E, Shinawi M, Zimmerman H, Amiel J, Faivre L, Colleaux L, Lory P, Cantagrel V (2018) De novo mutation screening in childhood-onset cerebellar atrophy identifies gain-of-function mutations in the CACNA1G calcium channel gene. Brain 141:1998–2013. https://doi.org/10.1093/brain/awy145

Curtis D (2021) Analysis of 200,000 exome-sequenced UK Biobank subjects implicates genes involved in increased and decreased risk of hypertension. Pulse (Basel) 9:17–29. https://doi.org/10.1159/000517419

D'Arco M, Dolphin AC (2012) L-type calcium channels: on the fast track to nuclear signaling. Sci Signal 5:pe34. https://doi.org/10.1126/scisignal.2003355. scisignal.2003355 [pii]

De Mingo Alemany MC, Mifsud Grau L, Moreno Macian F, Ferrer Lorente B, Leon Carinena S (2020) A de novo Cacna1d missense mutation in a patient with congenital hyperinsulinism, primary hyperaldosteronism and hypotonia. Channels (Austin). https://doi.org/10.1080/19336950.2020.1761171

De Rubeis S, He X, Goldberg AP, Poultney CS, Samocha K, Cicek AE, Kou Y, Liu L, Fromer M, Walker S, Singh T, Klei L, Kosmicki J, Shih-Chen F, Aleksic B, Biscaldi M, Bolton PF, Brownfeld JM, Cai J, Campbell NG, Carracedo A, Chahrour MH, Chiocchetti AG, Coon H, Crawford EL, Curran SR, Dawson G, Duketis E, Fernandez BA, Gallagher L, Geller E, Guter SJ, Hill RS, Ionita-Laza J, Jimenz Gonzalez P, Kilpinen H, Klauck SM, Kolevzon A, Lee I, Lei I, Lei J, Lehtimaki T, Lin CF, Ma'ayan A, Marshall CR, McInnes AL, Neale B, Owen MJ, Ozaki N, Parellada M, Parr JR, Purcell S, Puura K, Rajagopalan D, Rehnstrom K, Reichenberg A, Sabo A, Sachse M, Sanders SJ, Schafer C, Schulte-Ruther M, Skuse D, Stevens C, Szatmari P, Tammimies K, Valladares O, Voran A, Li-San W, Weiss LA, Willsey AJ, Yu TW, Yuen RK, Study DDD, Homozygosity Mapping Collaborative for Autism, Consortium UK, Cook EH, Freitag CM, Gill M, Hultman CM, Lehner T, Palotie A, Schellenberg GD, Sklar P, State MW, Sutcliffe JS, Walsh CA, Scherer SW, Zwick ME, Barett JC, Cutler DJ, Roeder K, Devlin B, Daly MJ, Buxbaum JD (2014) Synaptic, transcriptional and chromatin genes disrupted in autism. Nature 515:209–215. https://doi.org/10.1038/nature13772

Deciphering Developmental Disorders S (2015) Large-scale discovery of novel genetic causes of developmental disorders. Nature 519:223–228. https://doi.org/10.1038/nature14135

Dick IE, Tadross MR, Liang H, Tay LH, Yang W, Yue DT (2008) A modular switch for spatial Ca2+ selectivity in the calmodulin regulation of CaV channels. Nature 451:830–834. https://doi.org/10.1038/nature06529

Dodson PD, Forsythe ID (2004) Presynaptic K+ channels: electrifying regulators of synaptic terminal excitability. Trends Neurosci 27:210–217. https://doi.org/10.1016/j.tins.2004.02.012

El Ghaleb Y, Schneeberger PE, Fernandez-Quintero ML, Geisler SM, Pelizzari S, Polstra AM, van Hagen JM, Denecke J, Campiglio M, Liedl KR, Stevens CA, Person RE, Rentas S, Marsh ED, Conlin LK, Tuluc P, Kutsche K, Flucher BE (2021) CACNA1I gain-of-function mutations differentially affect channel gating and cause neurodevelopmental disorders. Brain 144:2092–2106. https://doi.org/10.1093/brain/awab101

Fernandes-Rosa FL, Williams TA, Riester A, Steichen O, Beuschlein F, Boulkroun S, Strom TM, Monticone S, Amar L, Meatchi T, Mantero F, Cicala MV, Quinkler M, Fallo F, Allolio B, Bernini G, Maccario M, Giacchetti G, Jeunemaitre X, Mulatero P, Reincke M, Zennaro MC (2014) Genetic spectrum and clinical correlates of somatic mutations in aldosterone-producing adenoma. Hypertension 64:354–361. https://doi.org/10.1161/HYPERTENSIONAHA.114.03419

Flanagan SE, Vairo F, Johnson MB, Caswell R, Laver TW, Lango Allen H, Hussain K, Ellard S (2017) A CACNA1D mutation in a patient with persistent hyperinsulinaemic hypoglycaemia, heart defects, and severe hypotonia. Pediatr Diabetes. https://doi.org/10.1111/pedi.12512

Fletcher CF, Lutz CM, O'Sullivan TN, Shaughnessy JD Jr, Hawkes R, Frankel WN, Copeland NG, Jenkins NA (1996) Absence epilepsy in tottering mutant mice is associated with calcium channel defects. Cell 87:607–617. https://doi.org/10.1016/s0092-8674(00)81381-1

Gagnon N, Caceres-Gorriti KY, Corbeil G, El Ghoyareb N, Ludwig N, Latour M, Lacroix A, Bourdeau I (2018) Genetic characterization of GnRH/LH-responsive primary aldosteronism. J Clin Endocrinol Metab 103:2926–2935. https://doi.org/10.1210/jc.2018-00087

Garza-Lopez E, Lopez JA, Hagen J, Sheffer R, Meiner V, Lee A (2018) Role of a conserved glutamine in the function of voltage-gated Ca(2+) channels revealed by a mutation in human CACNA1D. J Biol Chem 293:14444–14454. https://doi.org/10.1074/jbc.RA118.003681

Geisler S, Schopf CL, Stanika R, Kalb M, Campiglio M, Repetto D, Traxler L, Missler M, Obermair GJ (2019) Presynaptic alpha2delta-2 Calcium Channel subunits regulate postsynaptic GABAA receptor abundance and axonal wiring. J Neurosci 39:2581–2605. https://doi.org/10.1523/JNEUROSCI.2234-18.2019

Glasscock E, Qian J, Yoo JW, Noebels JL (2007) Masking epilepsy by combining two epilepsy genes. Nat Neurosci 10:1554–1558. https://doi.org/10.1038/nn1999

Guan F, Li L, Qiao C, Chen G, Yan T, Li T, Zhang T, Liu X (2015) Evaluation of genetic susceptibility of common variants in CACNA1D with schizophrenia in Han Chinese. Sci Rep 5:12935. https://doi.org/10.1038/srep12935

Guo Z, Nanba K, Udager A, McWhinney BC, Ungerer JPJ, Wolley M, Thuzar M, Gordon RD, Rainey WE, Stowasser M (2020) Biochemical, histopathological, and genetic characterization of posture-responsive and unresponsive APAs. J Clin Endocrinol Metab 105. https://doi.org/10.1210/clinem/dgaa367

Guzman JN, Sanchez-Padilla J, Chan CS, Surmeier DJ (2009) Robust pacemaking in substantia nigra dopaminergic neurons. J Neurosci 29:11011–11019. https://doi.org/10.1523/JNEUROSCI.2519-09.2009

Hacini I, De Sousa K, Boulkroun S, Meatchi T, Amar L, Zennaro MC, Fernandes-Rosa FL (2021) Somatic mutations in adrenals from patients with primary aldosteronism not cured after adrenalectomy suggest common pathogenic mechanisms between unilateral and bilateral disease. Eur J Endocrinol 185:405–412. https://doi.org/10.1530/EJE-21-0338

Helbig KL, Lauerer RJ, Bahr JC, Souza IA, Myers CT, Uysal B, Schwarz N, Gandini MA, Huang S, Keren B, Mignot C, Afenjar A, Billette de Villemeur T, Heron D, Nava C, Valence S, Buratti J, Fagerberg CR, Soerensen KP, Kibaek M, Kamsteeg EJ, Koolen DA, Gunning B, Schelhaas HJ, Kruer MC, Fox J, Bakhtiari S, Jarrar R, Padilla-Lopez S, Lindstrom K, Jin SC, Zeng X, Bilguvar K, Papavasileiou A, Xin Q, Zhu C, Boysen K, Vairo F, Lanpher BC, Klee EW, Tillema JM, Payne ET, Cousin MA, Kruisselbrink TM, Wick MJ, Baker J, Haan E, Smith N, Corbett MA, MacLennan AH, Gecz J, Biskup S, Goldmann E, Rodan LH, Kichula E, Segal E, Jackson KE, Asamoah A, Dimmock D, McCarrier J, Botto LD, Filloux F, Tvrdik T, Cascino GD, Klingerman S, Neumann C, Wang R, Jacobsen JC, Nolan MA, Snell RG, Lehnert K, Sadleir LG, Anderlid BM, Kvarnung M, Guerrini R, Friez MJ, Lyons MJ, Leonhard J, Kringlen G, Casas K, El Achkar CM, Smith LA, Rotenberg A, Poduri A, Sanchis-Juan A, Carss KJ, Rankin J, Zeman A, Raymond FL, Blyth M, Kerr B, Ruiz K, Urquhart J, Hughes I, Banka S, Deciphering Developmental Disorders S, Hedrich UBS, Scheffer IE, Helbig I, Zamponi GW et al (2018) De novo pathogenic variants in CACNA1E cause developmental and epileptic encephalopathy with contractures, macrocephaly, and dyskinesias. Am J Hum Genet. https://doi.org/10.1016/j.ajhg.2018.09.006

Hofer NT, Tuluc P, Ortner NJ, Nikonishyna YV, Fernandes-Quintero ML, Liedl KR, Flucher BE, Cox H, Striessnig J (2020) Biophysical classification of a CACNA1D de novo mutation as a high-risk mutation for a severe neurodevelopmental disorder. Mol Autism 11:4. https://doi.org/10.1186/s13229-019-0310-4

Hofer NT, Pinggera A, Nikonishyna YV, Tuluc P, Fritz EM, Obermair GJ, Striessnig J (2021) Stabilization of negative activation voltages of Cav1.3 L-type Ca(2+)-channels by alternative splicing. Channels (Austin) 15:38–52. https://doi.org/10.1080/19336950.2020.1859260

Huang H, Ng CY, Yu D, Zhai J, Lam Y, Soong TW (2014) Modest Cav1.3_{42}-selective inhibition by compound 8 is β-subunit dependent. Nat Commun 5:4481. https://doi.org/10.1038/ncomms5481. ncomms5481 [pii]

Imbrici P, Jaffe SL, Eunson LH, Davies NP, Herd C, Robertson R, Kullmann DM, Hanna MG (2004) Dysfunction of the brain calcium channel CaV2.1 in absence epilepsy and episodic ataxia. Brain 127:2682–2692. https://doi.org/10.1093/brain/awh301

Iossifov I, Ronemus M, Levy D, Wang Z, Hakker I, Rosenbaum J, Yamrom B, Lee YH, Narzisi G, Leotta A, Kendall J, Grabowska E, Ma B, Marks S, Rodgers L, Stepansky A, Troge J, Andrews P, Bekritsky M, Pradhan K, Ghiban E, Kramer M, Parla J, Demeter R, Fulton LL, Fulton RS, Magrini VJ, Ye K, Darnell JC, Darnell RB, Mardis ER, Wilson RK, Schatz MC, McCombie WR, Wigler M (2012) De novo gene disruptions in children on the autistic spectrum. Neuron 74:285–299. https://doi.org/10.1016/j.neuron.2012.04.009

Jiang MC, Birch DV, Heckman CJ, Tysseling VM (2021) The involvement of CaV1.3 channels in prolonged root reflexes and its potential as a therapeutic target in spinal cord injury. Front Neural Circuits 15:642111. https://doi.org/10.3389/fncir.2021.642111

Jiao J, Zhang M, Yang P, Huang Y, Hu X, Cai J, Yang C, Situ M, Zhang H, Fu L, Guo K, Huang Y (2020) Identification of de novo JAK2 and MAPK7 mutations related to autism spectrum disorder using whole-exome sequencing in a Chinese child and adolescent trio-based sample. J Mol Neurosci 70:219–229. https://doi.org/10.1007/s12031-019-01456-z

Joensuu M, Lanoue V, Hotulainen P (2018) Dendritic spine actin cytoskeleton in autism spectrum disorder. Prog Neuropsychopharmacol Biol Psychiatry 84:362–381. https://doi.org/10.1016/j.pnpbp.2017.08.023

Johnson BA, Javors MA, Lam YW, Wells LT, Tiouririne M, Roache JD, Ait-Daoud N, Lawson K (2005) Kinetic and cardiovascular comparison of immediate-release isradipine and sustained-release isradipine among non-treatment-seeking, cocaine-dependent individuals. Prog Neuropsychopharmacol Biol Psychiatry 29:15–20. https://doi.org/10.1016/j.pnpbp.2004.08.014

Kamide K, Yang J, Matayoshi T, Takiuchi S, Horio T, Yoshii M, Miwa Y, Yasuda H, Yoshihara F, Nakamura S, Nakahama H, Miyata T, Kawano Y (2009) Genetic polymorphisms of L-type calcium channel alpha1C and alpha1D subunit genes are associated with sensitivity to the antihypertensive effects of L-type dihydropyridine calcium-channel blockers. Circ J 73:732–740. https://doi.org/10.1253/circj.cj-08-0761

Kang S, Cooper G, Dunne SF, Dusel B, Luan CH, Surmeier DJ, Silverman RB (2012) Cav1.3-selective L-type calcium channel antagonists as potential new therapeutics for Parkinson's disease. Nat Commun 3:1146. https://doi.org/10.1038/ncomms2149. ncomms2149 [pii]

Kaur G, Pinggera A, Ortner NJ, Lieb A, Sinnegger-Brauns MJ, Yarov-Yarovoy V, Obermair GJ, Flucher BE, Striessnig J (2015) A polybasic plasma membrane binding motif in the I-II linker stabilizes voltage-gated Cav1.2 calcium channel function. J Biol Chem. https://doi.org/10.1074/jbc.M115.645671

Kitamoto T, Suematsu S, Yamazaki Y, Nakamura Y, Sasano H, Matsuzawa Y, Saito J, Omura M, Nishikawa T (2016) Clinical and steroidogenic characteristics of aldosterone-producing adenomas with ATPase or CACNA1D gene mutations. J Clin Endocrinol Metab 101:494–503. https://doi.org/10.1210/jc.2015-3284

Kitamoto T, Omura M, Suematsu S, Saito J, Nishikawa T (2018) KCNJ5 mutation as a predictor for resolution of hypertension after surgical treatment of aldosterone-producing adenoma. J Hypertens 36:619–627. https://doi.org/10.1097/HJH.0000000000001578

Klassen T, Davis C, Goldman A, Burgess D, Chen T, Wheeler D, McPherson J, Bourquin T, Lewis L, Villasana D, Morgan M, Muzny D, Gibbs R, Noebels J (2011) Exome sequencing of ion channel genes reveals complex profiles confounding personal risk assessment in epilepsy. Cell 145:1036–1048. https://doi.org/10.1016/j.cell.2011.05.025

Koschak A, Reimer D, Huber I, Grabner M, Glossmann H, Engel J, Striessnig J (2001) Alpha 1D (Cav1.3) subunits can form l-type Ca^{2+} channels activating at negative voltages. J Biol Chem 276:22100–22106. https://doi.org/10.1074/jbc.M101469200

Li B, Tadross MR, Tsien RW (2016) Sequential ionic and conformational signaling by calcium channels drives neuronal gene expression. Science 351:863–867. https://doi.org/10.1126/science.aad3647

Liaqat K, Schrauwen I, Raza SI, Lee K, Hussain S, Chakchouk I, Nasir A, Acharya A, Abbe I, Umair M, Ansar M, Ullah I, Shah K, University of Washington Center for Mendelian Genomics, Bamshad MJ, Nickerson DA, Ahmad W, Leal SM (2019) Identification of CACNA1D variants associated with sinoatrial node dysfunction and deafness in additional Pakistani families reveals a clinical significance. J Hum Genet 64:153–160. https://doi.org/10.1038/s10038-018-0542-8

Limpitikul WB, Dick IE, Ben-Johny M, Yue DT (2016) An autism-associated mutation in CaV1.3 channels has opposing effects on voltage- and Ca(2+)-dependent regulation. Sci Rep 6:27235. https://doi.org/10.1038/srep27235

Liss B, Striessnig J (2019) The potential of L-type calcium channels as a drug target for neuroprotective therapy in Parkinson's disease. Annu Rev Pharmacol Toxicol 59:263–289. https://doi.org/10.1146/annurev-pharmtox-010818-021214

Long S, Zhou H, Li S, Wang T, Ma Y, Li C, Zhou Y, Zhou S, Wu B, Wang Y (2019) The clinical and genetic features of co-occurring epilepsy and autism Spectrum disorder in Chinese children. Front Neurol 10:505. https://doi.org/10.3389/fneur.2019.00505

Lu X, Wang L, Lin X, Huang J, Charles GC, He M, Shen H, He J, Zhu J, Li H, Hixson JE, Wu T, Dai J, Lu L, Shen C, Chen S, He L, Mo Z, Hao Y, Mo X, Yang X, Li J, Cao J, Chen J, Fan Z, Li Y, Zhao L, Li H, Lu F, Yao C, Yu L, Xu L, Mu J, Wu X, Deng Y, Hu D, Zhang W, Ji X, Guo D, Guo Z, Zhou Z, Yang Z, Wang R, Yang J, Zhou X, Yan W, Sun N, Gao P, Gu D (2015) Genome-wide association study in Chinese identifies novel loci for blood pressure and hypertension. Hum Mol Genet 24:865–874. https://doi.org/10.1093/hmg/ddu478

Lule SA, Mentzer AJ, Namara B, Muwenzi AG, Nassanga B, Kizito D, Akurut H, Lubyayi L, Tumusiime J, Zziwa C, Akello F, Gurdasani D, Sandhu M, Smeeth L, Elliott AM, Webb EL (2019) A genome-wide association and replication study of blood pressure in Ugandan early adolescents. Mol Genet Genomic Med 7:e00950. https://doi.org/10.1002/mgg3.950

Mangoni ME, Couette B, Bourinet E, Platzer J, Reimer D, Striessnig J, Nargeot J (2003) Functional role of L-type Cav1.3 Ca^{2+} channels in cardiac pacemaker activity. Proc Natl Acad Sci U S A 100:5543–5548. https://doi.org/10.1073/pnas.0935295100. 0935295100 [pii]

Marcantoni A, Baldelli P, Hernandez-Guijo JM, Comunanza V, Carabelli V, Carbone E (2007) L-type calcium channels in adrenal chromaffin cells: role in pace-making and secretion. Cell Calcium 42:397–408. https://doi.org/10.1016/j.ceca.2007.04.015. S0143-4160(07)00101-7 [pii]

Marcantoni A, Vandael DH, Mahapatra S, Carabelli V, Sinnegger-Brauns MJ, Striessnig J, Carbone E (2010) Loss of Cav1.3 channels reveals the critical role of L-type and BK channel coupling in pacemaking mouse adrenal chromaffin cells. J Neurosci 30:491–504. https://doi.org/10.1523/JNEUROSCI.4961-09.2010. 30/2/491 [pii]

Marcantoni M, Fuchs A, Low P, Bartsch D, Kiehn O, Bellardita C (2020) Early delivery and prolonged treatment with nimodipine prevents the development of spasticity after spinal cord injury in mice. Sci Transl Med 12. https://doi.org/10.1126/scitranslmed.aay0167

Martinez-Rivera A, Hao J, Tropea TF, Giordano TP, Kosovsky M, Rice RC, Lee A, Huganir RL, Striessnig J, Addy NA, Han S, Rajadhyaksha AM (2017) Enhancing VTA Cav1.3 L-type Ca (2+) channel activity promotes cocaine and mood-related behaviors via overlapping AMPA receptor mechanisms in the nucleus accumbens. Mol Psychiatry 22:1735–1745. https://doi.org/10.1038/mp.2017.9

Monteleone S, Lieb A, Pinggera A, Negro G, Fuchs JE, Hofer F, Striessnig J, Tuluc P, Liedl KR (2017) Mechanisms responsible for omega-pore currents in Cav calcium channel voltage-sensing domains. Biophys J 113:1485–1495. https://doi.org/10.1016/j.bpj.2017.08.010

Murakami M, Yoshimoto T, Nakabayashi K, Nakano Y, Fukaishi T, Tsuchiya K, Minami I, Bouchi R, Okamura K, Fujii Y, Hashimoto K, Hata KI, Kihara K, Ogawa Y (2017) Molecular characteristics of the KCNJ5 mutated aldosterone-producing adenomas. Endocr Relat Cancer 24:531–541. https://doi.org/10.1530/ERC-17-0117

Murakami M, Rhayem Y, Kunzke T, Sun N, Feuchtinger A, Ludwig P, Strom TM, Gomez-Sanchez C, Knosel T, Kirchner T, Williams TA, Reincke M, Walch AK, Beuschlein F (2019)

In situ metabolomics of aldosterone-producing adenomas. JCI Insight 4. https://doi.org/10.1172/jci.insight.130356

Namkung Y, Skrypnyk N, Jeong MJ, Lee T, Lee MS, Kim HL, Chin H, Suh PG, Kim SS, Shin HS (2001) Requirement for the L-type Ca(2+) channel alpha(1D) subunit in postnatal pancreatic beta cell generation. J Clin Invest 108:1015–1022. https://doi.org/10.1172/JCI13310

Nanba K, Chen AX, Omata K, Vinco M, Giordano TJ, Else T, Hammer GD, Tomlins SA, Rainey WE (2016) Molecular heterogeneity in aldosterone-producing adenomas. J Clin Endocrinol Metab 101:999–1007. https://doi.org/10.1210/jc.2015-3239

Nanba K, Omata K, Else T, Beck PCC, Nanba AT, Turcu AF, Miller BS, Giordano TJ, Tomlins SA, Rainey WE (2018) Targeted molecular characterization of aldosterone-producing adenomas in White Americans. J Clin Endocrinol Metab 103:3869–3876. https://doi.org/10.1210/jc.2018-01004

Nanba K, Omata K, Gomez-Sanchez CE, Stratakis CA, Demidowich AP, Suzuki M, Thompson LDR, Cohen DL, Luther JM, Gellert L, Vaidya A, Barletta JA, Else T, Giordano TJ, Tomlins SA, Rainey WE (2019) Genetic characteristics of aldosterone-producing adenomas in blacks. Hypertension 73:885–892. https://doi.org/10.1161/HYPERTENSIONAHA.118.12070

Nanba K, Yamazaki Y, Bick N, Onodera K, Tezuka Y, Omata K, Ono Y, Blinder AR, Tomlins SA, Rainey WE, Satoh F, Sasano H (2020) Prevalence of somatic mutations in aldosterone-producing adenomas in Japanese patients. J Clin Endocrinol Metab 105. https://doi.org/10.1210/clinem/dgaa595

Neale BM, Kou Y, Liu L, Ma'ayan A, Samocha KE, Sabo A, Lin CF, Stevens C, Wang LS, Makarov V, Polak P, Yoon S, Maguire J, Crawford EL, Campbell NG, Geller ET, Valladares O, Schafer C, Liu H, Zhao T, Cai G, Lihm J, Dannenfelser R, Jabado O, Peralta Z, Nagaswamy U, Muzny D, Reid JG, Newsham I, Wu Y, Lewis L, Han Y, Voight BF, Lim E, Rossin E, Kirby A, Flannick J, Fromer M, Shakir K, Fennell T, Garimella K, Banks E, Poplin R, Gabriel S, DePristo M, Wimbish JR, Boone BE, Levy SE, Betancur C, Sunyaev S, Boerwinkle E, Buxbaum JD, Cook EH Jr, Devlin B, Gibbs RA, Roeder K, Schellenberg GD, Sutcliffe JS, Daly MJ (2012) Patterns and rates of exonic de novo mutations in autism spectrum disorders. Nature 485:242–245. https://doi.org/10.1038/nature11011

Nishimoto K, Tomlins SA, Kuick R, Cani AK, Giordano TJ, Hovelson DH, Liu CJ, Sanjanwala AR, Edwards MA, Gomez-Sanchez CE, Nanba K, Rainey WE (2015) Aldosterone-stimulating somatic gene mutations are common in normal adrenal glands. Proc Natl Acad Sci U S A 112: E4591–E4599. https://doi.org/10.1073/pnas.1505529112

Niu X, Yang Y, Chen Y, Cheng M, Liu M, Ding C, Tian X, Yang Z, Jiang Y, Zhang Y (2022) Genotype-phenotype correlation of CACNA1A variants in children with epilepsy. Dev Med Child Neurol 64:105–111. https://doi.org/10.1111/dmcn.14985

Omata K, Anand SK, Hovelson DH, Liu CJ, Yamazaki Y, Nakamura Y, Ito S, Satoh F, Sasano H, Rainey WE, Tomlins SA (2017) Aldosterone-producing cell clusters frequently harbor somatic mutations and accumulate with age in normal adrenals. J Endocr Soc 1:787–799. https://doi.org/10.1210/js.2017-00134

Omata K, Satoh F, Morimoto R, Ito S, Yamazaki Y, Nakamura Y, Anand SK, Guo Z, Stowasser M, Sasano H, Tomlins SA, Rainey WE (2018) Cellular and genetic causes of idiopathic hyperaldosteronism. Hypertension 72:874–880. https://doi.org/10.1161/HYPERTENSIONAHA.118.11086

O'Roak BJ, Vives L, Girirajan S, Karakoc E, Krumm N, Coe BP, Levy R, Ko A, Lee C, Smith JD, Turner EH, Stanaway IB, Vernot B, Malig M, Baker C, Reilly B, Akey JM, Borenstein E, Rieder MJ, Nickerson DA, Bernier R, Shendure J, Eichler EE (2012) Sporadic autism exomes reveal a highly interconnected protein network of de novo mutations. Nature 485:246–250. https://doi.org/10.1038/nature10989. nature10989 [pii]

Ortner NJ, Bock G, Vandael DH, Mauersberger R, Draheim HJ, Gust R, Carbone E, Tuluc P, Striessnig J (2014) Pyrimidine-2,4,6-triones are a new class of voltage-gated L-type Ca^{2+} channel activators. Nat Commun 5:3897. https://doi.org/10.1038/ncomms4897. ncomms4897 [pii]

Ortner NJ, Bock G, Dougalis A, Kharitonova M, Duda J, Hess S, Tuluc P, Pomberger T, Stefanova N, Pitterl F, Ciossek T, Oberacher H, Draheim HJ, Kloppenburg P, Liss B, Striessnig J (2017) Lower affinity of isradipine for L-type Ca^{2+} channels during substantia nigra dopamine neuron-like activity: implications for neuroprotection in Parkinson's disease. J Neurosci. https://doi.org/10.1523/JNEUROSCI.2946-16.2017

Ortner NJ, Kaserer T, Copeland JN, Striessnig J (2020) De novo CACNA1D Ca(2+) channelopathies: clinical phenotypes and molecular mechanism. Pflugers Arch 472:755–773. https://doi.org/10.1007/s00424-020-02418-w

Page KM, Heblich F, Davies A, Butcher AJ, Leroy J, Bertaso F, Pratt WS, Dolphin AC (2004) Dominant-negative calcium channel suppression by truncated constructs involves a kinase implicated in the unfolded protein response. J Neurosci 24:5400–5409. https://doi.org/10.1523/JNEUROSCI.0553-04.2004

Panagiotakos G, Haveles C, Arjun A, Petrova R, Rana A, Portmann T, Pasca SP, Palmer TD, Dolmetsch RE (2019) Aberrant calcium channel splicing drives defects in cortical differentiation in Timothy syndrome. Elife 8. https://doi.org/10.7554/eLife.51037

Park JH, Park YS, Rhim SY, Jhee OH, Kim SH, Yang SC, Lee MH, Shaw LM, Kang JS (2009) Quantification of isradipine in human plasma using LC-MS/MS for pharmacokinetic and bioequivalence study. J Chromatogr B Analyt Technol Biomed Life Sci 877:59–64. https://doi.org/10.1016/j.jchromb.2008.11.021

Peloquin JB, Rehak R, Doering CJ, McRory JE (2007) Functional analysis of congenital stationary night blindness type-2 CACNA1F mutations F742C, G1007R, and R1049W. Neuroscience 150:335–345. https://doi.org/10.1016/j.neuroscience.2007.09.021

Pinggera A (2015) Functional impact of human disease mutations, protein-protein and lipid-protein interactions on L-type Ca2+ channels. PhD thesis

Pinggera A, Lieb A, Benedetti B, Lampert M, Monteleone S, Liedl KR, Tuluc P, Striessnig J (2015) CACNA1D De novo mutations in autism spectrum disorders activate Cav1.3 L-type calcium channels. Biol Psychiatry 77:816–822. https://doi.org/10.1016/j.biopsych.2014.11.020

Pinggera A, Mackenroth L, Rump A, Schallner J, Beleggia F, Wollnik B, Striessnig J (2017) New gain-of-function mutation shows CACNA1D as recurrently mutated gene in autism spectrum disorders and epilepsy. Hum Mol Genet. https://doi.org/10.1093/hmg/ddx175

Pinggera A, Negro G, Tuluc P, Brown MJ, Lieb A, Striessnig J (2018) Gating defects of disease-causing de novo mutations in Cav1.3 Ca(2+) channels. Channels (Austin) 12:388–402. https://doi.org/10.1080/19336950.2018.1546518

Platzer J, Engel J, Schrott-Fischer A, Stephan K, Bova S, Chen H, Zheng H, Striessnig J (2000) Congenital deafness and sinoatrial node dysfunction in mice lacking class D L-type Ca^{2+} channels. Cell 102:89–97. S0092-8674(00)00013-1 [pii]

Plotkin JL, Shen W, Rafalovich I, Sebel LE, Day M, Chan CS, Surmeier DJ (2013) Regulation of dendritic calcium release in striatal spiny projection neurons. J Neurophysiol 110:2325–2336. https://doi.org/10.1152/jn.00422.2013

Reinbothe TM, Alkayyali S, Ahlqvist E, Tuomi T, Isomaa B, Lyssenko V, Renstrom E (2013) The human L-type calcium channel Cav1.3 regulates insulin release and polymorphisms in CACNA1D associate with type 2 diabetes. Diabetologia 56:340–349. https://doi.org/10.1007/s00125-012-2758-z

Richards S, Aziz N, Bale S, Bick D, Das S, Gastier-Foster J, Grody WW, Hegde M, Lyon E, Spector E, Voelkerding K, Rehm HL, Committee ALQA (2015) Standards and guidelines for the interpretation of sequence variants: a joint consensus recommendation of the American College of Medical Genetics and Genomics and the Association for Molecular Pathology. Genet Med 17:405–424. https://doi.org/10.1038/gim.2015.30

Sanchez-Padilla J, Guzman JN, Ilijic E, Kondapalli J, Galtieri DJ, Yang B, Schieber S, Oertel W, Wokosin D, Schumacker PT, Surmeier DJ (2014) Mitochondrial oxidant stress in locus coeruleus is regulated by activity and nitric oxide synthase. Nat Neurosci. https://doi.org/10.1038/nn.3717. nn.3717 [pii]

Sanders SJ, Murtha MT, Gupta AR, Murdoch JD, Raubeson MJ, Willsey AJ, Ercan-Sencicek AG, DiLullo NM, Parikshak NN, Stein JL, Walker MF, Ober GT, Teran NA, Song Y, El-Fishawy P, Murtha RC, Choi M, Overton JD, Bjornson RD, Carriero NJ, Meyer KA, Bilguvar K, Mane SM, Sestan N, Lifton RP, Gunel M, Roeder K, Geschwind DH, Devlin B, State MW (2012) De novo mutations revealed by whole-exome sequencing are strongly associated with autism. Nature 485:237–241. https://doi.org/10.1038/nature10945

Scharinger A, Eckrich S, Vandael DH, Schonig K, Koschak A, Hecker D, Kaur G, Lee A, Sah A, Bartsch D, Benedetti B, Lieb A, Schick B, Singewald N, Sinnegger-Brauns MJ, Carbone E, Engel J, Striessnig J (2015) Cell-type-specific tuning of Cav1.3 Ca(2+)-channels by a C-terminal automodulatory domain. Front Cell Neurosci 9:309. https://doi.org/10.3389/fncel.2015.00309

Schierberl K, Hao J, Tropea TF, Ra S, Giordano TP, Xu Q, Garraway SM, Hofmann F, Moosmang S, Striessnig J, Inturrisi CE, Rajadhyaksha AM (2011) Cav1.2 L-type Ca(2)(+) channels mediate cocaine-induced GluA1 trafficking in the nucleus accumbens, a long-term adaptation dependent on ventral tegmental area Ca(v)1.3 channels. J Neurosci 31:13562–13575. https://doi.org/10.1523/JNEUROSCI.2315-11.2011

Schierberl K, Giordano T, Satpute S, Hao J, Kaur G, Hofmann F, Moosmang S, Striessnig J, Rajadhyaksha A (2012) Cav 1.3 L-type Ca (2+) channels mediate long-term adaptation in dopamine D2L-mediated GluA1 trafficking in the dorsal striatum following cocaine exposure. Channels (Austin) 6:11–17. https://doi.org/10.4161/chan.19324

Scholl UI (2022) Genetics of primary aldosteronism. Hypertension 79:887–897. https://doi.org/10.1161/HYPERTENSIONAHA.121.16498

Scholl UI, Goh G, Stolting G, de Oliveira RC, Choi M, Overton JD, Fonseca AL, Korah R, Starker LF, Kunstman JW, Prasad ML, Hartung EA, Mauras N, Benson MR, Brady T, Shapiro JR, Loring E, Nelson-Williams C, Libutti SK, Mane S, Hellman P, Westin G, Akerstrom G, Bjorklund P, Carling T, Fahlke C, Hidalgo P, Lifton RP (2013) Somatic and germline CACNA1D calcium channel mutations in aldosterone-producing adenomas and primary aldosteronism. Nat Genet 45:1050–1054. https://doi.org/10.1038/ng.2695. ng.2695 [pii]

Scholl UI, Healy JM, Thiel A, Fonseca AL, Brown TC, Kunstman JW, Horne MJ, Dietrich D, Riemer J, Kucukkoylu S, Reimer EN, Reis AC, Goh G, Kristiansen G, Mahajan A, Korah R, Lifton RP, Prasad ML, Carling T (2015) Novel somatic mutations in primary hyperaldosteronism are related to the clinical, radiological and pathological phenotype. Clin Endocrinol (Oxf) 83:779–789. https://doi.org/10.1111/cen.12873

Schopf CL, Ablinger C, Geisler SM, Stanika RI, Campiglio M, Kaufmann WA, Nimmervoll B, Schlick B, Brockhaus J, Missler M, Shigemoto R, Obermair GJ (2021) Presynaptic alpha2delta subunits are key organizers of glutamatergic synapses. Proc Natl Acad Sci U S A 118. https://doi.org/10.1073/pnas.1920827118

Schraders M, Ruiz-Palmero L, Kalay E, Oostrik J, del Castillo FJ, Sezgin O, Beynon AJ, Strom TM, Pennings RJ, Zazo Seco C, Oonk AM, Kunst HP, Dominguez-Ruiz M, Garcia-Arumi AM, del Campo M, Villamar M, Hoefsloot LH, Moreno F, Admiraal RJ, del Castillo I, Kremer H (2012) Mutations of the gene encoding otogelin are a cause of autosomal-recessive nonsyndromic moderate hearing impairment. Am J Hum Genet 91:883–889. https://doi.org/10.1016/j.ajhg.2012.09.012

Semenova NA, Ryzhkova OR, Strokova TV, Taran NN (2018) The third case report a patient with primary aldosteronism, seizures, and neurologic abnormalities (PASNA) syndrome de novo variant mutations in the CACNA1D gene. Zh Nevrol Psikhiatr Im S S Korsakova 118:49–52. https://doi.org/10.17116/jnevro201811812149

Servili E, Trus M, Sajman J, Sherman E, Atlas D (2020) Elevated basal transcription can underlie timothy channel association with autism related disorders. Prog Neurobiol 191:101820. https://doi.org/10.1016/j.pneurobio.2020.101820

Shenfield GM, Boutagy J, Stokes GS, Rumble F, Dunagan F (1990) The pharmokinetics of isradipine in hypertensive subjects. Eur J Clin Pharmacol 38:209–211

Simmler MC, Cohen-Salmon M, El-Amraoui A, Guillaud L, Benichou JC, Petit C, Panthier JJ (2000) Targeted disruption of otog results in deafness and severe imbalance. Nat Genet 24:139–143. https://doi.org/10.1038/72793

Singh A, Gebhart M, Fritsch R, Sinnegger-Brauns MJ, Poggiani C, Hoda JC, Engel J, Romanin C, Striessnig J, Koschak A (2008) Modulation of voltage- and Ca^{2+}-dependent gating of Cav1.3 L-type calcium channels by alternative splicing of a C-terminal regulatory domain. J Biol Chem 283:20733–20744. https://doi.org/10.1074/jbc.M802254200

Smart SL, Lopantsev V, Zhang CL, Robbins CA, Wang H, Chiu SY, Schwartzkroin PA, Messing A, Tempel BL (1998) Deletion of the K(V)1.1 potassium channel causes epilepsy in mice. Neuron 20:809–819. https://doi.org/10.1016/s0896-6273(00)81018-1

Splawski I, Timothy KW, Sharpe LM, Decher N, Kumar P, Bloise R, Napolitano C, Schwartz PJ, Joseph RM, Condouris K, Tager-Flusberg H, Priori SG, Sanguinetti MC, Keating MT (2004) Cav1.2 calcium channel dysfunction causes a multisystem disorder including arrhythmia and autism. Cell 119:19–31. https://doi.org/10.1016/j.cell.2004.09.011

Splawski I, Timothy KW, Decher N, Kumar P, Sachse FB, Beggs AH, Sanguinetti MC, Keating MT (2005) Severe arrhythmia disorder caused by cardiac L-type calcium channel mutations. Proc Natl Acad Sci U S A 102:8089–8096.; discussion 8086-8. https://doi.org/10.1073/pnas.0502506102

Stanika RI, Flucher BE, Obermair GJ (2015) Regulation of postsynaptic stability by the L-type calcium channel CaV1.3 and its interaction with PDZ proteins. Curr Mol Pharmacol 8:95–101

Stanika R, Campiglio M, Pinggera A, Lee A, Striessnig J, Flucher BE, Obermair GJ (2016) Splice variants of the CaV1.3 L-type calcium channel regulate dendritic spine morphology. Sci Rep 6: 34528. https://doi.org/10.1038/srep34528

Strauss KA, Gonzaga-Jauregui C, Brigatti KW, Williams KB, King AK, Van Hout C, Robinson DL, Young M, Praveen K, Heaps AD, Kuebler M, Baras A, Reid JG, Overton JD, Dewey FE, Jinks RN, Finnegan I, Mellis SJ, Shuldiner AR, Puffenberger EG (2018) Genomic diagnostics within a medically underserved population: efficacy and implications. Genet Med 20:31–41. https://doi.org/10.1038/gim.2017.76

Striessnig J (2021) Voltage-gated Ca(2+)-channel alpha1-subunit de novo missense mutations: gain or loss of function – implications for potential therapies. Front Synaptic Neurosci 13:634760. https://doi.org/10.3389/fnsyn.2021.634760

Sun BB, Kurki MI, Foley CN, Mechakra A, Chen CY, Marshall E, Wilk JB, Biogen Biobank T, Chahine M, Chevalier P, Christe G, FinnGen PA, Daly MJ, Runz H (2022) Genetic associations of protein-coding variants in human disease. Nature 603:95–102. https://doi.org/10.1038/s41586-022-04394-w

Takahashi Y, Yamazaki K, Kamatani Y, Kubo M, Matsuda K, Asai S (2021) A genome-wide association study identifies a novel candidate locus at the DLGAP1 gene with susceptibility to resistant hypertension in the Japanese population. Sci Rep 11:19497. https://doi.org/10.1038/s41598-021-98144-z

Tan BZ, Jiang F, Tan MY, Yu D, Huang H, Shen Y, Soong TW (2011) Functional characterization of alternative splicing in the C terminus of L-type CaV1.3 channels. J Biol Chem 286:42725–42735. https://doi.org/10.1074/jbc.M111.265207

Tan GC, Negro G, Pinggera A, Tizen Laim NMS, Mohamed Rose I, Ceral J, Ryska A, Chin LK, Kamaruddin NA, Mohd Mokhtar N, Jamal ARA, Sukor N, Solar M, Striessnig J, Brown MJ, Azizan EA (2017) Aldosterone-producing adenomas: histopathology-genotype correlation and identification of a novel CACNA1D mutation. Hypertension 70:129–136. https://doi.org/10.1161/HYPERTENSIONAHA.117.09057

Tang ZZ, Liang MC, Lu S, Yu D, Yu CY, Yue DT, Soong TW (2004) Transcript scanning reveals novel and extensive splice variations in human l-type voltage-gated calcium channel, Cav1.2 alpha1 subunit. J Biol Chem 279:44335–44343. https://doi.org/10.1074/jbc.M407023200. M407023200 [pii]

Tanouye MA, Ferrus A, Fujita SC (1981) Abnormal action potentials associated with the Shaker complex locus of Drosophila. Proc Natl Acad Sci U S A 78:6548–6552. https://doi.org/10.1073/pnas.78.10.6548

Tisch M, De Mingo C, Alemany M, Suarez-Cubero M, Fauth C, Defrancesco M, Zschocke J, Gunther K, Edenhofer F (2022) Generation of the human induced pluripotent stem cell line (IBKMOLi002-A) from PBMCs of a patient carrying the heterozygous L271H mutation of the voltage-gated calcium channel subunit Cav1.3-encoding CACNA1D gene. Stem Cell Res 61: 102784. https://doi.org/10.1016/j.scr.2022.102784

Török F, Tezcan K, Filippini L, Fernández-Quintero ML, Zanetti L, Liedl KR, Drexel RS, Striessnig J, Ortner NJ (2022) Germline de novo variant F747S extends the phenotypic spectrum of CACNA1D Ca2+ channelopathies. Hum Mol Genet. https://doi.org/10.1093/hmg/ddac248

Tuluc P, Molenda N, Schlick B, Obermair GJ, Flucher BE, Jurkat-Rott K (2009) A CaV1.1 Ca2+ channel splice variant with high conductance and voltage-sensitivity alters EC coupling in developing skeletal muscle. Biophys J 96:35–44. https://doi.org/10.1016/j.bpj.2008.09.027

Tuluc P, Theiner T, Jacobo-Piqueras N, Geisler SM (2021) Role of high voltage-gated Ca(2+) channel subunits in pancreatic beta-cell insulin release. From structure to function. Cells 10. https://doi.org/10.3390/cells10082004

Tumiene B, Maver A, Writzl K, Hodzic A, Cuturilo G, Kuzmanic-Samija R, Culic V, Peterlin B (2018) Diagnostic exome sequencing of syndromic epilepsy patients in clinical practice. Clin Genet 93:1057–1062. https://doi.org/10.1111/cge.13203

Uchida S, Yamada S, Nagai K, Deguchi Y, Kimura R (1997) Brain pharmacokinetics and in vivo receptor binding of 1,4-dihydropyridine calcium channel antagonists. Life Sci 61:2083–2090. S0024320597008813 [pii]

Vandael DH, Zuccotti A, Striessnig J, Carbone E (2012) Ca(V)1.3-driven SK channel activation regulates pacemaking and spike frequency adaptation in mouse chromaffin cells. J Neurosci 32: 16345–16359. https://doi.org/10.1523/JNEUROSCI.3715-12.2012. 32/46/16345 [pii]

Vandael DH, Marcantoni A, Carbone E (2015) Cav1.3 channels as key regulators of neuron-like firings and catecholamine release in chromaffin cells. Curr Mol Pharmacol 8:149–161

Wang Y, Jin Z, Wang Z, Jiang X, Wang L (2013) Pharmacokinetic properties of isradipine after single-dose and multiple-dose oral administration in Chinese volunteers: a randomized, open-label, parallel-group phase I study. Biomed Chromatogr 27:1664–1670. https://doi.org/10.1002/bmc.2977

Wang B, Li X, Zhang X, Ma X, Chen L, Zhang Y, Lyu X, Tang Y, Huang Q, Gao Y, Fan Y, Ouyang J (2015) Prevalence and characterization of somatic mutations in chinese aldosterone-producing adenoma patients. Medicine (Baltimore) 94:e708. https://doi.org/10.1097/MD.0000000000000708

Wang C, Horigane SI, Wakamori M, Ueda S, Kawabata T, Fujii H, Kushima I, Kimura H, Ishizuka K, Nakamura Y, Iwayama Y, Ikeda M, Iwata N, Okada T, Aleksic B, Mori D, Yoshida T, Bito H, Yoshikawa T, Takemoto-Kimura S, Ozaki N (2022a) Identification of ultra-rare disruptive variants in voltage-gated calcium channel-encoding genes in Japanese samples of schizophrenia and autism spectrum disorder. Transl Psychiatry 12:84. https://doi.org/10.1038/s41398-022-01851-y

Wang H, Zhu JK, Cheng L, Mao G, Chen H, Wu X, Hong H, Wang C, Lin P, Chen J, Maboh RN, Chen H (2022b) Dominant role of CACNA1D exon mutations for blood pressure regulation. J Hypertens 40:819–834. https://doi.org/10.1097/HJH.0000000000003085

Wu J, Yan Z, Li Z, Yan C, Lu S, Dong M, Yan N (2015) Structure of the voltage-gated calcium channel Cav1.1 complex. Science 350:aad2395. https://doi.org/10.1126/science.aad2395

Wu J, Yan Z, Li Z, Qian X, Lu S, Dong M, Zhou Q, Yan N (2016) Structure of the voltage-gated calcium channel Cav1.1 at 3.6 A resolution. Nature 537:191–196. https://doi.org/10.1038/nature19321

Wu CH, Peng KY, Hwang DY, Lin YH, Wu VC, Chueh JS (2021) Novel mutations detection with next-generation sequencing and its association with clinical outcome in unilateral primary aldosteronism. Biomedicine 9. https://doi.org/10.3390/biomedicines9091167

Xie CB, Haris Shaikh L, Garg S, Tanriver G, Teo AE, Zhou J, Maniero C, Zhao W, Kang S, Silverman RB, Azizan EA, Brown MJ (2016) Regulation of aldosterone secretion by Cav1.3. Sci Rep 6:24697. https://doi.org/10.1038/srep24697

Xu W, Lipscombe D (2001) Neuronal Cav1.3α1 L-type channels activate at relatively hyperpolarized membrane potentials and are incompletely inhibited by dihydropyridines. J Neurosci 21:5944–5951. https://doi.org/10.1523/JNEUROSCI.21-16-05944.2001

Yamazaki Y, Nakamura Y, Omata K, Ise K, Tezuka Y, Ono Y, Morimoto R, Nozawa Y, Gomez-Sanchez CE, Tomlins SA, Rainey WE, Ito S, Satoh F, Sasano H (2017) Histopathological classification of cross-sectional image-negative hyperaldosteronism. J Clin Endocrinol Metab 102:1182–1192. https://doi.org/10.1210/jc.2016-2986

Yang YD, Song JY, Wang S, Wang Y, Song QY, Dong YH, Li CX, Wang HJ, Ma J (2019) Combined effects of the rs9810888 polymorphism in calcium voltage-gated channel subunit alpha1 D (CACNA1D) and lifestyle behaviors on blood pressure level among Chinese children. PLoS One 14:e0216950. https://doi.org/10.1371/journal.pone.0216950

Yao X, Gao S, Yan N (2022) Structural basis for pore blockade of human voltage-gated calcium channel Cav1.3 by motion sickness drug cinnarizine. Cell Res. https://doi.org/10.1038/s41422-022-00663-5

Zamponi GW, Striessnig J, Koschak A, Dolphin AC (2015) The physiology, pathology, and pharmacology of voltage-gated calcium channels and their future therapeutic potential. Pharmacol Rev 67:821–870. https://doi.org/10.1124/pr.114.009654

Zheng FF, Zhu LM, Nie AF, Li XY, Lin JR, Zhang K, Chen J, Zhou WL, Shen ZJ, Zhu YC, Wang JG, Zhu DL, Gao PJ (2015) Clinical characteristics of somatic mutations in Chinese patients with aldosterone-producing adenoma. Hypertension 65:622–628. https://doi.org/10.1161/HYPERTENSIONAHA.114.03346

Zhou J, Azizan EAB, Cabrera CP, Fernandes-Rosa FL, Boulkroun S, Argentesi G, Cottrell E, Amar L, Wu X, O'Toole S, Goodchild E, Marker A, Senanayake R, Garg S, Akerstrom T, Backman S, Jordan S, Polubothu S, Berney DM, Gluck A, Lines KE, Thakker RV, Tuthill A, Joyce C, Kaski JP, Karet Frankl FE, Metherell LA, Teo AED, Gurnell M, Parvanta L, Drake WM, Wozniak E, Klinzing D, Kuan JL, Tiang Z, Gomez Sanchez CE, Hellman P, Foo RSY, Mein CA, Kinsler VA, Bjorklund P, Storr HL, Zennaro MC, Brown MJ (2021) Somatic mutations of GNA11 and GNAQ in CTNNB1-mutant aldosterone-producing adenomas presenting in puberty, pregnancy or menopause. Nat Genet 53:1360–1372. https://doi.org/10.1038/s41588-021-00906-y

Zhu X, Feng T, Tayo BO, Liang J, Young JH, Franceschini N, Smith JA, Yanek LR, Sun YV, Edwards TL, Chen W, Nalls M, Fox E, Sale M, Bottinger E, Rotimi C, Consortium CB, Liu Y, McKnight B, Liu K, Arnett DK, Chakravati A, Cooper RS, Redline S (2015) Meta-analysis of correlated traits via summary statistics from GWASs with an application in hypertension. Am J Hum Genet 96:21–36. https://doi.org/10.1016/j.ajhg.2014.11.011

CACNA1A-Related Channelopathies: Clinical Manifestations and Treatment Options

Elisabetta Indelicato and Sylvia Boesch

Contents

Abstract

In the last decade, variants in the Ca^{2+} channel gene *CACNA1A* emerged as a frequent aetiology of rare neurological phenotypes sharing a common denominator of variable paroxysmal manifestations and chronic cerebellar dysfunction.

E. Indelicato (✉) · S. Boesch
Center for Rare Movement Disorders Innsbruck, Department of Neurology, Medical University of Innsbruck, Innsbruck, Austria
e-mail: elisabetta.indelicato@i-med.ac.at

J. Striessnig (ed.), *Voltage-gated Ca^{2+} Channels: Pharmacology, Modulation and their Role in Human Disease*, Handbook of Experimental Pharmacology 279,
https://doi.org/10.1007/164_2022_625

The spectrum of paroxysmal manifestations encompasses migraine with hemiplegic aura, episodic ataxia, epilepsy and paroxysmal non-epileptic movement disorders. Additional chronic neurological symptoms range from severe developmental phenotypes in early-onset cases to neurobehavioural disorders and chronic cerebellar ataxia in older children and adults.

In the present review we systematically approach the clinical manifestations of *CACNA1A* variants, delineate genotype–phenotype correlations and elaborate on the emerging concept of an age-dependent phenotypic spectrum in *CACNA1A* disease. We furthermore reflect on different therapy options available for paroxysmal symptoms in *CACNA1A* and address open issues to prioritize in the future clinical research.

Keywords

Acetazolamide · *CACNA1A* · Cerebellar ataxia · Developmental delay · Epilepsy · Episodic ataxia · Familial hemiplegic migraine · Genetic testing · Therapy

1 Introduction

The advent of the genetic era unveiled the critical role of ion channel dysfunction as substrate of rare neurological disorders with paroxysmal manifestations (Graves and Hanna 2005). Variants in the Ca^{2+} channel gene *CACNA1A* emerged as a particularly frequent aetiology of such rare phenotypes (Indelicato and Boesch 2021).

The *CACNA1A* gene encodes the pore-forming subunit α1 of the neuronal voltage-gated Ca^{2+} channel P/Q. In the central nervous system (CNS), P/Q channels are ubiquitously expressed and particularly abundant in cerebellum granules and Purkinje cells. At the presynaptic terminal, P/Q channels activation upon depolarization results in a Ca^{2+} inflow which in turn triggers the vesicular neurotransmitter release (Catterall 2011). Beyond the regulation of synaptic transmission and related plasticity processes, P/Q Ca^{2+} currents influence the gating of K^+ channels, transcriptional activity as well as intracellular signalling pathways (Zamponi et al. 2015).

The first association between *CACNA1A* and human diseases dates to 1996, as its mutations were linked to two autosomal dominant neurological disorders: familial hemiplegic migraine type 1 (FHM1) and episodic ataxia type 2 (EA2) (Ophoff et al. 1996). FHM1 and EA2 are classical channelopathies featuring peculiar recurring neurological symptoms, such as migraine with motor aura and attacks of paroxysmal vertigo and ataxia. Shortly after, a third autosomal dominant disorder, the spinocerebellar ataxia type 6 (SCA6), was mapped at *CACNA1A* locus (Zhuchenko et al. 1997). SCA6 may present with paroxysmal vertigo as inaugural symptom, but primarily features a late-onset progressive ataxic syndrome, which does not fit with the phenotype of a classical channelopathy.

In the last decade, cumulative reports delineated an even broader spectrum of clinical manifestations of *CACNA1A* variants, which in children can predate the classical hemiplegic migraine/episodic ataxia. On the one hand, the episodic manifestations expanded to epilepsy as well as to additional paroxysmal non-epileptic movement disorders (Indelicato and Boesch 2021). On the other hand, a role of *CACNA1A* in determining early-onset, severe neurodevelopmental and neuropsychiatric phenotypes is emerging (Indelicato and Boesch 2021).

The paradigm change from three defined allelic disorders FHM1, EA2 and SCA6, to an evolving age-dependent spectrum in non-polyglutamine *CACNA1A* mutations is slowing getting its way across the scientific community. Though, the pathophysiological basis of this clinical variability and evolution is still largely unexplored.

Herein, we aim at offering an exhaustive, updated overview on the clinical spectrum of *CACNA1A* mutations, highlighting genotype–phenotype correlations, and focusing on age-dependent clinical features as well as on established and emerging therapeutic concepts.

2 Molecular Genetics of *CACNA1A* Disorders: Brief Overview

The *CACNA1A* locus resides on chromosome 19. The gene contains 47 exons and can through alternative splicing be translated in myriad isoforms with different regional distribution and kinetic properties (Lipscombe et al. 2013). The final structure of the α1 subunit recapitulates the canonical organization shared by several ion channels as illustrated in Fig. 1. Six transmembrane domains (S1–S6) are expressed in a repeated fashion and organized in a tertiary structure in which the domain S4, the voltage sensor, lines the pore opening (Catterall 2011). The S4 domain is enriched of positively charged amino acids. Several disease-associated missense variants lead to substitution of positively charged arginine in this segment (Indelicato and Boesch 2021).

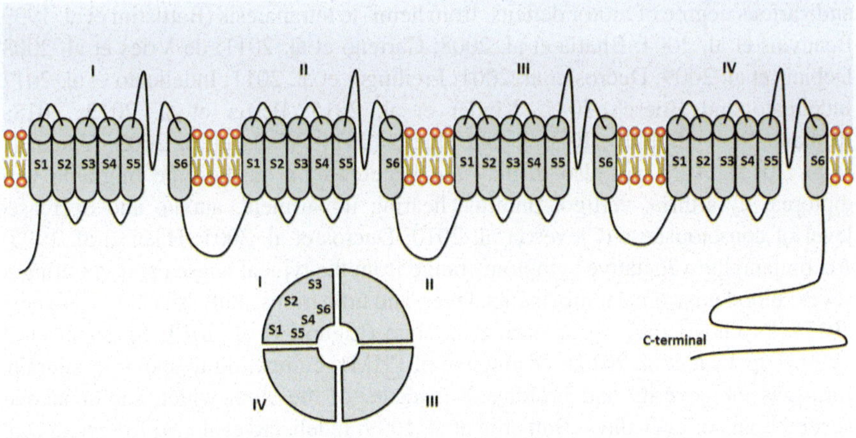

Fig. 1 Schematic representation of the α1 subunit of the voltage-gated Ca^{2+} channel P/Q

CACNA1A scores among the 2% most "intolerant" genes (Petrovski et al. 2013), a notion that underlines the need of a careful interpretation of new variants of – currently – unknown significance. Disease-associated genotypes span from missense mutations, usually with gain-of-function net effect, to loss-of-function truncating mutations and small deletions. The expansion of a CAG repeat coding for polyglutamine in the C-terminus of the gene represents a further disease genotype as we will address later.

Irrespective of the type of mutation, classical *CACNA1A* disorders feature an autosomal dominant pattern of inheritance. Few reports described exceptional cases of biallelic pathogenic *CACNA1A* variants. Individuals bearing biallelic pathogenic variants display an early-onset and a growing severity of phenotype from biallelic missense mutation (Ko et al. 2021) to one missense plus one truncating (Reinson et al. 2016) to a biallelic truncating mutation (Arteche-Lopez et al. 2021) which led to death few months after birth. Additional dysmorphic features as well as cerebral malformations may be evident in these cases.

3 One Side of the Clinical Phenotype: Paroxysmal Features

3.1 Hemiplegic Migraine

Migraine is a well-known episodic feature of *CACNA1A* disorders and migraine with hemiplegic aura represents the key feature of FHM1 (OMIM # 141500), the first genetically framed form of migraine with aura (Ophoff et al. 1996). Up to date, three further monogenic forms of familial hemiplegic migraine have been identified (Riant et al. 2022; Russell and Ducros 2011).

Migraine spells in FHM1 usually start in the first two decades and display a variety of positive and negative symptoms during the aura. These include positive visual phenomena, hemianopsia, tingling/numbness of the extremities, dysphasia, and various degree of motor deficits, from hemi- to tetraparesis (Battistini et al. 1999; Beauvais et al. 2004; Bhatia et al. 2008; Carreno et al. 2011; de Vries et al. 2008; Debiais et al. 2009; Ducros et al. 2001; Freilinger et al. 2011; Indelicato et al. 2017; Indelicato and Boesch 2021; Kinder et al. 2015; Pelzer et al. 2013, 2018a; Romaniello et al. 2010; Russell and Ducros 2011; Wada et al. 2002; Yamazaki et al. 2011). Aura can also display the features of a basilar-type migraine with diplopia, dysarthria, vertigo, tinnitus, hearing impairment, ataxia, and decreased level of consciousness (Cleves et al. 2010; Ducros et al. 2001; Haan et al. 1995). Accompanying vegetative symptoms range from the typical nausea and vomiting to severe blood pressure derangements, fever, and urticarious skin rash (Indelicato et al. 2017). Psychosis may occur during the aura (Liguori et al. 2013; Spranger et al. 1999; Topakian et al. 2014). Distinctive of FHM1, comparing to sporadic migraine forms, is the severity and prolonged character of the aura, which can span over several hours or even days (Battistini et al. 1999; Indelicato et al. 2017; Vahedi et al. 2000) and largely outlasts the burden of the subsequent headache. Co-occurring disturbance of consciousness is especially disabling and may range from somnolence to coma (Curtain et al. 2006; Debiais et al. 2009; Ducros et al. 2001; Echenne et al. 1999; Fitzsimons and Wolfenden 1985; Kors et al. 2001; Malpas et al. 2010; Stam et al. 2009; Vahedi et al. 2000; Wada et al. 2002). Cumulative reports highlighted that severe migraine attacks may present as encephalitic state with additional fever and seizures, as well as severe hemiplegia (Battistini et al. 1999; Chan et al. 2008; Stam et al. 2009; Yamazaki et al. 2011). Although these dramatic clinical pictures may resolve without sequelae, in the setting of specific mutations, of which the S218L is the prototype, arising complications may be lethal (Debiais et al. 2009; Fitzsimons and Wolfenden 1985; Kors et al. 2001; Malpas et al. 2010; Stam et al. 2009).

Classical migraine triggers such as stress may be reported also by FHM1 patients. Peculiar triggers in FHM1 are represented by cerebral angiography (Ducros et al. 2001; Indelicato et al. 2017), as well as by bagatelle trauma in the setting of specific mutations, such as S218L (Curtain et al. 2006; Kors et al. 2001; Stam et al. 2009). In this setting, a progressive cerebral oedema may develop and lead to the above-mentioned severe attacks.

Hemiplegic migraine results in various degrees of slowing on routine scalp EEG, which can be visualized also several days after resolution of the symptoms (Indelicato et al. 2017, 2021b).

The frequency and course of hemiplegic migraine may be highly variable. Attack frequency spans from 1 to 2 spells in a lifespan to several episodes per month. In several cases in the literature and in our clinical experience, hemiplegic migraine shows a biphasic course (Indelicato et al. 2017). Indeed, a florid phase with recurring spells in the youth may transit in a long-lasting symptom-free interval (Ducros et al. 2001), which is eventually followed by flaring up at the end of 40s/beginning of 50s. Notably, patients may not recall their migraine attack (Indelicato et al. 2017; Stam et al. 2011).

FHM1 patients may suffer also from attacks of "common" migraine without hemiplegia.

FHM1 is typically caused by missense mutations with gain-of-function effect on Ca^{2+} currents, which are believed to facilitate a cascade of events that results in cortical spreading depression classically seen in migraine aura (de Vries et al. 2009; Rajakulendran et al. 2012; Striessnig 2020; van den Maagdenberg et al. 2010). Malignant mutations such as S218L display a particularly accentuated effect on the channel kinetics (Loonen et al. 2019; van den Maagdenberg et al. 2010).

3.2 Episodic Ataxia

Recurring spells of dysarthria, vertigo and coordination disturbances, a spectrum of symptoms defined as "episodic ataxia", represent the cardinal feature of EA2 (OMIM #108500). As for hemiplegic migraine, the genetic background of episodic ataxia is manifold, but the subtype due to *CACNA1A* mutations is the most frequent (Jen et al. 2007).

Episodic ataxia spells usually start in the first two life decades (Jen et al. 2004, 2007; Nachbauer et al. 2014; Subramony et al. 2003), but late-onset cases have been described (Imbrici et al. 2005). Reported symptoms include diplopia, blurred vision, tinnitus, dysarthria, coordination problems including marked balance disturbances and nausea. Cervical dystonia occasionally occurs as further ictal symptom (Hu et al. 2013; Spacey et al. 2005). Headache may also accompany ataxia spells (Nachbauer et al. 2014). Furthermore, EA2 patients may suffer from migraine occurring independently from spells of episodic ataxia (Jen et al. 2004; Subramony et al. 2003; Verriello et al. 2021).

Typically, episodic ataxia spells are shorter than migraine episodes, may even last few minutes, but are more frequent, recurring also several times per day. Physical and emotional stress are typical triggers (Jen et al. 2004). Furthermore, phenytoin, alcohol and caffeine consumption may precipitate the attacks (Kipfer and Strupp 2014; Nachbauer et al. 2014).

EA2 is typically associated with loss-of-function variants, usually mutations which interrupt the reading frame (deletion, in-frame insertion, truncating mutations) (Imbrici et al. 2005; Jen et al. 2004, 2007; Nachbauer et al. 2014; Ophoff et al. 1996; Riant et al. 2010).

3.3 Epilepsy

Shortly after the genetic allocation of FHM1 and EA2, several reports described epilepsy as additional features in *CACNA1A* families (Imbrici et al. 2004; Jouvenceau et al. 2001; Kors et al. 2004; Rajakulendran et al. 2010; Stam et al. 2008). Since *CACNA1A* diseases are channelopathies, this finding is not surprising, though the clinical picture and the underlying pathophysiological mechanisms are manifold and still unclear.

Three main patterns of epileptic comorbidities may be delineated (Brunklaus 2021; Indelicato and Boesch 2021):

- In FHM1 families, seizures mostly occur concomitantly with severe migraine attacks accompanied by fever and coma state. Particularly frequent is the description of seizures following trivial head trauma in patients bearing the S218L mutations (Stam et al. 2009; Yamazaki et al. 2011). Additional mutations recurrently associated with epileptic seizures are the R1349Q (Malpas et al. 2010; Sanchez-Albisua et al. 2013; Tantsis et al. 2016), and Y1385C (Ducros et al. 2001; Vahedi et al. 2000). Epileptic seizures in the interval are rare. All in all, seizures in FHM1 are considered "symptomatic", thus comparable to seizures caused by various non-specific major brain insults.
- Conversely, in EA2 families epilepsy is frequently described as "independent" manifestation usually occurring in children before the onset of paroxysmal ataxia (Choi et al. 2013; Damaj et al. 2015; Indelicato and Boesch 2021; Indelicato et al. 2021a, b; Riant et al. 2010; Strupp et al. 2005). Absence epilepsy and its correlating EEG changes (3 Hz spike-waves complexes) are the most common

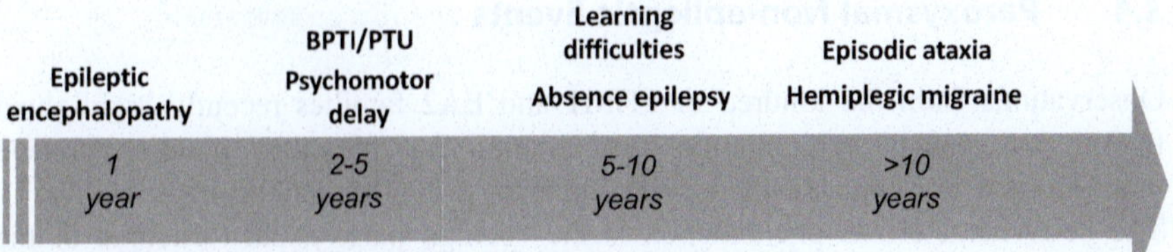

Fig. 2 The age-dependent phenotype of *CACNA1A* disorders (From Indelicato and Boesch 2021)

findings (Imbrici et al. 2004; Indelicato et al. 2021a, b; Jung et al. 2010; Labrum et al. 2009; Mantuano et al. 2010; Rajakulendran et al. 2010; Stendel et al. 2020), suggesting a susceptibility mechanism for generalized seizures in the setting of loss-of-function state of P/Q channels (Indelicato et al 2021b; Llinas et al. 2007). Description of epilepsy/seizure in adults with EA2 phenotypes is rare. In adults focal seizures as well as well-defined localized EEG foci have been described (Cuenca-Leon et al. 2009; Verriello et al. 2021).

- A third pattern emerged with the advent of whole exome sequencing studies in large paediatric cohorts affected by severe, therapy-resistant seizures accompanied by developmental delay/regression. These studies unravelled that de novo missense *CACNA1A* mutations recurrently cause severe early-onset epilepsy which phenotypically may present as classical developmental-epileptic encephalopathies such as Dravet syndrome, Lennox-Gastaut syndrome and epilepsy of infancy with migrating focal seizures (Epi et al. 2013; Epi 2016; Le Roux et al. 2021; Mitta et al. 2020). The term "developmental and epileptic encephalopathy-42" (DEE42) (OMIM #601011) has been coined to describe such new disease entities. Carriers of de novo missense mutations often present with early episodes of status epilepticus, especially triggered by fever (Guerin et al. 2008; Le Roux et al. 2021; Niu et al. 2022). Together with novel missense mutations, other well-known *CACNA1A* variants have been associated with these severe phenotypes, first of all the S218L (Epi 2016). In some children both severe epilepsy and episodes of flaccid hemiparesis can be noticed (Le Roux et al. 2021), a phenotype that overlaps with earlier reports of FHM1 families with a particularly severe disease in the offspring.

Taken together older and newer reports on carriers of *CACNA1A* mutations suggest that the nature and severity of symptoms may be at least partially age-dependent, an aspect overlooked by a majority because of its transient nature and absence later in life (see Fig. 2 as well as Indelicato and Boesch (2021); Indelicato et al. (2021b)).

3.4 Paroxysmal Non-epileptic Events

Observations from the kindred of FHM1 and EA2 families recently highlighted further paroxysmal, non-epileptic manifestation that in young children predate migraine and episodic ataxia (Gur-Hartman et al. 2020; Indelicato et al. 2017). The best characterized phenomenon is the "benign paroxysmal torticollis of the infancy" (BPTI) (Cuenca-Leon et al. 2008; Roubertie et al. 2008). This term indicated a form of paroxysmal cervical dystonia accompanied by pallor, nausea and malaise (Rosman et al. 2009). The strong association between BPTI and later onset of migraine in the general population contributed to the inclusion of this condition in the list of the "periodic syndromes" considered as precursors of migraine (Headache Classification Committee of the International Headache, S. 2013). While BPTI occurs isolated in other children (Danielsson et al. 2018), in *CACNA1A* mutation carriers, further subtle or overt neurological signs are usually detected at the examination (Gur-Hartman et al. 2020; Humbertclaude et al. 2020; Mantuano et al. 2010).

Similar to BPTI, the so-called paroxysmal tonic upward gaze (PTU) describes a condition of sudden tonic upwards deviation of the bulbi which is slowly emerging as a frequent manifestation in young children with *CACNA1A* mutations. As BPTI, the PTU spells predate the classical paroxysmal motor phenotypes (Blumkin et al. 2010; Chang et al. 2021; Gur-Hartman et al. 2020; Humbertclaude et al. 2020; Quade et al. 2020; Tantsis et al. 2016; Zhang et al. 2021).

4 The Other Side of the Clinical Phenotype: Chronic Neurological Manifestations

4.1 Cerebellar Ataxia and Further Chronic Motor Signs

Although some FHM1 and EA2 families present with an exquisite paroxysmal disorder, the majority of patients eventually develop some degree of chronic cerebellar signs in the course of their disease (Ducros et al. 2001; Indelicato and Boesch 2021; Jen et al. 2007).

Cerebellar features in FHM1 and EA2 patients are usually mild. The most frequent interictal sign is nystagmus, which frequently shows a major down-beat component (Indelicato et al. 2019, 2021a; Marti et al. 2008; Strupp et al. 2007; Yabe et al. 2008). The second most common chronic feature is gait ataxia. In our experience and as it has been shown through objective gait analysis (Indelicato et al. 2021a), the gait in *CACNA1A* patient is not constantly broadened as in other ataxic disorders, though instability upon external perturbances may be prominent and falls occur early in disease course (Indelicato et al. 2021a). Conversely, dysmetria of the extremities is usually mild, if present at all. Chronic cerebellar signs usually do not display a relevant progression, but exceptions are also encountered. Indeed, in some cases paroxysmal symptoms soon abate and a rather fast progressive gait ataxia dominates the clinical picture. An isolated "SCA6-like" chronic progressive cerebellar syndrome may also be the only presentation (Coutelier et al. 2017; Marti et al.

2008; Nikonishyna et al. 2022) in carriers of conventional, non-polyglutamine, *CACNA1A* mutations. In severe cases, further motor signs such as spasticity and positive Babinski signs may be evident. Additional dystonic features, especially cervical dystonia, are recurrently reported (Cuenca-Leon et al. 2009; Fuerte-Hortigon et al. 2020; Spacey et al. 2005).

The time of onset of chronic cerebellar signs is highly variable. In young children, cerebellar ataxia is not easy to diagnose. Indeed, the typical signs of a cerebellar syndrome may not be evident before the second-third year of life (Bertini et al. 2018). Unspecific signs such as generalized hypotonia or delay in the acquisition of motor milestones offer a clinical clue (Izquierdo-Serra et al. 2020) and can be later recalled by the parents of young adults upon clinical interview (Indelicato et al. 2019). Severe motor features in infants usually frame within a general developmental delay phenotype as described in the following paragraph. Mild cerebellar signs can be noticed for the first time in young adults during an acute referral because of a migraine or episodic ataxia spell.

4.2 Developmental Delay, Cognitive and Behavioural Features

The parents of young adults with FHM1 and EA2 can often recall some delay in the acquisition of early milestones, described as a slight delay in walking or speaking (Indelicato et al. 2019). Later, learning problems may become evident and lead to the inclusion in special education programs (Indelicato et al. 2019). In schoolchildren, prominent behavioural problems may arise, compatible with attention-deficit hyperactivity disorder or autistic spectrum disorders (Damaj et al. 2015; Indelicato et al. 2019; Nardello et al. 2020; Tantsis et al. 2016). In the literature, reports of such prominent neurobehavioural features are limited to children. This observation, along with our clinical experience, suggests that in children arising from FHM1 and EA2 families behavioural abnormalities may ameliorate spontaneously (Indelicato et al. 2017). Though, subtle cognitive dysfunction may be detectable in later life (Indelicato et al. 2019; Karner et al. 2010, 2012). Several authors reported that adult patients are mostly employed in unskilled labour, have poor education level and often psychiatric comorbidities, as well as impulsive/sociopathic behaviour (Freilinger et al. 2008; Indelicato et al. 2019; Nachbauer et al. 2014). Few studies applied formal neuropsychological testing in adults with genetically confirmed *CACNA1A* disease (Indelicato et al. 2019; Karner et al. 2010, 2012). These works revealed deficits in several cognitive domains, first of all in executive functions, figural memory and visuo-constructive abilities. Late-onset progressive cognitive decline consistent with dementia is rare (Freilinger et al. 2011).

Cumulative reports highlighted that a severe neurodevelopmental phenotype, isolated or along with severe epilepsy can dominate the clinical picture in infants with de novo *CACNA1A* mutations (Gur-Hartman et al. 2020; Humbertclaude et al. 2020; Luo et al. 2017; Ohba et al. 2013; Weyhrauch et al. 2016). A developmental regression after the normal milestones acquisition has also been reported (Guerin et al. 2008), often following severe hemiplegic attacks with encephalitic features, a clinical picture which can overlap with that of mitochondrial diseases such as MELAS (Knierim et al. 2011).

4.3 Imaging Findings

The typical imaging findings in FHM1 and EA2 are represented by an isolated cerebellar atrophy with solely – or prevalent – involvement of the vermis (Indelicato et al. 2019). The degree of atrophy is usually mild. Parallel to the variable clinical manifestations, more severe atrophy pattern may appear. Cerebellar atrophy may not be evident in infants, but eventually developed in a majority of cases later in the disease course (Gur-Hartman et al. 2020). Further findings in early-onset severe cases encompass cerebral undersulcation, mild cerebral atrophy and thinning of the corpus callosum (Ko et al. 2021; Reinson et al. 2016; Weyhrauch et al. 2016). During severe hemiplegic migraine attacks, imaging techniques can capture a hypermetabolism or oedema in the hemisphere contralateral to the aura symptoms which can later evolve in a hemicerebral atrophy (Indelicato et al. 2017; Pelzer et al. 2018b; Tashiro et al. 2014).

5 Spinocerebellar Ataxia Type 6: A Polyglutamine *CACNA1A* Disorder

SCA6 is caused by a pathologically expanded CAG repeat, coding for a polyglutamine stretch, at the C-terminus of *CACNA1A* (Zhuchenko et al. 1997). Normal alleles host up to 18 repeats, while disease causing alleles contain >19 repeats (19–33 repeats). As in other polyglutamine disorders, the length of the pathologically expanded allele is inversely correlated to the age at onset and the disease severity (Geschwind et al. 1997; Schols et al. 1998; Stevanin et al. 1997). Phenotypically, SCA6 resembles other polyglutamine-related spinocerebellar ataxias, by featuring a chronic progressive cerebellar syndrome with adult onset (Geschwind et al. 1997; Schols et al. 1998; Stevanin et al. 1997). Somehow differently from other SCAs, it presents with a pure cerebellar phenotype, without major features of brainstem or cerebral involvement, and it usually manifests later in life, in the fifth to sixth decade. Generally, the phenotype of SCA6 is milder than that of other SCAs, which usually also bear longer CAG-repeats. Natural history data revealed that the lifespan in SCA6 is not shortened (Jacobi et al. 2015). Differently from other SCAs and in line with its allelic disorders, SCA6 patients may display some marked clinical fluctuations at the beginning of disease, which sometimes resemble the paroxysmal ataxia episodes of EA2 (Jen et al. 1998).

From a pathophysiological point of view, SCA6 is not primary a channelopathy. Indeed, α1-subunits bearing expanded CAG do not seem to affect calcium currents (Giunti et al. 2015). Ten years ago, the research group of Gomez et al. discovered a novel internal ribosomal entry site at *CACNA1A* locus, implicating that gene encodes not only the α1-subunit, but also a second shorter protein, carrying the

CAG-expansion (Du et al. 2013). The newly described peptide α1ACT translocates in the nucleus, where it acts as transcription factor, regulating the expression of several genes with a pivotal role in neuronal growth (Du et al. 2019). Dysfunction of the polyglutamine-bearing α1ACT protein is thus believed to be the main driver of SCA6 pathophysiology.

6 Treatment Options in *CACNA1A* Disorders

The availability of effective pharmacological treatments makes *CACNA1A* disorders a not-to-be-missed diagnosis in the genetic work-up of suggestive clinical phenotypes. Despite the lack of controlled trials, cumulative reports and the clinical experience undoubtedly proved the effectiveness of several therapeutics in the prevention of disabling paroxysmal manifestations in FHM1 and EA2. Acetazolamide, aminopyridines, flunarizine and topiramate are drugs with an established effect as interval prophylaxis (Indelicato and Boesch 2021; Pelzer et al. 2013; Strupp et al. 2007). Up to date, no clear recommendations are available concerning the treatment of *CACNA1A* related epilepsy. Since seizures are also paroxysmal manifestations, an effect of the established effective prophylaxes could be assumed.

Over the time, paroxysmal manifestations subside, spontaneously or upon medical treatment, and chronic neurological symptoms, first of all gait ataxia, become the major burden of disease. Although the interval therapy does not have major impact on chronic balance disorder, it may still result in some stabilization, by smoothening clinical fluctuations. In this setting, rehabilitation represents the main long-term therapy. Neurodevelopmental and neuropsychiatric issues similarly await a targeted pharmacological treatment.

In the following paragraphs, we summarize the evidence supporting the application of several pharmacological treatments in *CACNA1A* disorders, address open issues and priorities for future clinical research.

6.1 Acetazolamide: A Serendipitous Discovery

In 1978, Griggs and colleagues reported on a large family with autosomal dominant inherited paroxysmal ataxia which displayed a dramatic response to a serendipitous trial of acetazolamide (Griggs et al. 1978). The confirmation of the dramatic effect of acetazolamide in following reports contributed to delineate the disease entity "acetazolamide-responsive paroxysmal ataxia" (Neufeld et al. 1996). This clinically defined disorder received its genetic allocation as EA2 after the description of *CACNA1A* mutations by Ophoff and colleagues.

Acetazolamide is a carbonic anhydrase inhibitor. Carbonic anhydrase catalyses the interconversion between water and carbon dioxide to carbonic acid, which in the body fluids solubilizes as hydrogen and bicarbonate ions. The inhibition of carbonic anhydrase in the kidney hampers the reabsorption of bicarbonate as well as of sodium and chloride, which goes along with excretion of water along with ions resulting in a diuretic effect (2012). The consequence at a systemic level is a decrease in blood pressure, intracranial and intraocular pressures. Furthermore, bicarbonate excretion increases the pH of the blood as well as of the cerebrospinal fluid.

The clinical application of acetazolamide spans from the altitude sickness to glaucoma, intracranial hypertension to epilepsy (Van Berkel and Elefritz 2018). Up to date, the beneficial effect of acetazolamide in *CACNA1A* disorders is believed to directly derive from the pH alterations elicited by the drug. Earlier studies applying phosphorous magnetic resonance spectroscopy in small non-genetically confirmed cohorts revealed that pH in the cerebellum is higher in patients compared to controls and normalizes upon treatment with acetazolamide (Bain et al. 1992; Sappey-Marinier et al. 1999). Changes in extra- and intracellular pH influence the potassium currents as well as the opening of sodium and calcium channels and may thus mitigate the effect of an abnormal opening kinetics in the setting of *CACNA1A* mutations (Strupp et al. 2007). This positive effect is not restricted to P/Q channel mutations, as acetazolamide improves the manifestations of further ion channel disorders, such as hypokalemic periodic paralysis. Notably, the influence of pH fluctuations induced by acetazolamide may be more marked in the setting of a mutated ion channel than in the wild-type (Strupp et al. 2007).

Acetazolamide was primarily applied in EA2, but it has been also proven effective in the prevention of hemiplegic migraine (Battistini et al. 1999; Indelicato et al. 2017), as well as in the treatment of paroxysmal dizziness that predates chronic ataxia in SCA6 (Jen et al. 1998). Sparse reports suggested that acetazolamide may also mitigate some chronic features. Our recommendation is a trial of acetazolamide also in primary chronic courses, as a positive effect on fluctuations may also contribute to disease stabilization. In our clinical experience, acetazolamide contributed also to the stabilization of psychopathological status in *CACNA1A* patients with concomitant schizophrenia (Mechtcheriakov et al. 2003).

Several side effects limit the application of acetazolamide. The most frequent is represented by distal paraesthesia, a non-dangerous symptom that may however be bothersome for patients and lead to early withdrawal. In our clinical experience, paraesthesias are both an indicator of therapy compliance and of expected clinical benefit. Further side effects, probably induced by the metabolic acidosis, are fatigue, muscle stiffness, gastrointestinal disturbances as well as nephrolithiasis (Strupp et al. 2007). Shifts in blood electrolytes represent a caveat for treatment in older patients or in the setting of impaired renal function. Blood count alterations may also occur.

Up to very recent time, no randomized trials investigated the effect of acetazolamide in *CACNA1A* disorders. Despite this and the moderate tolerability, acetazolamide represents the first line therapy in the setting of paroxysmal symptoms due to *CACNA1A* mutations. The usual daily dosage ranges from 250 to 1,000 mg. In the clinical routine, a trial of acetazolamide is always performed in absence of contraindications. Clinical experience (personal observation as well as of Strupp et al. (2007)) shows that in some patients acetazolamide may lose efficacy over the time. In this case or in the setting of adverse events/contraindication, further pharmacological interventions are available (see Table 1).

Table 1 Established therapeutic compounds and their applications in *CACNA1A* disorder

Compound	Dosage	Mechanisms of action	Application
Acetazolamide	250–1,000 mg/daily	Carbon anhydrase inhibition	First line therapy for EA2, as well as prevention of paroxysmal manifestation in FHM1
4-aminopyridine	10–15 mg/daily	K^+ channel blocker	EA2, interictal nystagmus, episodic ataxia in other settings (e.g. SCA6)
Fampridine	20 mg/daily	K^+ channel blocker (prolonged-release form of 4-AP)	As 4-AP
Flunarizine	5–10 mg/daily	Ca^{2+} channel blocker	First line migraine prophylaxis in FHM1
Topiramate	50–100 mg/daily	Na^+ channel blocker, AMPA inhibition, GABA enhancer, weak carbonic anhydrase inhibition	Overlap phenotypes episodic ataxia/hemiplegic migraine as well as overlap with epileptic manifestations

6.2 Aminopyridines in the Treatment of Episodic Ataxia

Aminopyridines are potassium channel blockers which have been originally applied in the treatment of down-beat nystagmus (Strupp et al. 2003). The beneficial effect on nystagmus may derive from an increasing inhibitory firing of the Purkinje cells. An earlier report described a beneficial effect of 4-aminopyridines in two patients with EA2 who no longer responded to acetazolamide (Strupp et al. 2004). This observation was corroborated in an observational study (Strupp et al. 2004) as well as in a randomized, double-blind, placebo-controlled, crossover trial, comparing 4-aminopyridine with placebo (Strupp et al. 2011). Based on these data, a treatment with 4-aminopyridine 15 mg/die is recommended in patients with EA2. However, 4-aminopyridine is not licensed for other indications and its availability may be limited. Instead, the prolonged-release form fampridine (Fampyra, Biogen) holds an approval for the symptomatic treatment of gait disturbances in multiple sclerosis and has been recently investigated in a crossover trial versus acetazolamide and placebo in a group of patients with episodic ataxia with and without confirmed *CACNA1A* mutation (Muth et al. 2021). Fampridine was effective in reducing the number of attacks to 63% compared to placebo. In comparison, acetazolamide appeared to be likely more effective (reduction of attacks to 52%), though it was, as expected, far less well tolerated. Aminopyridines can be applied also to treat the interictal nystagmus, which in both EA2 and SCA6 patients may be highly disabling and contribute to balance disorder.

Concerning safety issues, aminopyridines may increase seizure risk and are contraindicated in the setting of prolonged QT interval as well as in renal insufficiency.

6.3 Migraine Prophylaxis in FHM1: Flunarizine, Topiramate and Other Compounds

FHM1 patients may experience severe migraine attacks. Especially in the setting of disorders of consciousness, the establishment of an effective prophylaxis is mandatory since there is no treatment available to arrest the aura.

The best characterized migraine prophylaxis in FHM1 is flunarizine (Karsan et al. 2018; Peer Mohamed et al. 2012). Flunarizine is a non-selective Ca^{2+} channel blocker which is approved also in the treatment of common migraine and has been proven to be safe and effective in children (Peer Mohamed et al. 2012). Experimental evidence suggests that blocking of Ca^{2+} and Na^{+} currents induced by flunarizine raises the threshold for the cortical spreading depression phenomenon underlying migraine (Eikermann-Haerter et al. 2012; Ye et al. 2011). Flunarizine may be particularly beneficial in the treatment of migraine with aura and several reports confirmed its effectiveness in FHM1 (Pelzer et al. 2013). The usual dosage is 10 mg daily. Common side effects include sedation as well as depressive mood and weight gain. Flunarizine also increases the risk of developing a pharmacologically induced parkinsonism after long-term use (Karsan et al. 2018). Considering this risk profile, evening schedule in the administration as well as caution in the treatment of specific patient group (elderly, known depression) should be paid.

Sparse reports highlight a relevant effect of the Ca^{2+} channel blocker verapamil in migraine with hemiplegic aura (Yu and Horowitz 2001; Yu and Horowitz 2003). Though, to the best of our knowledge, only one genetically confirmed FHM1 case with good response to verapamil is reported, notably with severe attacks induced by trivial trauma within an R1349Q mutation (Knierim et al. 2011).

As previously mentioned, acetazolamide is effective also in the prevention of migraine in FHM1 while this effect cannot be generalized to other forms of migraine (Vahedi et al. 2002). Interestingly, topiramate is also an established migraine prophylaxis which bears carbon anhydrase inhibition properties and has been reported to be effective in sparse reports in both FHM1 and EA2 (Gonzalez-Mingot et al. 2022; Indelicato et al. 2017). In the clinical experience of ours and others (Pelzer et al. 2013) topiramate is generally better tolerated in *CACNA1A* patients as in common migraineurs. Side effects of topiramate partly overlap with those of acetazolamide, due to clear overlapping mechanisms. Furthermore, topiramate may have central side effects such as impaired concentration, speech disorders and cognitive disturbances which are reversible upon drug withdrawal. Topiramate and to less extent acetazolamide also have an anorexiant effect which may be advantageous in the clinical practice.

Controversially, poor evidence is available for other classical migraine prophylaxes such as beta blockers (propranolol, metoprolol) (Pelzer et al. 2013). We personally used metoprolol with benefit as add-on prophylaxis in a patient experiencing severe hypertensive crisis concomitant to hemiplegic attacks (Indelicato et al. 2017). Reports from migraine with aura suggest a possible beneficial effect of valproate and lamotrigine (Pelzer et al. 2013), which could be used as reserve treatments in the setting of specific comorbidities such as epilepsy.

6.4 The Treatment of *CACNA1A* Related Epilepsy: Where Do We Stand?

Data on therapy of *CACNA1A* related epilepsy are controversial and systematic analysis are lacking. Generally, early-onset severe phenotypes present with the stigmata of a therapy refractory epilepsy, while later onset epileptic syndromes appear to be prone to a good seizure control upon classical antiepileptic drugs (AED) (Le Roux et al. 2021; Niu et al. 2022; Verriello et al. 2021).

Interestingly, acetazolamide bears also anticonvulsive properties and in the past it was applied in the treatment of catamenial aggravation of epilepsy (Lim et al. 2001). The pharmacodynamically related drug topiramate is an approved AED. Interestingly, recent findings point towards a favourable response to topiramate in *CACNA1A* related epilepsy (Le Roux et al. 2021). Zhang et al. reported three seizure free children after adding up topiramate to AED polytherapy (Zhang et al. 2020). Le Roux et al. similarly reported that add-on therapy with topiramate reduced seizure frequency in eight out of nine patients (one patient seizure free after topiramate) (Le Roux et al. 2021). In the latter study, acetazolamide was less effective than topiramate in seizure control. Contrasting findings have been reported for levetiracetam (Le Roux et al. 2021). Finally, additional therapy with valproate has been described to be beneficial both in epilepsy and migraine (Niu et al. 2022; Pelzer et al. 2013; Zhang et al. 2020). In the same report of Le Roux (Le Roux et al. 2021), ketogenic diet resulted in clinical improvement in one out of four patients; vagus nerve stimulation yielded no benefit in four implanted patients. The singular description of a "dramatic" improvement of therapy resistant absence seizure to pyridoxine in one child (Du et al. 2017) was not replied up to now in further reports. Finally, cumulative literature suggest that seizure may spontaneously improve over the time independently from AED administration (Indelicato et al. 2021b).

6.5 Acute Treatment of EA2 and FHM1 Spells

Up to date, no specific treatment is available which effectively abates ataxia or hemiplegic migraine spells. One single report described abating aura manifestations upon administration of intranasal ketamine (Kaube et al. 2000). The mechanism of action is unclear and this finding has not been replicated. Therefore, and considering the safety profile of ketamine, its application is not recommended. In the acute treatment of severe hemiplegic attacks with brain oedema, an empirical symptomatic management with corticosteroid or hypertonic saline should be considered. The treatment of the migraine related headache relies on the same therapeutics available for common migraine, with a controversial caveat for triptans. Since triptans are powerful vasoconstricting agents, their application in FHM was believed to possibly aggravate neurological deficits and leading to migrainous strokes. Currently, the US Food and Drug Administration approved package labelling states that triptans are contraindicated in patients with hemiplegic migraine. Some retrospective case series showed a safe profile for triptans in the treatment of hemiplegic migraine (Artto et al. 2007; Mathew et al. 2016). In our clinical practice, we do not apply triptans in FHM1.

7 Conclusion and Future Directions

Advancements in clinical research in *CACNA1A* disorders are gaining a faster pace. The recent creation of consortia of clinicians and patients advocacy representatives in Europe and North America is further contributing to these developments. These advancements bring along relevant implications for the clinical routine and set priorities for the future lines of research.

In the clinical routine, the achievement of a genetic diagnosis in case of paroxysmal or hemiplegic manifestation should become mandatory to stratify carriers with severe mutations, rule out differential diagnosis (such as *PRRT2*, *ATP1A2*, etc.) and guide the choice of an interval prophylaxis. Furthermore, we recommend the introduction of regular cognitive and behavioural testing with standardized instruments to detect and address the often neglected non-motor symptoms of *CACNA1A* disorders.

CACNA1A disease spectrum is manifold and evolves over the time. Paediatricians and adult neurologists deal only with a limited side of it. Efforts should be made to develop clinical registers for the collection of longitudinal data to guide rigorous clinical and treatment follow-up and establish valid diagnostic and therapeutic algorithms. In turn, this will translate in an optimized management transition from child to adult neurology. Natural history studies will also help us understanding the response of further symptoms, such as epilepsy, to available pharmacological therapy as well as elucidating their impact on cognitive and motor development.

Understanding how the phenotype evolves across life will be pivotal in directing basic research in the quest for underlying pathophysiological mechanisms. Long-standing efforts in the SCA6 research culminated in the seminal discovery of the bicistronic nature of *CACNA1A* and set the target for the development of oriented therapeutic intervention (Pastor et al. 2018). Such a key mechanism is still to be unveiled in paroxysmal *CACNA1A* disorders. The definition of suitable experimental models, for example induced pluripotent stem cells derived systems, represents the first essential step towards this target (Bavassano et al. 2017).

Future advancements in clinical and basic research will bridge the gaps in our scattered knowledge on *CACNA1A* disorders and ameliorate the management and outcome of these still underdiagnosed clinical entities.

References

Arteche-Lopez A et al (2021) Biallelic variants in genes previously associated with dominant inheritance: CACNA1A, RET and SLC20A2. Eur J Hum Genet 29:1520–1526

Artto V et al (2007) Treatment of hemiplegic migraine with triptans. Eur J Neurol 14:1053–1056
Bain PG et al (1992) Familial periodic cerebellar ataxia: a problem of cerebellar intracellular pH homeostasis. Ann Neurol 31:147–154
Battistini S et al (1999) A new CACNA1A gene mutation in acetazolamide-responsive familial hemiplegic migraine and ataxia. Neurology 53:38–43
Bavassano C et al (2017) Bicistronic CACNA1A gene expression in neurons derived from spinocerebellar ataxia type 6 patient-induced pluripotent stem cells. Stem Cells Dev 26:1612–1625
Beauvais K et al (2004) New CACNA1A gene mutation in a case of familial hemiplegic migraine with status epilepticus. Eur Neurol 52:58–61
Bertini E, Zanni G, Boltshauser E (2018) Nonprogressive congenital ataxias. Handb Clin Neurol 155:91–103
Bhatia R et al (2008) Sporadic hemiplegic migraine: report of a case with clinical and radiological features. J Headache Pain 9:385–388
Blumkin L et al (2010) Congenital ataxia, mental retardation, and dyskinesia associated with a novel CACNA1A mutation. J Child Neurol 25:892–897
Brunklaus A (2021) Advances in genotype-phenotype associations for CACNA1A-related epilepsies. Eur J Paediatr Neurol 33:A2
Carreno O et al (2011) Acute striatal necrosis in hemiplegic migraine with de novo CACNA1A mutation. Headache 51:1542–1546
Catterall WA (2011) Voltage-gated calcium channels. Cold Spring Harb Perspect Biol 3:a003947
Chan YC et al (2008) Electroencephalographic changes and seizures in familial hemiplegic migraine patients with the CACNA1A gene S218L mutation. J Clin Neurosci 15:891–894
Chang MY et al (2021) Paroxysmal tonic upgaze (PTU) associated with CACNA1A mutation and gross motor delay. J AAPOS 25:353–353.e1
Choi KD et al (2013) Possible anticipation associated with a novel splice site mutation in episodic ataxia type 2. Neurol Sci 34:1629–1632
Cleves C et al (2010) Link between confusional migraine, hemiplegic migraine and episodic ataxia type 2: hypothesis, family genealogy, gene typing and classification. Cephalalgia 30:740–743
Coutelier M et al (2017) A panel study on patients with dominant cerebellar ataxia highlights the frequency of channelopathies. Brain 140:1579–1594
Cuenca-Leon E et al (2008) Genetic analysis of 27 Spanish patients with hemiplegic migraine, basilar-type migraine and childhood periodic syndromes. Cephalalgia 28:1039–1047
Cuenca-Leon E et al (2009) Late-onset episodic ataxia type 2 associated with a novel loss-of-function mutation in the CACNA1A gene. J Neurol Sci 280:10–14
Curtain RP et al (2006) Minor head trauma-induced sporadic hemiplegic migraine coma. Pediatr Neurol 34:329–332
Damaj L et al (2015) CACNA1A haploinsufficiency causes cognitive impairment, autism and epileptic encephalopathy with mild cerebellar symptoms. Eur J Hum Genet 23:1505–1512
Danielsson A et al (2018) Benign paroxysmal torticollis of infancy does not lead to neurological sequelae. Dev Med Child Neurol 60:1251–1255
de Vries B et al (2008) CACNA1A mutation linking hemiplegic migraine and alternating hemiplegia of childhood. Cephalalgia 28:887–891
de Vries B et al (2009) Molecular genetics of migraine. Hum Genet 126:115–132
Debiais S et al (2009) The FHM1 mutation S218L: a severe clinical phenotype? A case report and review of the literature. Cephalalgia 29:1337–1339
(2012) Diuretics. In: LiverTox: clinical and research information on drug-induced liver injury, Bethesda
Du X et al (2013) Second cistron in CACNA1A gene encodes a transcription factor mediating cerebellar development and SCA6. Cell 154:118–133
Du X et al (2017) Dramatic response to pyridoxine in a girl with absence epilepsy with ataxia caused by a de novo CACNA1A mutation. Seizure 45:189–191

Du X et al (2019) alpha1ACT is essential for survival and early cerebellar programming in a critical neonatal window. Neuron 102:770–785.e7
Ducros A et al (2001) The clinical spectrum of familial hemiplegic migraine associated with mutations in a neuronal calcium channel. N Engl J Med 345:17–24
Echenne B et al (1999) Recurrent episodes of coma: an unusual phenotype of familial hemiplegic migraine with linkage to chromosome 1. Neuropediatrics 30:214–217
Eikermann-Haerter K, Can A, Ayata C (2012) Pharmacological targeting of spreading depression in migraine. Expert Rev Neurother 12:297–306
Epi KC (2016) De novo mutations in SLC1A2 and CACNA1A are important causes of epileptic encephalopathies. Am J Hum Genet 99:287–298
Epi KC et al (2013) De novo mutations in epileptic encephalopathies. Nature 501:217–221
Fitzsimons RB, Wolfenden WH (1985) Migraine coma. Meningitic migraine with cerebral oedema associated with a new form of autosomal dominant cerebellar ataxia. Brain 108(Pt 3):555–577
Freilinger T et al (2008) Expansion of the phenotypic spectrum of the CACNA1A T666M mutation: a family with familial hemiplegic migraine type 1, cerebellar atrophy and mental retardation. Cephalalgia 28:403–407
Freilinger T et al (2011) A novel mutation in CACNA1A associated with hemiplegic migraine, cerebellar dysfunction and late-onset cognitive decline. J Neurol Sci 300:160–163
Fuerte-Hortigon A et al (2020) Novel CACNA1A variant may cause cervical dystonia and cerebellar ataxia syndrome. J Neurol Sci 415:116909
Geschwind DH et al (1997) Spinocerebellar ataxia type 6. Frequency of the mutation and genotype-phenotype correlations. Neurology 49:1247–1251
Giunti P et al (2015) Molecular mechanism of spinocerebellar ataxia type 6: glutamine repeat disorder, channelopathy and transcriptional dysregulation. The multifaceted aspects of a single mutation. Front Cell Neurosci 9:36
Gonzalez-Mingot C, Lopez-Ortega R, Brieva-Ruiz L (2022) The efficacy of combining topiramate and 4-aminopyridine to reduce relapses and interictal progression in two cases of episodic ataxia type 2. Neurol Sci:1–3
Graves TD, Hanna MG (2005) Neurological channelopathies. Postgrad Med J 81:20–32
Griggs RC et al (1978) Hereditary paroxysmal ataxia: response to acetazolamide. Neurology 28: 1259–1264
Guerin AA et al (2008) Stepwise developmental regression associated with novel CACNA1A mutation. Pediatr Neurol 39:363–364
Gur-Hartman T, Berkowitz O, Yosovich K et al (2020) Clinical phenotypes of infantile onset CACNA1A-related disorder. Eur J Neurol Paedriatr Neurol 30:144–154
Haan J et al (1995) Is familial hemiplegic migraine a hereditary form of basilar migraine? Cephalalgia 15:477–481
Headache Classification Committee of the International Headache, S (2013) The international classification of headache disorders, 3rd edition (beta version). Cephalalgia 33:629–808
Hu Y et al (2013) Identification of a novel nonsense mutation p.Tyr1957Ter of CACNA1A in a Chinese family with episodic ataxia 2. PLoS One 8:e56362
Humbertclaude V et al (2020) Cognitive impairment in children with CACNA1A mutations. Dev Med Child Neurol 62:330–337
Imbrici P et al (2004) Dysfunction of the brain calcium channel CaV2.1 in absence epilepsy and episodic ataxia. Brain 127:2682–2692
Imbrici P et al (2005) Late-onset episodic ataxia type 2 due to an in-frame insertion in CACNA1A. Neurology 65:944–946
Indelicato E, Boesch S (2021) From genotype to phenotype: expanding the clinical spectrum of CACNA1A variants in the era of next generation sequencing. Front Neurol 12:639994
Indelicato E et al (2017) Ten years of follow-up in a large family with familial hemiplegic migraine type 1: clinical course and implications for treatment. Cephalalgia 38(6):1167–1176
Indelicato E et al (2019) The neuropsychiatric phenotype in CACNA1A mutations: a retrospective single center study and review of the literature. Eur J Neurol 26:66–e7

Indelicato E et al (2021a) Instrumented gait analysis defines the walking signature of CACNA1A disorders. J Neurol 269(6):2941–2947

Indelicato E, Unterberger I, Nachbauer W, Eigentler A, Amprosi M, Zeiner F, Haberlandt E, Kaml M, Boesch S (2021b) The electrophysiological footprint of *CACNA1A* disorders. J Neurol 268(7):2493–2505

Izquierdo-Serra M, Fernandez-Fernandez JM, Serrano M (2020) Rare CACNA1A mutations leading to congenital ataxia. Pflugers Arch 472:791–809

Jacobi H et al (2015) Long-term disease progression in spinocerebellar ataxia types 1, 2, 3, and 6: a longitudinal cohort study. Lancet Neurol 14:1101–1108

Jen JC et al (1998) Spinocerebellar ataxia type 6 with positional vertigo and acetazolamide responsive episodic ataxia. J Neurol Neurosurg Psychiatry 65:565–568

Jen J, Kim GW, Baloh RW (2004) Clinical spectrum of episodic ataxia type 2. Neurology 62:17–22

Jen JC et al (2007) Primary episodic ataxias: diagnosis, pathogenesis and treatment. Brain 130: 2484–2493

Jouvenceau A et al (2001) Human epilepsy associated with dysfunction of the brain P/Q-type calcium channel. Lancet 358:801–807

Jung J et al (2010) Phenotypic variability of episodic ataxia type 2 mutations: a family study. Eur Neurol 64:114–116

Karner E et al (2010) Cognitive functions, emotional behavior, and quality of life in familial hemiplegic migraine. Cogn Behav Neurol 23:106–111

Karner E et al (2012) Long-term outcome of cognitive functions, emotional behavior, and quality of life in a family with familial hemiplegic migraine. Cogn Behav Neurol 25:85–92

Karsan N et al (2018) Flunarizine in migraine-related headache prevention: results from 200 patients treated in the UK. Eur J Neurol 25:811–817

Kaube H et al (2000) Aura in some patients with familial hemiplegic migraine can be stopped by intranasal ketamine. Neurology 55:139–141

Kinder S et al (2015) Novel frameshift mutation in the CACNA1A gene causing a mixed phenotype of episodic ataxia and familiar hemiplegic migraine. Eur J Paediatr Neurol 19:72–74

Kipfer S, Strupp M (2014) The clinical spectrum of autosomal-dominant episodic ataxias. Mov Disord Clin Pract 1:285–290

Knierim E et al (2011) Recurrent stroke due to a novel voltage sensor mutation in Cav2.1 responds to verapamil. Stroke 42:e14–e17

Ko PY et al (2021) Two missense CACNA1A variants in a single family with variable neurobehavioral, cerebellar, epileptic, and oculomotor features. Neuropediatrics 52:186–191

Kors EE et al (2001) Delayed cerebral edema and fatal coma after minor head trauma: role of the CACNA1A calcium channel subunit gene and relationship with familial hemiplegic migraine. Ann Neurol 49:753–760

Kors EE et al (2004) Childhood epilepsy, familial hemiplegic migraine, cerebellar ataxia, and a new CACNA1A mutation. Neurology 63:1136–1137

Labrum RW et al (2009) Large scale calcium channel gene rearrangements in episodic ataxia and hemiplegic migraine: implications for diagnostic testing. J Med Genet 46:786–791

Le Roux M et al (2021) CACNA1A-associated epilepsy: electroclinical findings and treatment response on seizures in 18 patients. Eur J Paediatr Neurol 33:75–85

Liguori C et al (2013) May a suspicious psychiatric disorder hide sporadic hemiplegic migraine? Genetic test as prompting factor for diagnosis. Neurol Sci 34:1845–1846

Lim LL et al (2001) Acetazolamide in women with catamenial epilepsy. Epilepsia 42:746–749

Lipscombe D, Allen SE, Toro CP (2013) Control of neuronal voltage-gated calcium ion channels from RNA to protein. Trends Neurosci 36:598–609

Llinas RR et al (2007) Gamma-band deficiency and abnormal thalamocortical activity in P/Q-type channel mutant mice. Proc Natl Acad Sci U S A 104:17819–17824

Loonen ICM et al (2019) Brainstem spreading depolarization and cortical dynamics during fatal seizures in Cacna1a S218L mice. Brain 142:412–425

Luo X et al (2017) Clinically severe CACNA1A alleles affect synaptic function and neurodegeneration differentially. PLoS Genet 13:e1006905
Malpas TJ et al (2010) Sporadic hemiplegic migraine and delayed cerebral oedema after minor head trauma: a novel de novo CACNA1A gene mutation. Dev Med Child Neurol 52:103–104
Mantuano E et al (2010) Identification of novel and recurrent CACNA1A gene mutations in fifteen patients with episodic ataxia type 2. J Neurol Sci 291:30–36
Marti S et al (2008) Progressive cerebellar ataxia with variable episodic symptoms–phenotypic diversity of R1668W CACNA1A mutation. Eur Neurol 60:16–20
Mathew PG et al (2016) A retrospective analysis of triptan and dhe use for basilar and hemiplegic migraine. Headache 56(5):841–848
Mechtcheriakov S et al (2003) Schizophrenia and episodic ataxia type 2. J Neurol Neurosurg Psychiatry 74:688–689
Mitta N et al (2020) Genotype-phenotype correlates of infantile-onset developmental & epileptic encephalopathy syndromes in South India: a single centre experience. Epilepsy Res 166:106398
Muth C et al (2021) Fampridine and acetazolamide in EA2 and related familial EA: a prospective randomized placebo-controlled trial. Neurol Clin Pract 11:e438–e446
Nachbauer W et al (2014) Episodic ataxia type 2: phenotype characteristics of a novel CACNA1A mutation and review of the literature. J Neurol 261:983–991
Nardello R et al (2020) Two distinct phenotypes, hemiplegic migraine and episodic ataxia type 2, caused by a novel common CACNA1A variant. BMC Neurol 20:155
Neufeld MY et al (1996) The electroencephalogram in acetazolamide-responsive periodic ataxia. Mov Disord 11:283–288
Nikonishyna YV et al (2022) Novel CACNA1A variant p.Cys256Phe disrupts disulfide bonds and causes spinocerebellar ataxia. Mov Disord 37:401–404
Niu X et al (2022) Genotype-phenotype correlation of CACNA1A variants in children with epilepsy. Dev Med Child Neurol 64:105–111
Ohba C et al (2013) Diagnostic utility of whole exome sequencing in patients showing cerebellar and/or vermis atrophy in childhood. Neurogenetics 14:225–232
Ophoff RA et al (1996) Familial hemiplegic migraine and episodic ataxia type-2 are caused by mutations in the Ca2+ channel gene CACNL1A4. Cell 87:543–552
Pastor PDH et al (2018) Targeting the CACNA1A IRES as a treatment for spinocerebellar ataxia type 6. Cerebellum 17:72–77
Peer Mohamed B, Goadsby PJ, Prabhakar P (2012) Safety and efficacy of flunarizine in childhood migraine: 11 years' experience, with emphasis on its effect in hemiplegic migraine. Dev Med Child Neurol 54:274–277
Pelzer N et al (2013) Familial and sporadic hemiplegic migraine: diagnosis and treatment. Curr Treat Options Neurol 15:13–27
Pelzer N et al (2018a) Clinical spectrum of hemiplegic migraine and chances of finding a pathogenic mutation. Neurology 90:e575–e582
Pelzer N et al (2018b) Brain atrophy following hemiplegic migraine attacks. Cephalalgia 38:1199–1202
Petrovski S et al (2013) Genic intolerance to functional variation and the interpretation of personal genomes. PLoS Genet 9:e1003709
Quade A et al (2020) Paroxysmal tonic upgaze: a heterogeneous clinical condition responsive to carbonic anhydrase inhibition. Eur J Paediatr Neurol 25:181–186
Rajakulendran S et al (2010) Genetic and functional characterisation of the P/Q calcium channel in episodic ataxia with epilepsy. J Physiol 588:1905–1913
Rajakulendran S, Kaski D, Hanna MG (2012) Neuronal P/Q-type calcium channel dysfunction in inherited disorders of the CNS. Nat Rev Neurol 8:86–96
Reinson K et al (2016) Biallelic CACNA1A mutations cause early onset epileptic encephalopathy with progressive cerebral, cerebellar, and optic nerve atrophy. Am J Med Genet A 170:2173–2176

Riant F et al (2010) Identification of CACNA1A large deletions in four patients with episodic ataxia. Neurogenetics 11:101–106
Riant F et al (2022) Hemiplegic migraine associated with PRRT2 variations: a clinical and genetic study. Neurology 98:e51–e61
Romaniello R et al (2010) A wide spectrum of clinical, neurophysiological and neuroradiological abnormalities in a family with a novel CACNA1A mutation. J Neurol Neurosurg Psychiatry 81: 840–843
Rosman NP et al (2009) The neurology of benign paroxysmal torticollis of infancy: report of 10 new cases and review of the literature. J Child Neurol 24:155–160
Roubertie A et al (2008) Benign paroxysmal tonic upgaze, benign paroxysmal torticollis, episodic ataxia and CACNA1A mutation in a family. J Neurol 255:1600–1602
Russell MB, Ducros A (2011) Sporadic and familial hemiplegic migraine: pathophysiological mechanisms, clinical characteristics, diagnosis, and management. Lancet Neurol 10:457–470
Sanchez-Albisua I et al (2013) Possible effect of corticoids on hemiplegic attacks in severe hemiplegic migraine. Pediatr Neurol 49:286–288
Sappey-Marinier D et al (1999) Phosphorus and proton magnetic resonance spectroscopy in episodic ataxia type 2. Ann Neurol 46:256–259
Schols L et al (1998) Spinocerebellar ataxia type 6: genotype and phenotype in German kindreds. J Neurol Neurosurg Psychiatry 64:67–73
Spacey SD et al (2005) Two novel CACNA1A gene mutations associated with episodic ataxia type 2 and interictal dystonia. Arch Neurol 62:314–316
Spranger M et al (1999) Familial hemiplegic migraine with cerebellar ataxia and paroxysmal psychosis. Eur Neurol 41:150–152
Stam AH et al (2008) CACNA1A R1347Q: a frequent recurrent mutation in hemiplegic migraine. Clin Genet 74:481–485
Stam AH et al (2009) Early seizures and cerebral oedema after trivial head trauma associated with the CACNA1A S218L mutation. J Neurol Neurosurg Psychiatry 80:1125–1129
Stam AH et al (2011) A long-term follow-up study of 18 patients with sporadic hemiplegic migraine. Cephalalgia 31:199–205
Stendel C et al (2020) Association of a novel splice site mutation in P/Q-type calcium channels with childhood epilepsy and late-onset slowly progressive non-episodic cerebellar ataxia. Int J Mol Sci 21:3810
Stevanin G et al (1997) Clinical and molecular features of spinocerebellar ataxia type 6. Neurology 49:1243–1246
Striessnig J (2020) Voltage-gated Ca^{2+}-channel de novo missense mutations: gain or loss of function – implications for potential therapies. Front Synaptic Neurosci 13:634760
Strupp M et al (2003) Treatment of downbeat nystagmus with 3,4-diaminopyridine: a placebo-controlled study. Neurology 61:165–170
Strupp M et al (2004) Treatment of episodic ataxia type 2 with the potassium channel blocker 4-aminopyridine. Neurology 62:1623–1625
Strupp M et al (2005) Dysfunction of the brain calcium channel CaV2.1 in absence epilepsy and episodic ataxia–a comment. Brain 128:E32. Author reply E33
Strupp M, Zwergal A, Brandt T (2007) Episodic ataxia type 2. Neurotherapeutics 4:267–273
Strupp M et al (2011) A randomized trial of 4-aminopyridine in EA2 and related familial episodic ataxias. Neurology 77:269–275
Subramony SH et al (2003) Novel CACNA1A mutation causes febrile episodic ataxia with interictal cerebellar deficits. Ann Neurol 54:725–731
Tantsis EM et al (2016) Eye movement disorders are an early manifestation of CACNA1A mutations in children. Dev Med Child Neurol 58:639–644
Tashiro Y et al (2014) Repeated encephalopathy and hemicerebral atrophy in a patient with familial hemiplegic migraine type 1. Intern Med 53:2245–2250
Topakian R et al (2014) Rare clinical findings in a patient with sporadic hemiplegic migraine: FDG-PET provides diminished brain metabolism at 10-year follow-up. Cephalalgia 34:392–396

Vahedi K et al (2000) CACNA1A gene de novo mutation causing hemiplegic migraine, coma, and cerebellar atrophy. Neurology 55:1040–1042

Vahedi K et al (2002) Efficacy and tolerability of acetazolamide in migraine prophylaxis: a randomised placebo-controlled trial. J Neurol 249:206–211

Van Berkel MA, Elefritz JL (2018) Evaluating off-label uses of acetazolamide. Am J Health Syst Pharm 75:524–531

van den Maagdenberg AM et al (2010) High cortical spreading depression susceptibility and migraine-associated symptoms in Ca(v)2.1 S218L mice. Ann Neurol 67:85–98

Verriello L et al (2021) Epilepsy and episodic ataxia type 2: family study and review of the literature. J Neurol 268:4296–4302

Wada T et al (2002) Wide clinical variability in a family with a CACNA1A T666m mutation: hemiplegic migraine, coma, and progressive ataxia. Pediatr Neurol 26:47–50

Weyhrauch DL et al (2016) Whole exome sequencing and heterologous cellular electrophysiology studies elucidate a novel loss-of-function mutation in the CACNA1A-encoded neuronal P/Q-type calcium channel in a child with congenital hypotonia and developmental delay. Pediatr Neurol 55:46–51

Yabe I et al (2008) Downbeat positioning nystagmus is a common clinical feature despite variable phenotypes in an FHM1 family. J Neurol 255:1541–1544

Yamazaki S et al (2011) Hemiconvulsion-hemiplegia-epilepsy syndrome associated with CACNA1A S218L mutation. Pediatr Neurol 45:193–196

Ye Q et al (2011) Flunarizine blocks voltage-gated Na(+) and Ca(2+) currents in cultured rat cortical neurons: a possible locus of action in the prevention of migraine. Neurosci Lett 487: 394–399

Yu W, Horowitz SH (2001) Familial hemiplegic migraine and its abortive therapy with intravenous verapamil. Neurology 57:1732–1733

Yu W, Horowitz SH (2003) Treatment of sporadic hemiplegic migraine with calcium-channel blocker verapamil. Neurology 60:120–121

Zamponi GW et al (2015) The physiology, pathology, and pharmacology of voltage-gated calcium channels and their future therapeutic potential. Pharmacol Rev 67:821–870

Zhang L et al (2020) CACNA1A gene variants in eight Chinese patients with a wide range of phenotypes. Front Pediatr 8:577544

Zhang LP, Jia Y, Wang YP (2021) Identification of two de novo variants of CACNA1A in pediatric Chinese patients with paroxysmal tonic upgaze. Front Pediatr 9:722105

Zhuchenko O et al (1997) Autosomal dominant cerebellar ataxia (SCA6) associated with small polyglutamine expansions in the alpha 1A-voltage-dependent calcium channel. Nat Genet 15: 62–69

$Ca_V3.2$ (*CACNA1H*) in Primary Aldosteronism

Hoang An Dinh, Gabriel Stölting, and Ute I. Scholl

Contents

Abstract

Aldosterone is a steroid hormone produced in the zona glomerulosa (ZG) of the adrenal cortex. The most prominent function of aldosterone is the control of electrolyte homeostasis and blood pressure via the kidneys. The primary factors regulating aldosterone synthesis are the serum concentrations of angiotensin II and potassium. The T-type voltage-gated calcium channel $Ca_V3.2$ (encoded by *CACNA1H*) is an important component of electrical as well as intracellular

Hoang An Dinh and Gabriel Stölting contributed equally to this work.

H. A. Dinh · G. Stölting
Berlin Institute of Health at Charité – Universitätsmedizin Berlin, Center of Functional Genomics, Berlin, Germany

U. I. Scholl (✉)
Berlin Institute of Health at Charité – Universitätsmedizin Berlin, Center of Functional Genomics, Berlin, Germany

Department of Nephrology and Medical Intensive Care, Charité – Universitätsmedizin Berlin, Corporate Member of Freie Universität Berlin and Humboldt-Universität zu Berlin, Berlin, Germany
e-mail: ute.scholl@bih-charite.de

J. Striessnig (ed.), *Voltage-gated Ca^{2+} Channels: Pharmacology, Modulation and their Role in Human Disease*, Handbook of Experimental Pharmacology 279,
https://doi.org/10.1007/164_2023_660

calcium oscillations, which govern aldosterone production in the ZG. Excessive aldosterone production that is (partially) uncoupled from physiological stimuli leads to primary aldosteronism, the most common cause of secondary hypertension. Germline gain-of-function mutations in *CACNA1H* were identified in familial hyperaldosteronism, whereas somatic mutations are a rare cause of aldosterone-producing adenomas. In this review, we summarize these findings, put them in perspective, and highlight missing knowledge.

Keywords

Aldosterone · *CACNA1H* · Calcium channel · $Ca_V3.2$ · Primary aldosteronism

1 Introduction

The mineralocorticoid hormone aldosterone is synthesized from its precursor cholesterol in the zona glomerulosa, the outermost layer of the adrenal cortex (Fig. 1a). Aldosterone regulates blood pressure and electrolyte homeostasis by binding to the mineralocorticoid receptor, leading to increased renal NaCl reabsorption alongside with water and increased potassium secretion (Simpson et al. 1953).

The main stimuli of aldosterone synthesis are elevated serum potassium (hyperkalemia) and volume depletion (Fig. 1b). Hyperkalemia directly depolarizes ZG cells, activating voltage-gated calcium channels (VGCC). Volume depletion activates the renin-angiotensin system, leading to the synthesis of angiotensin II (Ang II). In the ZG, Ang II binds to its cognate G protein-coupled receptor, depolarizing ZG cells via inhibition of potassium channels. Additionally, the formation of IP_3 may release calcium from intracellular stores.

Among other pathways, this influx of calcium increases the expression of the aldosterone synthase (*CYP11B2*), which catalyzes the last step of aldosterone synthesis (Fig. 1b) (Yagci and Müller 1996; Clyne et al. 1996).

Primary aldosteronism (PA) occurs when aldosterone is produced despite low renin (and thus Ang II) and low or normal potassium levels. PA is likely the most common cause of secondary hypertension (Funder 2017) but identification of cases is hampered by low screening rates (Jaffe et al. 2020).

Common causes of PA are diffuse hyperplasia of the adrenal gland, aldosterone-producing (micro)nodules or aldosterone-producing adenomas (APAs). APAs are benign tumors at least 10 mm in diameter, whereas aldosterone-producing (micro)-nodules are less than 10 mm in size. While nodules are recognizable in traditional hematoxylin-eosin staining, micronodules are only visible after immunohistochemistry for aldosterone synthase (Williams et al. 2021). Over the past years, somatic mutations in ion channels, ion pumps, and other genes have been identified as causes of >95% of APAs. Heterozygous gain-of function mutations in VGCC genes *CACNA1D* (encodes L-type calcium channel α-subunit $Ca_V1.3$) (Azizan et al.

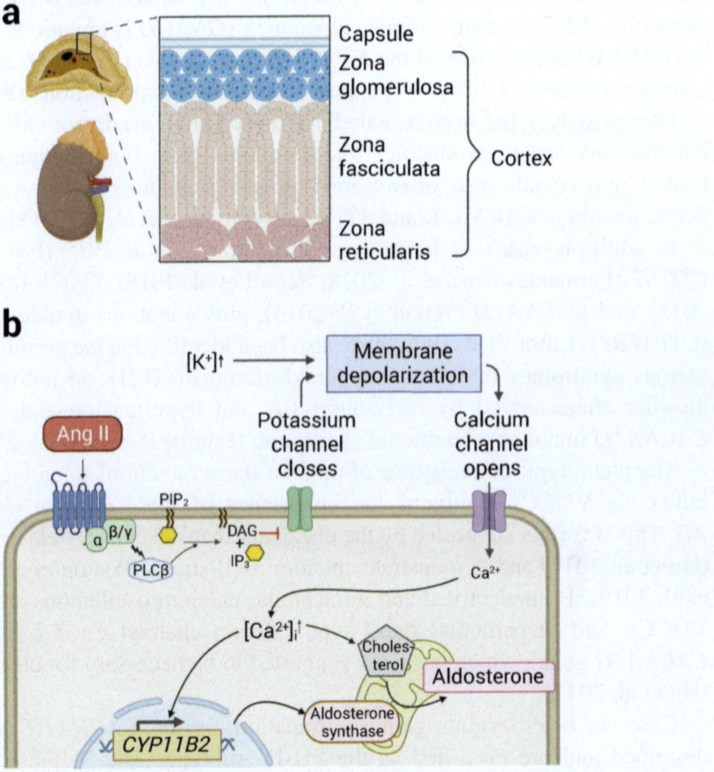

Fig. 1 (**a**) The adrenal glands are located apical to the kidneys. The human adrenal cortex consists of three cell layers, the zona glomerulosa (producing aldosterone), zona fasciculata (producing cortisol) and the zona reticularis (producing androgens). (**b**) The two main stimuli of aldosterone production, angiotensin II (Ang II) and potassium (K^+) facilitate calcium influx via depolarization of ZG cells. Ang II binds to its cognate GPCR, leading to the activation of phospholipase C (PLCß), cleaving phosphatidylinositol 4,5-bisphosphate (PIP_2) into diacylglycerol (DAG) and inositol-1,4,5-trisphosphate (IP_3). DAG can inhibit background potassium channels, facilitating depolarization. The calcium influx through voltage-gated calcium channels is required for several steps of aldosterone synthesis including transcription of *CYP11B2* (encoding aldosterone synthase), and as a co-factor for the transport of cholesterol for the synthesis of aldosterone

2013; Scholl et al. 2013) and *CACNA1H* (encodes T-type calcium channel α-subunit $Ca_V3.2$) (Nanba et al. 2020) have been demonstrated to directly lead to increased calcium influx. On the other hand, mutations in other ion channels such as *KCNJ5* (encodes the inward rectifier potassium channel Kir3.4) (Choi et al. 2011) and *CLCN2* (encodes the chloride channel ClC-2) (Dutta et al. 2019) as well as in ion pump genes *ATP1A1* (encodes α_1- subunit of the Na^+/K^+-ATPase) (Beuschlein et al. 2013; Azizan et al. 2013), *ATP2B3* (encodes plasma membrane Ca^{2+} ATPase) (Beuschlein et al. 2013) and zinc efflux transporter ZnT1 (*SLC30A1*) (Rege et al. 2022) are assumed to indirectly increase intracellular calcium levels by depolarizing

ZG cells and activating VGCCs. Mutations in β-catenin (*CTNNB1*) activate Wnt signaling, and G-proteins subunits alpha q/11 (*GNAQ/11*) mutations activate signaling pathways downstream of the Ang II receptor (Zhou et al. 2021). Mutations in cell adhesion molecule 1 (*CADM1*) appear to modulate gap junctions (Wu et al. 2020).

Interestingly, it has been revealed that these mutations do not only occur in APAs but that aldosterone-producing (micro)nodules from both, patients with PA and healthy individuals also often carry somatic mutations in known APA disease genes, mostly in *CACNA1D* and *ATP1A1* (Nishimoto et al. 2015; Omata et al. 2018).

In addition, cases of heterozygous mutations in *KCNJ5* (Choi et al. 2011), *CLCN2* (Fernandes-Rosa et al. 2018; Scholl et al. 2018), *CACNA1H* (Scholl et al. 2015), and *CACNA1D* (Scholl et al. 2013), plus mutations in aldosterone synthase (*CYP11B2*) (Lifton et al. 1992) have also been identified in the germline and result in various syndromes of familial hyperaldosteronism (FH), an autosomal-dominant disorder characterized by early-onset PA and hypertension and, in the case of *CACNA1D* mutations, additional syndromic features (Scholl et al. 2013).

The phenotypic convergence of most of these mutations to an increased calcium influx via VGCCs highlights the importance of these calcium channels for the ZG. This is further supported by the discovery that ZG cells are electrically excitable (Hu et al. 2012) and demonstrate calcium oscillations (Penton et al. 2012; Schewe et al. 2019). Both electrical and intracellular calcium oscillations are dependent on VGCCs, and in particular the T-type calcium channel $Ca_V3.2$ (encoded by the *CACNA1H* gene), which has been suggested to be necessary for electrical signaling (Hu et al. 2012).

Cases of heterozygous germline mutations in the *CACNA1H* gene have been described and are classified as the FH-IV subtype. The M1549V mutation was identified in five unrelated families presenting with early-onset PA (Scholl et al. 2015). Furthermore, an individual with a mutation at the same position, M1549I, also had early-onset PA and an additional multiplex developmental disorder (Daniil et al. 2016). We demonstrated that the M1549V mutation in *CACNA1H* leads to gain of channel function due to reduced inactivation and a shift of activation to more negative potentials resulting in increased calcium influx and aldosterone production (Scholl et al. 2015). Further mutations with slightly different or yet unknown biophysical phenotype have been discovered since (Table 1; Fig. 2a), including a somatic mutation in aldosterone-producing adenomas (I1430T) (Nanba et al. 2020).

Within this chapter, we would like to take a closer look at the role of $Ca_V3.2$ in the healthy ZG, the current knowledge about mutations in PA as well as missing parts in our understanding of ZG physiology and pharmacological interventions regarding T-type calcium channels.

2 The $Ca_V3.2$ Calcium Channel

The $Ca_V3.2$ is a low-voltage activated calcium channel from the T-type (Ca_V3) family encoded by the *CACNA1H* gene on chromosome 16 (Cribbs et al. 1998). The reference $Ca_V3.2$ protein comprises 2,353 amino acids and has a calculated mass of

Table 1 Published *CACNA1H* variants associated with primary aldosteronism

Mutation	Germline/ somatic	Position	# of cases	Age at dx	Abnormalities in clinical imaging	Electrophysiological phenotype
M1549V Scholl et al. (2015)	Germline	S6, DIII	8 (6m, 2f)	2m – 9y	No	Delayed inactivation, left-shifted activation
M1549I Daniil et al. (2016)	Germline	S6, DIII	1 (1m)	2m	No	Delayed inactivation, left-shifted activation
S196L Daniil et al. (2016)	Germline	S4, DI	2 (1m, 1f)	>30y	Nodule visible in CT	Delayed inactivation
V1951E Daniil et al. (2016)	Germline	C-terminus	1 (1f)	>50y	Left adrenal nodule	Faster recovery from inactivation
P2083L Daniil et al. (2016)	Germline	C-terminus	2 (2m)	>30y	Slight hyperplasia	Faster recovery from inactivation
R890H Wulczyn et al. (2019)	Germline	S4, DII	1 (2f)	>30y	No	Loss-of-function; smaller whole cell currents in HEK293
I1430T Nanba et al. (2020)	Somatic	S5, DIII	3 (1f, 2m)	>40y	2 APA, 1 not visible	Unknown, but increased aldosterone synthesis in HAC15 cells
P277S Wu et al. (2021)	Somatic	Extracellular between S5 and S6, DI	1	?	Unilateral hyperplasia or APA	Unknown
W482X Wu et al. (2021)	Somatic	Cytoplasmic between S6, DI and S1, DII	1	?	Unilateral hyperplasia or APA	Unknown
T615P Wu et al. (2021)	Somatic	Cytoplasmic between S6, DI and S1, DII	1	?	Unilateral hyperplasia or APA	Unknown
Q875R Wu et al. (2021)	Somatic	Extracellular, after S3, DII	1	?	Unilateral hyperplasia or APA	Unknown
Q904X Wu et al. (2021)	Somatic	S4, DII	1	?	Unilateral hyperplasia or APA	Unknown
V1937M Tseng et al. (2022)	Somatic	C-terminus	1	43	0.6 cm left nodule	Unknown, but increased aldosterone synthesis in HAC15 cells

Dx, diagnosis

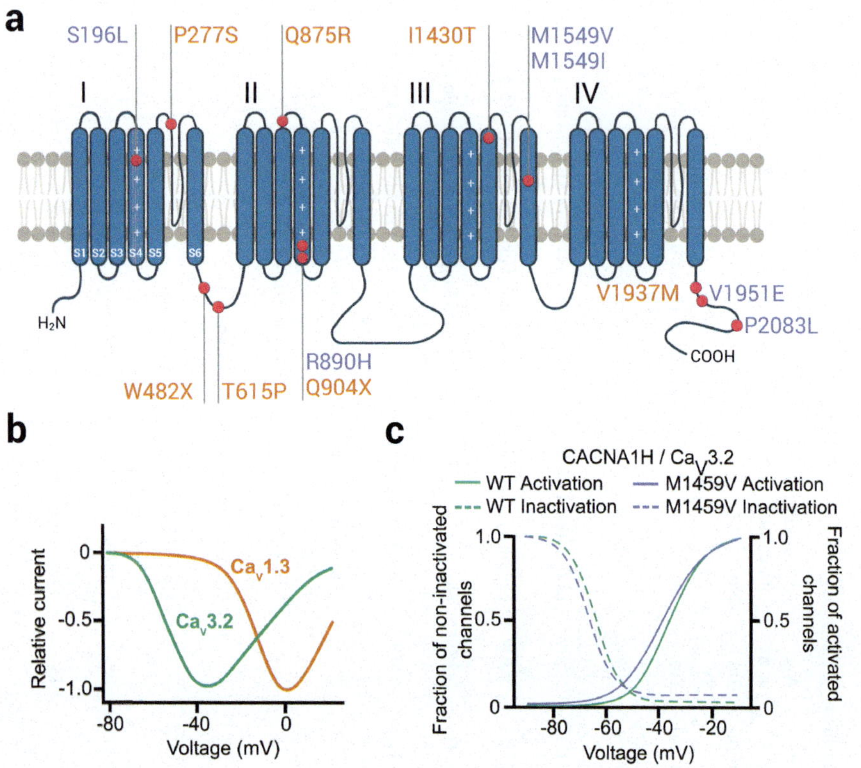

Fig. 2 (**a**) Transmembrane topology model of $Ca_V3.2$. Domains I–IV and segments S1–S6 are indicated. All currently published mutations are plotted at their approximate positions with red circles. Purple mutations were found in the germline, red mutations as somatic mutations within the adrenal gland. (**b**) Exemplary current-voltage relationships of $Ca_V3.2$ (green) and $Ca_V1.3$ (orange) demonstrating the difference in voltage dependence. Plots are based on data from Scholl et al. (2013, 2015). **c** The M1459V mutation in $Ca_V3.2$ results in a small shift of the activation and inactivation curves toward more negative potentials. This plot is based on Scholl et al. (2015)

259 kDa (UniProtKB: O95180-1). Other members of the Ca_V3 family are $Ca_V3.1$ (*CACNA1G*) and $Ca_V3.3$ (*CACNA1I*). Each Ca_V3 isoform can undergo alternative splicing resulting in multiple variants, which exhibit varied electrophysiological characteristics and drug sensitivity (Zhong et al. 2006). For $Ca_V3.2$, the major alternative splice variant replaces 7 amino acids from the reference sequence in exon 26 with a single lysine (UniProtKB: O95180-2).

All three Ca_V3 genes encode a single polypeptide chain organized in four largely homologous domains making up a pore-forming α_1-subunit (Fig. 2a). In contrast to the α_1-subunits from the Ca_V1 and Ca_V2 families, which require binding of an auxiliary β-subunit and are physiologically associated with an $\alpha_2\delta$-subunit to form a functional channel protein (Neely and Hidalgo 2014), Ca_V3 channels lack β-subunit binding sites (Waard et al. 1994; Lambert et al. 1997) and appear as functional α_1-subunit monomers.

Ca_V3 VGCCs differ from Ca_V1 and Ca_V2 channels in their threshold of activation. The Ca_V3 channels are activated by smaller membrane depolarizations (low-voltage activated, LVA; Fig. 2b) (Carbone and Lux 1984; Perez-Reyes et al. 1998). Activation typically starts already at voltages as low as −70 mV. Ca_V1 and Ca_V2 members require larger depolarizations (high-voltage activated, HVA; Fig. 2b), typically above −40 mV. Furthermore, Ca_V3 channels are characterized by their fast inactivation resulting in fast decaying, transient Ca^{2+} currents and were therefore termed T-type channels. In contrast, $Ca_V1.3$ channels were categorized as L-type channels due to their slow inactivation kinetics that lead to long-lasting inward currents (Nilius et al. 1985).

The inactivation of Ca_V3 channels, including Cav3.2, is strongly voltage dependent as well. Steady-state inactivation curves typically range from no inactivation at −80 mV to full inactivation at −40 mV (Fig. 2c). The opposing voltage dependences of activation and inactivation result in a range of voltages where channels can be activated while not becoming completely inactivated (approximately between −65 and −40 mV for $Ca_V3.2$). Within this range, Ca_V3 channels do not mediate transient but rather steady-state calcium influx, the so-called "window current" (Fig. 2c) (Perez-Reyes et al. 1998; Cribbs et al. 1998).

Each of the four domains of the $Ca_V3.2$ α_1-subunit comprises six transmembrane helices (S1–S6) and a membrane loop (P-loop), which links together the S5 and S6 segment. The S1–S4 segments form the voltage sensing domain with the S4 comprising positively charged residues crucial for voltage sensitivity. This sequence is highly conserved among voltage-gated cation channels. The P-loop and the S5 and S6 helices from each domain constitute the pore of the channel (Fig. 2b) (Cribbs et al. 1998).

The four domains are connected via three intracellular loops, each linking the S6 and S1 segments of the respective upstream and downstream domain. These loops as well as the amino- and carboxy-terminus harbor several interaction sites for Ca_V3 channel regulation, mediated by various second messengers and proteins (Wolfe et al. 2003; Blesneac et al. 2015).

For example, phosphorylation of $Ca_V3.2$, carried out by kinases CaMKII, PKC or PKA within the II-III linker region, enhances its activity (Blesneac et al. 2015). G-protein β_2-subunits also interact with the II-III domain linker, instead reducing the channel open probability (Wolfe et al. 2003). In addition, other posttranslational modifications, including glycosylation (Weiss et al. 2013) and ubiquitination (García-Caballero et al. 2018), were found to regulate $Ca_V3.2$ channel activity and expression at the surface.

3 $Ca_V3.2$ Calcium Channels in the Healthy Zona Glomerulosa

In ZG cells, Cav3.2 channels are the most expressed subtype among the Ca_V3 family (Scholl et al. 2013). It appears that exclusively the long, reference isoform is expressed in human ZG cells. Among T-type channels, besides $Ca_V3.2$, only $Ca_V3.1$ was reported to show meaningful expression in the ZG (Scholl et al. 2013; Thuesen et al. 2019).

The threshold of activation of $Ca_V3.2$ (~−70 mV) is very close to the resting membrane potential of ZG cells (~−80 mV). Small depolarizations may thus already activate $Ca_V3.2$ and further depolarize ZG cells. This mechanism has been postulated to be the reason why voltage oscillations of ZG cells subsided when $Ca_V3.2$ was blocked (Hu et al. 2012). Individual depolarizations are very uniform in appearance and action potential-like. Their frequency was reported to depend on the intensity of the extracellularly applied stimuli of aldosterone synthesis such as Ang II or K^+ (Hu et al. 2012).

Several laboratories, including our own, have now demonstrated that oscillations are not just visible in the membrane voltage but also as an oscillatory influx of calcium (Penton et al. 2012; Schewe et al. 2019; Guagliardo et al. 2020). As mentioned in the introduction, calcium is required for the regulation of expression of the aldosterone synthase (*CYP11B2*) but also at various other steps along the pathway towards aldosterone synthesis (Spät and Hunyady 2004). Baseline calcium levels in the ZG are already rather high (several 100 nM), and oscillatory calcium influx can increase mean concentrations even further (Seidel et al. 2021).

Unlike for the electrical signals, calcium oscillations occur in bursts of calcium spikes with pauses in between. Within these bursts, the frequency of spiking does not vary with increasing stimuli. Rather the number and length of bursts are regulated (Schewe et al. 2019; Guagliardo et al. 2020). It is therefore currently unclear whether electrical and calcium signals are linked 1:1. The broadly similar frequency and involvement of VGCCs in both make this likely, however. The observed discrepancies may be due to the intracellular calcium-buffering substances typically included in patch-clamp experiments that may impair the physiological function of calcium-dependent potassium channels, which may be required for the repolarization of ZG cells.

Inhibition of $Ca_V3.2$ with Ni^{2+} led to a complete loss of electrical oscillations in the original study (Hu et al. 2012). The authors concluded that $Ca_V3.2$ is a crucial component of ZG voltage oscillations, likely being necessary for proper regulation of aldosterone synthesis. However, we and others observed normal serum aldosterone levels in mice lacking $Ca_V3.2$ under a normal salt diet (Thuesen et al. 2019; Seidel et al. 2021). This unexpected result indicates that, at least under basal conditions, $Ca_V3.2$ might be dispensable for aldosterone synthesis. Our own observations also suggest that calcium oscillations are still present in mice lacking $Ca_V3.2$ (unpublished observation). However, further investigations are required to elucidate the effects on the cellular level and whether other Ca_V3 channels may be upregulated in response to the knock-out of $Ca_V3.2$.

4 $Ca_V3.2$ Calcium Channels in Primary Aldosteronism

To date, several cases of primary aldosteronism have been associated with mutations in the *CACNA1H* gene (Fig. 2a, Table 1). Germline variants cause FH-IV. The first identified mutation, M1549V, was found by exome sequencing of 40 unrelated individuals who had hypertension and PA by 10 years of age (Scholl et al. 2015). Five of these subjects carried the M1549V mutations. Of the identified mutations, one occurred de novo in an affected index case, one de novo in the affected mother of another index case. Interestingly, two parents who carried the mutation were normotensive in adulthood, suggesting incomplete penetrance. Where available, adrenal imaging did not reveal adrenal masses or enlargement, except for the mother of one index case, who showed mild left adrenal fullness and underwent unilateral adrenalectomy without cure at age 29 years. Histology demonstrated microscopic glomerulosa hyperplasia (Scholl et al. 2015). A second case with de novo mutation at the same residue (M1549I) was reported to have early-onset PA without abnormalities on adrenal imaging and a developmental disorder (Daniil et al. 2016). In the same publication, two additional germline variants (S196L and P2083L) were found in patients with familial aggregation of primary aldosteronism. The pathogenicity of the V1951E variant found in the germline of a patient with sporadic APA and cured by unilateral adrenalectomy is somewhat more doubtful. After the discovery of *CACNA1H* mutations in FH-IV, a somatic I1430T mutation was discovered in two APAs (Nanba et al. 2020), followed by the description of five (P277S, W482X, T615P, Q875R, Q904X) and one (V1937M) somatic APA mutation(s) in two subsequent publications (Wu et al. 2021; Tseng et al. 2022), respectively.

PA associated mutations do not appear to cluster around specific protein domains but are spread across the entire $Ca_V3.2$ channel (Fig. 2a). Functional information on the impact of the mutations on channel function is only available for some of the mutations (Table 1). With one exception, all the mutations studied by patch-clamp electrophysiology exhibited changes in inactivation kinetics, either delaying inactivation or having a faster recovery from inactivation. Mutations at M1549 appear to also cause a shift of the voltage dependence of activation and inactivation to more negative potentials (Scholl et al. 2015). Interestingly, patients with these mutations were typically diagnosed at a younger age than those with mutations that result in more subtle changes to inactivation (Daniil et al. 2016). Given that the window current of $Ca_V3.2$ is already close to the resting membrane potential, it may be speculated that the left-shift of the activation leads to higher constitutive calcium influx into ZG cells, thereby increasing aldosterone synthesis. Overall, a gain-of-function effect has been suggested as underlying mechanism of *CACNA1H* mutations in PA.

Studies in the aldosterone-producing H295R or HAC15 human adrenocortical carcinoma cell lines demonstrated increased *CYP11B2* expression and aldosterone production in cells transfected with M1549V compared to WT *CACNA1H* (Reimer et al. 2016). For the M1549I, S196L, P2083L, and V1951E variants, aldosterone levels and *CYP11B2* expression were not significantly elevated under basal conditions but only after stimulation with Ang II or K^+ (Daniil et al. 2016). Heterologous expression of the I1430T mutant increased *CYP11B2* expression and aldosterone production over baseline, but a comparison with WT overexpression was not performed (Nanba et al. 2020). V2937M led to increased *CYP11B2* expression and aldosterone production (Tseng et al. 2022). No functional analysis was reported for the P277S, W482X, T615P, Q875R, and Q904X variants. Of note, two of these variants are nonsense variants leading to loss of a significant part of the protein, for which gain of function appears less likely, although it cannot be excluded (Wu et al. 2021). Also, for the R890H mutation (Wulczyn et al. 2019), a loss-of-function has been described in patch clamp experiments, which is currently difficult to reconcile with the pathomechanism of the other mutations and may require further investigation if more such mutations are found.

These results are further corroborated by findings from our recently published mouse model carrying the M1560V (homologous to the human M1549V) mutation (Seidel et al. 2021). This model showed a mild phenotype of primary aldosteronism with normal adrenal histology, increased aldosterone-to-renin ratios (also under high salt diet), and slightly elevated blood pressure. Calcium imaging of acutely prepared adrenal gland slices revealed increased basal calcium levels, supporting the hypothesis of higher constitutive calcium influx caused by the M1549V mutation. Unlike the effect of mutations in mouse models of the $Ca_V1.3$ calcium (unpublished observation) or ClC-2 chloride channel (Schewe et al. 2019), however, M1560V mice exhibited slower calcium oscillations under stimulation while showing higher mean calcium levels overall. While this may be the result of the slower inactivation of mutant $Ca_V3.2$ channels, it may also be caused by an increased activation of calcium-dependent potassium channels due to the higher calcium influx. Their activation would counteract the increased excitability expected from the left-shifted activation curve of mutant $Ca_V3.2$ channels and explain the decrease in calcium oscillation frequency. However, the combined activation of hyperpolarizing potassium and depolarizing calcium conductances may keep the membrane potential within the range of the mutant $Ca_V3.2$ window current, maintaining higher basal calcium levels in ZG cells.

5 Pharmacology of $Ca_V3.2$ Calcium Channels in Primary Aldosteronism

Since the discovery of $Ca_V3.2$ as a critical component of calcium influx in the ZG, it has been suggested that specific inhibition of this channel may be a good strategy to attenuate aldosterone synthesis. This hypothesis was further supported when it was discovered that $Ca_V3.2$ is a critical component of the voltage oscillations observed in murine ZG cells. In Hu et al. (2012), it was demonstrated that the application of 50 μM Ni^{2+}, which is (somewhat) specific for the inhibition of $Ca_V3.2$ among VGCCs, resulted in a complete inhibition of voltage oscillations in patch clamped murine ZG cells. In an inversion of this concept, the authors could elicit voltage oscillations in, otherwise non-oscillating, H295R cells by the transfection of

CACNA1H. As discussed above, the equivalence of voltage and calcium oscillations is currently still unclear, but contrary to this concept, mice deficient for $Ca_V3.2$ have normal blood pressure, aldosterone and renin serum levels (Thuesen et al. 2019; Seidel et al. 2021). Mibefradil, thought to specifically block T-type calcium channels such as $Ca_V3.2$, inhibits aldosterone production in vitro, but does not show lasting in vivo effects (Schmitt et al. 1992; Rossier et al. 1998; Ragueneau et al. 2001). Newer results from our lab demonstrate that ZG cells from *Cacna1h* KO mice still exhibit calcium oscillations with only slightly altered properties compared to WT, suggesting that calcium oscillations in the ZG can be maintained by both T- and L-type VGCCs (unpublished results).

These results indicate that the inhibition of T-type calcium channels may only transiently decrease calcium influx into ZG cells. Long-term inhibition, such as mimicked in *Cacna1h* KO mice, does not sustain the reduced synthesis of aldosterone and may therefore not constitute a useful general strategy in treating aldosteronism. On the other hand, a dual blockade of T- and L-type such as with efonidipine or benidipine may provide an interesting strategy to pursue as both drugs are approved for clinical use in some countries. An attenuation of aldosterone release from H295R cells through these compounds has been demonstrated (Imagawa et al. 2006; Akizuki et al. 2008) even though these cells apparently lack $Ca_V3.2$ (Hu et al. 2012). However, such unspecific inhibition might increase the likelihood of unwanted side effects due to the important role of these channels in various organs.

In the case of primary aldosteronism caused by mutations in *CACNA1H*, it remains to be seen if specific inhibitors may reduce influx through mutated channels sufficiently to reduce aldosterone synthesis back to baseline levels. Specific inhibitory drugs of T-type calcium channels are currently not clinically approved, making this at present a theoretical consideration only.

6 Summary

Overall, the contribution of $Ca_V3.2$ to the physiology of aldosterone synthesis and the role of *CACNA1H* variants in the pathophysiology of the ZG are well established. With the increased interest in the genetics of primary aldosteronism and the advent of personalized medicine, it is to be expected that more mutations in this channel will be found. It would be desirable to use this knowledge for further translation to the bedside with more specific therapeutic strategies as the current first-line therapy with mineralocorticoid receptor antagonists is prone to adverse effects. Such developments could be based on the recently published mouse models of primary aldosteronism and used to initiate further clinical studies.

Acknowledgments This work was supported by grants from the Fritz Thyssen Stiftung (10.16.1.027MN), the Ministerium für Kultur und Wissenschaft der Landes Nordrhein-Westfalen (Rückkehrprogramm and Junges Kolleg) and the Stiftung Charité (BIH Johanna Quandt Professorship) (all to UIS) and the Deutsche Forschungsgemeinschaft (CRC 1453 and CRC 1365 to UIS; STO 1260/1-1 to GS). Some of the figures were created with BioRender.com.

References

Akizuki O, Inayoshi A, Kitayama T et al (2008) Blockade of T-type voltage-dependent Ca2+ channels by benidipine, a dihydropyridine calcium channel blocker, inhibits aldosterone production in human adrenocortical cell line NCI-H295R. Eur J Pharmacol 584:424–434. https://doi.org/10.1016/j.ejphar.2008.02.001

Azizan EAB, Poulsen H, Tuluc P et al (2013) Somatic mutations in ATP1A1 and CACNA1D underlie a common subtype of adrenal hypertension. Nat Genet 45:1055–1060. https://doi.org/10.1038/ng.2716

Beuschlein F, Boulkroun S, Osswald A et al (2013) Somatic mutations in ATP1A1 and ATP2B3 lead to aldosterone-producing adenomas and secondary hypertension. Nat Genet 45:440–444. 444e1-2

Blesneac I, Chemin J, Bidaud I et al (2015) Phosphorylation of the Cav3.2 T-type calcium channel directly regulates its gating properties. Proc Natl Acad Sci U S A 112:13705–13710

Carbone E, Lux HD (1984) A low voltage-activated, fully inactivating ca channel in vertebrate sensory neurones. Nature 310:501–502. https://doi.org/10.1038/310501a0

Choi M, Scholl UI, Yue P et al (2011) K+ channel mutations in adrenal aldosterone-producing adenomas and hereditary hypertension. Science 331:768–772

Clyne CD, White PC, Rainey WE (1996) Calcium regulates human CYP11B2 transcription. Endocr Res 22:485–492

Cribbs LL, Lee J-H, Yang J et al (1998) Cloning and characterization of α1H from human heart, a member of the T-type Ca2+ channel gene family. Circ Res 83:103–109

Daniil G, Fernandes-Rosa FL, Chemin J et al (2016) CACNA1H mutations are associated with different forms of primary aldosteronism. EBioMedicine 13:225–236

Dutta RK, Arnesen T, Heie A et al (2019) A somatic mutation in CLCN2 identified in a sporadic aldosterone-producing adenoma. Eur J Endocrinol 181:K37–K41. https://doi.org/10.1530/EJE-19-0377

Fernandes-Rosa FL, Daniil G, Orozco IJ et al (2018) A gain-of-function mutation in the CLCN2 chloride channel gene causes primary aldosteronism. Nat Genet 1

Funder J (2017) Primary aldosteronism: the next five years. Horm Metab Res 49:977–983. https://doi.org/10.1055/s-0043-119802

Garcia-Caballero A, Zhang F-X, Hodgkinson V, Huang J, Chen L, Souza IA, Cain S, Kass J, Alles S, Snutch TP, Zamponi GW (2018) T-type calcium channels functionally interact with spectrin (α/β) and ankyrin B. Mol Brain 11(1). https://doi.org/10.1186/s13041-018-0368-5

Guagliardo NA, Klein PM, Gancayco CA et al (2020) Angiotensin II induces coordinated calcium bursts in aldosterone-producing adrenal rosettes. Nat Commun 11:1–15

Hu C, Rusin CG, Tan Z et al (2012) Zona glomerulosa cells of the mouse adrenal cortex are intrinsic electrical oscillators. J Clin Investig 122:2046–2053

Imagawa K, Okayama S, Takaoka M et al (2006) Inhibitory effect of Efonidipine on aldosterone synthesis and secretion in human Adrenocarcinoma (H295R) cells. J Cardiovasc Pharmacol 47: 133. https://doi.org/10.1097/01.fjc.0000197539.12685.f5

Jaffe G, Gray Z, Krishnan G et al (2020) Screening rates for primary aldosteronism in resistant hypertension. Hypertension 75:650–659. https://doi.org/10.1161/HYPERTENSIONAHA.119.14359

Lambert RC, Maulet Y, Mouton J et al (1997) T-type Ca2+ current properties are not modified by Ca2+ channel β subunit depletion in Nodosus ganglion neurons. J Neurosci 17:6621–6628. https://doi.org/10.1523/JNEUROSCI.17-17-06621.1997

Lifton RP, Dluhy RG, Powers M et al (1992) A chimaeric llβ-hydroxylase/aldosterone synthase gene causes glucocorticoid-remediable aldosteronism and human hypertension. Nature 355: 262–265. https://doi.org/10.1038/355262a0

Nanba K, Blinder AR, Rege J et al (2020) Somatic CACNA1H mutation as a cause of aldosterone-producing adenoma. Hypertension 75:645–649. https://doi.org/10.1161/HYPERTENSIONAHA.119.14349

Neely A, Hidalgo P (2014) Structure-function of proteins interacting with the α1 pore-forming subunit of high-voltage-activated calcium channels. Front Physiol 5:209

Nilius B, Hess P, Lansman JB, Tsien RW (1985) A novel type of cardiac calcium channel in ventricular cells. Nature 316:443–446. https://doi.org/10.1038/316443a0

Nishimoto K, Tomlins SA, Kuick R et al (2015) Aldosterone-stimulating somatic gene mutations are common in normal adrenal glands. Proc Natl Acad Sci U S A 112:E4591–E4599

Omata K, Satoh F, Morimoto R et al (2018) Cellular and genetic causes of idiopathic hyperaldosteronism. Hypertension 72:874–880. https://doi.org/10.1161/HYPERTENSIONAHA.118.11086

Penton D, Bandulik S, Schweda F et al (2012) Task3 potassium channel gene invalidation causes low renin and salt-sensitive arterial hypertension. Endocrinology 153:4740–4748. https://doi.org/10.1210/en.2012-1527

Perez-Reyes E, Cribbs LL, Daud A et al (1998) Molecular characterization of a neuronal low-voltage-activated T-type calcium channel. Nature 391:896–900. https://doi.org/10.1038/36110

Ragueneau I, Sao AB, Démolis J-L et al (2001) Comparison of sympathetic modulation induced by single oral doses of mibefradil, amlodipine, and nifedipine in healthy volunteers. Clin Pharmacol Ther 69:122–129. https://doi.org/10.1067/mcp.2001.113406

Rege J, Nanba K, Bandulik S et al (2022) Zinc transporter somatic gene mutations cause primary aldosteronism. https://doi.org/10.1101/2022.07.25.501443

Reimer EN, Walenda G, Seidel E, Scholl UI (2016) CACNA1HM1549V mutant calcium channel causes autonomous aldosterone production in HAC15 cells and is inhibited by Mibefradil. Endocrinology 157:3016–3022. https://doi.org/10.1210/en.2016-1170

Rossier MF, Ertel EA, Vallotton MB, Capponi AM (1998) Inhibitory action of mibefradil on calcium signaling and aldosterone synthesis in bovine adrenal glomerulosa cells. J Pharmacol Exp Ther 287:824–831

Schewe J, Seidel E, Forslund S et al (2019) Elevated aldosterone and blood pressure in a mouse model of familial hyperaldosteronism with ClC-2 mutation. Nat Commun 10:5155

Schmitt R, Kleinbloesem CH, Belz GG et al (1992) Hemodynamic and humoral effects of the novel calcium antagonist Ro 40-5967 in patients with hypertension. Clin Pharmacol Ther 52:314–323. https://doi.org/10.1038/clpt.1992.147

Scholl UI, Goh G, Stölting G et al (2013) Somatic and germline CACNA1D calcium channel mutations in aldosterone-producing adenomas and primary aldosteronism. Nat Genet 45:1050–1054

Scholl UI, Stölting G, Nelson-Williams C et al (2015) Recurrent gain of function mutation in calcium channel CACNA1H causes early-onset hypertension with primary aldosteronism. eLife Sci 4:e06315

Scholl UI, Stölting G, Schewe J et al (2018) CLCN2 chloride channel mutations in familial hyperaldosteronism type II. Nat Genet 50:349–354

Seidel E, Schewe J, Zhang J et al (2021) Enhanced Ca2+ signaling, mild primary aldosteronism, and hypertension in a familial hyperaldosteronism mouse model (Cacna1hM1560V/+). Proc Natl Acad Sci U S A 118

Simpson SA, Tait JF, Wettstein A et al (1953) Isolierung eines neuen kristallisierten Hormons aus Nebennieren mit besonders hoher Wirksamkeit auf den Mineralstoffwechsel. Experientia 9: 333–335

Spät A, Hunyady L (2004) Control of aldosterone secretion: a model for convergence in cellular signaling pathways. Physiol Rev 84:489–539. https://doi.org/10.1152/physrev.00030.2003

Thuesen AD, Finsen SH, Rasmussen LL et al (2019) Deficiency of T-type ca $^{2+}$ channels ca $_v$ 3.1 and ca $_v$ 3.2 has no effect on angiotensin II-induced hypertension but differential effect on plasma aldosterone in mice. Am J Physiol Renal Physiol 317:F254–F263

Tseng C-S, Peng K-Y, Wang S-M et al (2022) A novel somatic mutation of CACNA1H p.V1937M in unilateral primary hyperaldosteronism. Front Endocrinol 13

Waard MD, Pragnell M, Campbell KP (1994) Ca2+ channel regulation by a conserved β subunit domain. Neuron 13:495–503. https://doi.org/10.1016/0896-6273(94)90363-8

Weiss N, Black SAG, Bladen C et al (2013) Surface expression and function of Cav3.2 T-type calcium channels are controlled by asparagine-linked glycosylation. Pflugers Arch Eur J Physiol 465:1159–1170. https://doi.org/10.1007/s00424-013-1259-3

Williams TA, Gomez-Sanchez CE, Rainey WE et al (2021) International histopathology consensus for unilateral primary aldosteronism. J Clin Endocrinol Metabol 106:42–54. https://doi.org/10.1210/clinem/dgaa484

Wolfe JT, Wang H, Howard J et al (2003) T-type calcium channel regulation by specific G-protein βγ subunits. Nature 424:209–213

Wu C-H, Peng K-Y, Hwang D-Y et al (2021) Novel mutations detection with next-generation sequencing and its association with clinical outcome in unilateral primary aldosteronism. Biomedicine 9:1167. https://doi.org/10.3390/biomedicines9091167

Wu X, Garg S, Cabrera CP et al (2020) OR34-02 somatic transmembrane domain mutations of a cell adhesion molecule, CADM1, cause primary aldosteronism by preventing gap junction communication between adrenocortical cells. J Endocrine Soc 4:OR34-02. https://doi.org/10.1210/jendso/bvaa046.492

Wulczyn K, Perez-Reyes E, Nussbaum RL, Park M (2019) Primary aldosteronism associated with a germline variant in CACNA1H. BMJ Case Rep CP 12:e229031. https://doi.org/10.1136/bcr-2018-229031

Yagci A, Müller J (1996) Induction of steroidogenic enzymes by potassium in cultured rat zona glomerulosa cells depends on calcium influx and intact protein synthesis. Endocrinology 137:4331–4338. https://doi.org/10.1210/endo.137.10.8828493

Zhong X, Liu JR, Kyle JW et al (2006) A profile of alternative RNA splicing and transcript variation of CACNA1H, a human T-channel gene candidate for idiopathic generalized epilepsies. Hum Mol Genet 15:1497–1512. https://doi.org/10.1093/hmg/ddl068

Zhou J, Azizan EAB, Cabrera CP et al (2021) Somatic mutations of GNA11 and GNAQ in CTNNB1-mutant aldosterone-producing adenomas presenting in puberty, pregnancy or menopause. Nat Genet 53:1360–1372

$Ca_V3.3$ Channelopathies

Yousra El Ghaleb and Bernhard E. Flucher

Contents

Abstract

$Ca_V3.3$ is the third member of the low-voltage-activated calcium channel family and the last to be recognized as disease gene. Previously, CACNA1I, the gene encoding $Ca_V3.3$, had been described as schizophrenia risk gene. More recently,

Y. El Ghaleb · B. E. Flucher (✉)
Institute of Physiology, Medical University Innsbruck, Innsbruck, Austria
e-mail: bernhard.e.flucher@i-med.ac.at

J. Striessnig (ed.), *Voltage-gated Ca^{2+} Channels: Pharmacology, Modulation and their Role in Human Disease*, Handbook of Experimental Pharmacology 279,
https://doi.org/10.1007/164_2022_631

de novo missense mutations in CACNA1I were identified in patients with variable degrees of neurodevelopmental disease with and without epilepsy. Their functional characterization indicated gain-of-function effects resulting in increased calcium load and hyperexcitability of neurons expressing $Ca_V3.3$. The amino acids mutated in the $Ca_V3.3$ disease variants are located in the vicinity of the channel's activation gate and thus are classified as gate-modifying channelopathy mutations. A persistent calcium leak during rest and prolonged calcium spikes due to increased voltage sensitivity of activation and slowed kinetics of channel inactivation, respectively, may be causal for the neurodevelopmental defects. The prominent expression of $Ca_V3.3$ in thalamic reticular nucleus neurons and its essential role in generating the rhythmic thalamocortical network activity are consistent with a role of the mutated channels in the etiology of epileptic seizures and thus suggest T-type channel blockers as a viable treatment option.

Keywords

CACNA1I · Epilepsy · Low-voltage-activated calcium channels · Neurodevelopmental disease · T-type calcium channels

1 Introduction

The $Ca_V3.3$ calcium channel is the third member of the of T-type calcium channel family. Together with $Ca_V3.1$ and $Ca_V3.2$, they constitute the group of low-voltage-activated (LVA) calcium channels with a threshold of activation below −50 mV, close to the typical resting potentials of many neurons (Perez-Reyes 2003). Therefore, T-type calcium channels are ideally equipped for functions in regulating cellular excitability. Ca_V3 channels are further characterized by their rapid activation and inactivation kinetics and small single channel conductance. That is why they were designated "T-type," for *transient* and *tiny*, respectively.

Voltage-gated calcium channels and voltage-gated sodium channels together form a single protein family, based on their common structure and presumed common ancestry. Both, their biophysical characteristics and their primary functions in cellular excitability position the T-type calcium channels in between the high-voltage gated calcium channels and voltage-gated sodium channels. Accordingly, T-type channels are primarily found in the central and peripheral nervous system, where they contribute to the processing of pain, sensory and motor functions, rhythm generation and the regulation of sleep (Cain and Snutch 2010; Carbone et al. 2014; Bourinet et al. 2016; Leresche and Lambert 2017; Harding and Zamponi 2022). However, they are also expressed in some non-neuronal tissues including neuro-endocrine cells, glial cells, the cardio-vascular system, and the kidney, where they play more traditional calcium channel roles, in regulating secretion, contractility, and transcriptional regulation (Latour et al. 2003; Hansen 2015; Yu et al. 2020).

Owing to their dual roles in cellular excitability and activity-dependent regulation of calcium-dependent cell functions, their involvement in the pathophysiology of disease also shows two faces. On the one hand, altered gating properties of Ca_V3 disease variants can have direct effects on neuronal information processing and network activity, resulting in hyperexcitability, pain, or seizures (Weiss and Zamponi 2020). On the other hand, aberrant calcium influx due to gain-of-function mutations can alter secretion of neurotransmitters and hormones, cause developmental defects or even the death of the affected cells (Alza et al. 2022). Thus, in the recent years, T-type calcium channels have been implicated in the pathology of a wide range of neurological, neurodevelopmental, and psychiatric disorders. At the same time, their critical position as regulators of excitability and calcium influx in specific subsets of neurons makes Ca_V3 channels promising drug targets for treating pathologies ranging from pain to epilepsy (Weiss and Zamponi 2019).

Since their cloning around the turn of the millennium, numerous genetic variants of the genes encoding $Ca_V3.1$ and $Ca_V3.2$ channel isoforms, CACNA1G and CACNA1H, respectively, have been linked to diseases including late onset cerebellar ataxia, childhood cerebellar atrophy, primary aldosteronism, autism spectrum disorder, and different forms of epilepsy. The evidence for the disease associations of these two channels has been comprehensively discussed in a recent review by Philippe Lory and colleagues (2020). At that time, the third member of this family, $Ca_V3.3$, had only been known as a risk gene for schizophrenia. However, since then, several de novo missense variants have been identified and functionally characterized, linking $Ca_V3.3$ to neurodevelopmental disease and epilepsy (El Ghaleb et al. 2021). Therefore, in this review, we focus on the emerging evidence showing that $Ca_V3.3$ can be causal for severe brain disease.

2 Structure and Modulation of $Ca_V3.3$

Ca_V3 channels have the domain structure typical of Ca_V and Na_V channels. Their pore-forming α_1 subunit is the product of a single gene with four homologous repeats (I–IV), each comprised of six transmembrane helices (S1-S6) (Fig. 1a). Helices S1-S4 of each repeat form an independent voltage-sensing domain. The S5 and S6 helices of all four repeats and their connecting pore loops together form the central channel pore with the selectivity filter just below the outer vestibule and the activation gate at the cytoplasmic end of the S6 helices. The four repeats are connected by cytoplasmic loops of different lengths and the channels contain a long cytoplasmic C-terminal tail including multiple modulation sites.

Notably, the T-type calcium channels differ from the high-voltage activated channels in that they do not require the auxiliary $\alpha_2\delta$, β, and γ subunits for functional expression, nor for modulation of their current properties. Although, few studies did report increased membrane expression and current density when T-type channels were co-expressed with auxiliary $\alpha_2\delta$ and β subunits in heterologous cells, even this had no effect on the biophysical gating properties (Dubel et al. 2004; Arteaga-Tlecuitl et al. 2018). In keeping with the absence of a classical regulation by β subunits, Ca_V3 channels lack the high-affinity binding site for β subunits (known as AID) in the cytoplasmic loop between repeats I and II. Instead, this I–II linker in

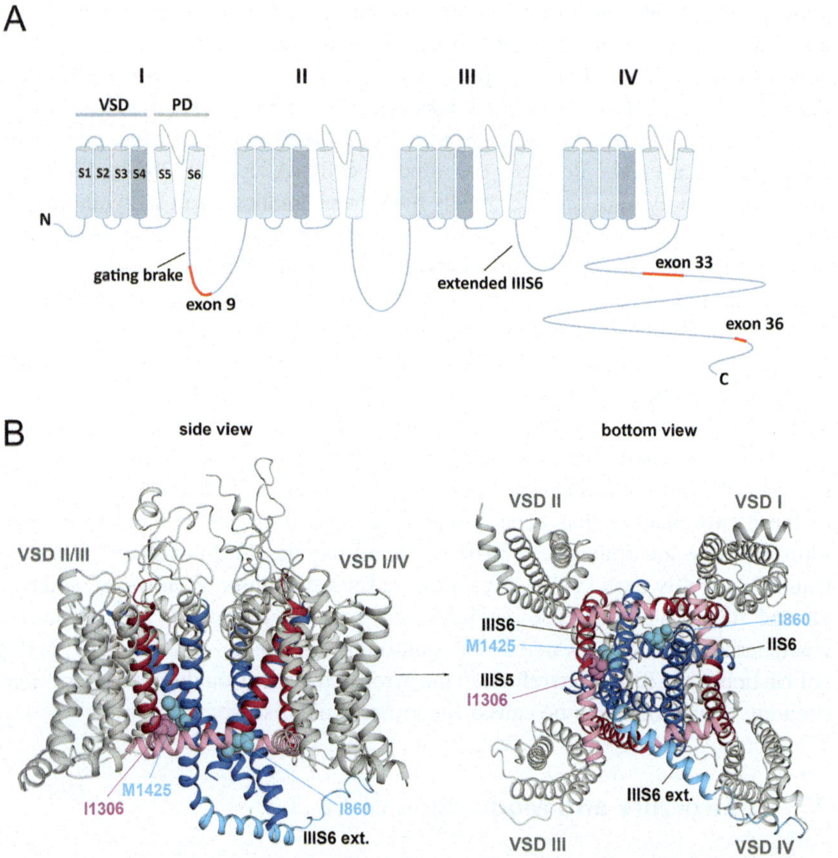

Fig. 1 Structure of $Ca_V3.3$ highlighting unique features and positions of gate-modifying disease mutations. (**a**) Domain structure of $Ca_V3.3$ comprising four domains, each with six transmembrane segments (S1-S6), of which S1-S4 form the voltage sensor domain (VSD) and S5 and S6 the pore domain (PD). $Ca_V3.3$ contains three alternatively spliced exons (marked in red): Exon 9 in the intracellular I–II linker, exon 33 and exon 36 in the C-terminus. The proximal end of the I-II loop contains the so-called gating brake. The S6 helix of the third PD extends into the cytoplasm. (**b**) Structure model of the human $Ca_V3.3$ channel (courtesy of Monica Fernández-Quintero). Shown in color are the PD helices composing the channel gate (S6 blue, S5 magenta, S4–S5 linker pink) and the residues mutated in the recently described disease-related missense variants (El Ghaleb et al. 2021). Left (side view) shows the S6 bundle crossing and the extended IIIS6 helix unique for the Ca_V3 channels. Right (bottom view): The four VSDs (gray) are symmetrically arranged around the common PD. The channel gate is composed of the four S6 helices, closely flanked by the S5 helices and enclosed by the cytoplasmic S4–S5 linkers, which connect the VSDs to the pore domain and presumably actuate the channel gate. The four described mutations linked to neurodevelopmental disorder and epilepsy reside at the bundle crossing on the cytoplasmic side of the S6 (I860 and M1425, blue) and S5 (I1306, red) helices

itself is important for surface expression and modulation of the channels, as it contains sequences important for regulating the gating properties of T-type channels (Perez-Reyes 2010).

2.1 The Gating Brake

The so-called *gating brake* constitutes a helix-loop-helix structure in the proximal I-II loop (Fig. 1a) that accelerates activation and inactivation kinetics and is thought to stabilize T-channels in the closed state, preventing the channels from opening at even more hyperpolarized potentials. Importantly, the gating brake in T-type channels can be functionally regulated by calcium/calmodulin and by phosphorylation. Calmodulin binds to the gating brake with high affinity in the apo-state, without calcium. Co-expression of a calmodulin variant with mutations in its four calcium binding sites abolishes the action of the gating brake and thus leads to channels with the runaway gating phenotype, similar to channels with deletions in the gating brake (Chemin et al. 2017). Furthermore, at the transition from the proximal helix to the loop sequence, the gating brake contains three functionally relevant phosphorylation sites (Blesneac et al. 2015). A triple mutation of these serines and threonine to alanine in $Ca_V3.2$ altered the kinetics and voltage-dependence of activation and inactivation, similar to the deletion of the gating brake itself. As these features of the gating brake are conserved between the three Ca_V3 isoforms, modulation of gating properties by calcium and phosphorylation/dephosphorylation of the cytoplasmic I-II loop may represent a general mechanism for the physiological regulation of T-type channel gating.

In contrast to deletions in the proximal part of the I-II loop containing the gating brake, deletions of sequences in its central regions did not alter channel gating, but increased surface expression several-fold. Therefore, the I-II loop of Ca_V3 channels is an important determinant of the channels´ gating properties and surface targeting, and a hot spot for their regulation by multiple signaling mechanism. Likewise, genetic variants in the I-II loop are likely to show functionally relevant effects on current properties. Thus, sequence variations in the I-II loop will contribute to the functional diversity of T-type currents in different tissues and cells, and potentially to pathologies related to aberrant current properties. Indeed, in $Ca_V3.2$ numerous missense variants linked to idiopathic generalized epilepsy that lead to gating or trafficking defects are located in the I-II loop. One of them, C456S, significantly increased spontaneous firing and lowered the threshold for rebound burst firing when expressed in hippocampal neurons, consistent with the importance of this sequence for regulating T-type current properties (Eckle et al. 2014).

2.2 The Selectivity Filter

Also, the selectivity filter presents a remarkable difference between LVA and HVA calcium channels. In HVA calcium channels the selectivity filter is formed by four glutamate residues, one in each pore domain (EEEE). In contrast, in LVA T-type channels the selectivity filter consists of two glutamates in domains I and II, whereas domains III and IV contain aspartates in the corresponding positions (EEDD). The result of this difference is that T-type channels are less selective for calcium than HVA. However, the alternative amino acid arrangement in the selectivity filter may also contribute to some of the differences of the gating property in LVA and HVA calcium channels (Talavera and Nilius 2006; He et al. 2022).

2.3 Domain-Specific Functions

Over the last two decades, several groups used a chimera approach to identify the structural determinants for the activation of Ca_V3 channels at low voltages that set them apart from HVA channels. However, the results are partially conflicting. Li and coworkers (2004) reported that domains I, III, and IV, but not domain II, were critical for low-voltage activation of $Ca_V3.1$, and that transferring only the S4 segments was not sufficient to confer high-voltage activation to this T-type channel. In contrast, a more recent study on $Ca_V3.3$ indicated that IIS4 and to a lesser degree IVS4, plus their respective S4–S5 linkers, determine the LVA properties of $Ca_V3.3$ (Sanchez-Sandoval et al. 2018). While these studies disagree about the importance of specific VSDs for determining the low-voltage activation of T-type channels, they consistently indicate that the four voltage sensors contribute differentially to the activation of Ca_V3 channels.

What about the structural basis of the functional differences among the Ca_V3 members? Although there is a high level of conservation between the sequences of T-type channels, $Ca_V3.3$ activates and inactivates remarkably slower and at less hyperpolarized potentials than $Ca_V3.1$ and $Ca_V3.2$. Using chimeras of $Ca_V3.1$ and $Ca_V3.3$, several regions in the protein were found to be important for the slower kinetics and relatively depolarized voltage-dependence of $Ca_V3.3$. Initially, Park et al. (2004) concluded from his chimera study in Xenopus oocytes that no single region is sufficient to explain the specific gating properties of $Ca_V3.1$ and $Ca_V3.3$. In contrast, Hamid's subsequent study in mammalian cells indicated that specifically domain IV is involved in producing the differences in activation and in the recovery from inactivation of $Ca_V3.1$ and $Ca_V3.3$ channels, whereas inactivation kinetics depended on the concerted action of structural determinants across domains I, II, and IV (Hamid et al. 2006). Finally, the study by Kang and coworkers (2008) highlighted a prominent role of the III-IV loop and domain I in determining the distinct activation and inactivation kinetics of $Ca_V3.1$ and $Ca_V3.3$, which is in line with recent structural analyses of T-type channels.

2.4 The IIIS6 Segment

The structure of $Ca_V3.3$ has been solved at 3.3 Å resolution using cryo-EM (He et al. 2022). Its general structure resembles the typical fold of Na_V and Ca_V channels with four independent voltage-sensing domains arranged around a common channel pore in a domain-swapped configuration. Compared to the high-voltage activated calcium channels, the structure of $Ca_V3.3$ shows some remarkable differences in the pore domain of the third repeat. Unlike the S6 helices of the other three repeats, the IIIS6 α-helix does not end shortly below the helix crossing at the gate, but extends into the cytoplasm for six additional turns (Fig. 1). This IIIS6 helix extension is almost entirely composed of charged amino acids; it bends toward the membrane and directly contacts the voltage-sensing domain IV. Although the corresponding domain is not resolved in the cryo-EM structure of $Ca_V3.1$ (Zhao et al. 2019), the sequence of this IIIS6 helix extension is highly conserved among all three Ca_V3 channels. Charge-neutralizing mutations or the disconnection of the cytoplasmic helix extension from IIIS6 caused a substantial shift of the voltage-dependence of activation to more depolarizing potentials (He et al. 2022). Thus, this distinguishing structural feature might contribute to the typical activation properties of the low-voltage activated channels. Because of its net positive charge and its direct contact with the voltage-sensing domain IV, the IIIS6 helix extension has been suggested to directly participate in voltage sensing of T-type channels. Interestingly, a role for IIIS6 in T-type channel gating was already exposed in 2001 by Marksteiner et al. (2001), who discovered a hotspot (M1510, F1511, V1512) of inactivation determinants in segment IIIS6, corresponding to the earlier identified inactivation determinants (IFV-motif) in IIIS6 of $Ca_V2.1$ (Hering et al. 1998). This shows that despite the many differences between LVA and HVA structure and function, the two channel sub-families apply distinct mechanism in the same regions to achieve analogous functional regulation.

2.5 The C-Terminus

Recently, it has been demonstrated that the initial proximal region of the C-terminus contains a sequence of 20 amino acids, which is highly conserved among the Ca_V3 channels and is a determining structure for shaping Cav3.3 gating properties (Jurkovicova-Tarabova et al. 2019). When deleting these 20 amino acids, a substantial hyperpolarizing shift of the voltage-dependence of activation and inactivation, slowed inactivation kinetics, and weaker voltage sensitivity of the channel were observed. The C-terminus of $Ca_V3.3$ is also known for its role in calcium-dependent regulation. An increase in the cytoplasmic calcium concentration shifts the activation and inactivation toward more hyperpolarized potentials and accelerates the kinetics of activation and inactivation (Lee et al. 2017). Notably, the shift in the inactivation curve was more robust than that of the activation curve, resulting in a decreased window current and calcium influx at rest. As explained in more detail below, the window current is particularly important for the physiology and pathophysiology of $Ca_V3.3$. Since $Ca_V3.3$ presents the slowest inactivation kinetics among the three T-type channels, its contribution to intracellular calcium concentrations is larger than that of $Ca_V3.1$ and $Ca_V3.2$, which could explain this negative feedback mechanism in $Ca_V3.3$ specifically. This effect seems to be mediated via a calcium/calmodulin

(CaM) interaction with the C-terminus of $Ca_V3.3$, reminiscent of calcium-dependent inactivation (CDI) in HVA calcium channels (Lee et al. 2017). Moreover, the calcium/CaM interaction also activates calmodulin-dependent protein kinase II c (CaMKIIc), which can in turn increase T-type channel expression (Wen et al. 2018).

2.6 Splice Variants

Finally, alternative splicing adds to the structural and functional diversity of T-type calcium channels. The human CACNA1I gene, coding for $Ca_V3.3$, consists of 37 exons with three sites of alternative splicing to regulate the inclusion or deletion of exon 9, 33, and 36. Exon 9 codes for 35 amino acids at the proximal end of the I–II linker, while exon 33 (13 amino acids) and 36 (one amino acid) code for extra sequences in the C-terminus (Fig. 1a) (Mittman et al. 1999; Chemin et al. 2001a). Inclusion of exon 9 in the +33 variation improved recovery from inactivation and increased the voltage sensitivity of steady-state inactivation. On the other hand, inclusion of exon 9 in the Δ33 variation slowed both activation and inactivation. Computer simulations of neuronal firing in the NEURON model indicated that separately exon 9 or Δ33 reduce burst firing, but in combination they enhance firing (Murbartián et al. 2004). This further highlights the role of the I–II linker and the C-terminus in the channel gating of $Ca_V3.3$ and suggests an interaction between these two intracellular structures, to adjust their respective effect on gating and providing $Ca_V3.3$ with a fine-tuning mechanism to shape their low-threshold calcium spikes.

3 $Ca_V3.3$ Physiology

3.1 Expression Patterns

Low-voltage-activated T-type calcium currents are present in most neurons. However, the T-type current characteristics vary considerably between neurons in different brain regions and nuclei (Broicher et al. 2008). Accordingly, the three subtypes of T-type calcium channels show distinct and complementary expression patterns in the central and peripheral nervous system (Talley et al. 1999). High levels of $Ca_V3.1$ are found in the inferior olive, the thalamus, and in cerebellar Purkinje cells. $Ca_V3.2$ is highly expressed in the sensory ganglia, pituitary, the thalamus, and the dentate gyrus; and $Ca_V3.3$ in the cerebral cortex and in the thalamus. Hippocampal pyramidal cells and olfactory granule cells express all three Ca_V3 subtypes. In the thalamus, $Ca_V3.2$ and $Ca_V3.3$ are co-expressed in the reticular nucleus (NRT), whereas $Ca_V3.1$ is preferentially expressed in the thalamic relay neurons (Talley et al. 1999; Kim et al. 2001) (Fig. 2). On the protein level it had been shown that virtually all NRT neurons in the thalamus express $Ca_V3.3$, whereas in the cerebral cortex its expression was limited to a subset of GABAergic neurons (Liu et al. 2011). In both, the cortical

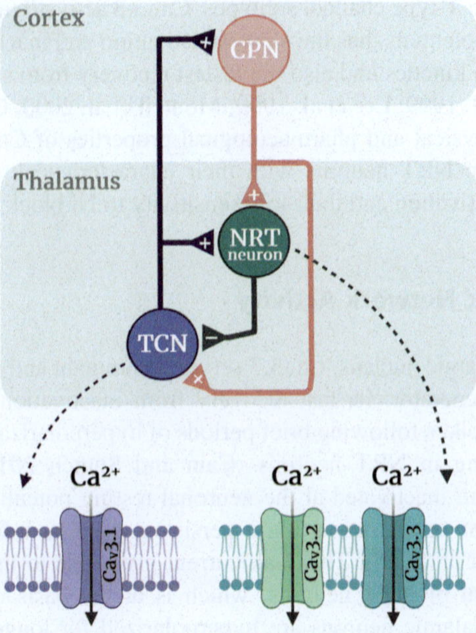

Fig. 2 Thalamocortical projections controlled by T-type channels. In the thalamus, thalamocortical relay neurons (TCNs) express CaV3.1 channels and have excitatory projections to reticular thalamic nucleus (NRT) neurons, which express both CaV3.2 and CaV3.3, and to cortical pyramidal neurons (CPNs). CPNs in turn extend excitatory projections to both NRT neurons and TCNs, whereas NRT neurons exert inhibitory hyperpolarizing input onto TCNs, allowing CaV3.1 channels to de-inactivate and respond with rhythmic bursting. Figure created with BioRender.com

and thalamic neurons, $Ca_V3.3$ was preferentially localized in the postsynaptic compartment, the dendrites and soma (Mckay et al. 2006).

$Ca_V3.3$ expression is not only spatially but also temporally regulated (Yunker et al. 2003). Its expression in the forebrain and hindbrain is considerably modified during development, much more than that of $Ca_V3.1$. By embryonic day 14 in situ, $Ca_V3.3$, but not $Ca_V3.1$, is already expressed in the brain and spinal cord. Overall, the fetal brain expresses about 25% more $Ca_V3.3$ than the adult brain. At birth, a unique long $Ca_V3.3$ isoform (~260 kDa) is expressed but gradually decreases in parallel with the increase of the shorter (~190 kDa) isoform (Yunker et al. 2003). The observed differences in size, as detected by Western blot analysis, possibly reflect truncations of the C-terminus as a result of posttranslational modifications (Murbartián et al. 2004; Perez-Reyes 2006). Several regulatory mechanisms likely shape the developmental expression of $Ca_V3.3$, including its many phosphorylation sites and glycosylation motifs (Lee et al. 1999; Mittman et al. 1999). Since many of the $Ca_V3.3$ pathologies are developmental disorders (see below), disruption of the spatial and temporal expression patterns of $Ca_V3.3$ might contribute to these channelopathies.

Among the three T-type channel subtypes, $Ca_V3.3$ activates and inactivates at the most depolarized potentials, has the slowest activation and inactivation kinetics, the fastest deactivation kinetics and also the fastest recovery from steady-state inactivation (Klöckner et al. 1999; Lee et al. 1999; Monteil et al. 2000; Gomora et al. 2002). The specific biophysical and pharmacological properties of $Ca_V3.3$ endow T-type calcium currents in NRT neurons with their characteristically slow decay, rapid recovery from inactivation and their low sensitivity to Ni block (Astori et al. 2011).

3.2 Thalamic Network Activity

In the reticular thalamic nucleus, $Ca_V3.3$ serves a dominant and essential function as oscillatory burst generator. Its fast recovery from inactivation enables $Ca_V3.3$ to generate calcium spikes following brief periods of hyperpolarization, thus triggering rebound burst firing in NRT neurons (Cain and Snutch 2010). Although most $Ca_V3.3$ channels are inactivated at the neuronal resting potential, a minor fraction of channels remains open, supporting a persistent calcium influx termed window current (Crunelli et al. 2005). This leak current gives rise to the bi-stability of the membrane potential of NRT neurons, which is at the basis of their burst firing behavior. When thalamic neurons are hyperpolarized for longer than 100 ms, for example by inhibitory synaptic input, $Ca_V3.3$ channels exit the inactivated state becoming ready for subsequent depolarization, the so-called calcium spikes, which in turn trigger bursts of sodium action potentials. The influx of calcium during the calcium spikes eventually activates small-conductance calcium-activated potassium channels, which hyperpolarize the cell again and terminate the burst of action potentials (Cueni et al. 2008). This characteristic firing pattern is termed rebound burst firing (Lee et al. 2014).

Such rhythmic action potential bursting of NRT neurons provides barrages of hyperpolarizing inhibitory synaptic input on thalamocortical relay neurons, which in turn respond with rhythmic bursting due to $Ca_V3.1$ activation (Fig. 2). Thus, $Ca_V3.3$ endows NRT neurons with their specific firing pattern that underlies the rhythmic activity of the thalamocortical network. This activity brings about sleep spindles, an EEG hallmark of natural non-rapid eye movement (NREM-) sleep, as well as spike-wave discharge in generalized absence epilepsy. While the origin and exact mode of rhythm generation in the thalamocortical network is still controversial, the importance of $Ca_V3.3$ and NRT burst firing for sleep spindles and spike-wave discharge is undisputed (Huguenard and McCormick 2007; Lee et al. 2014). Considering the central importance of $Ca_V3.3$ channels in the activity of NRT neurons, it is not surprising that their firing pattern is highly sensitive to any alteration of biophysical properties of $Ca_V3.3$, which result in a change of the voltage range or magnitude of the window current (Crunelli et al. 2005; Chevalier et al. 2006; Dreyfus et al. 2010).

3.3 Genetic Mouse Models

The central role of $Ca_V3.3$ in rhythm generation is reflected in the phenotype of $Ca_V3.3$ knockout mice. In the absence of $Ca_V3.3$, the GABAergic NRT neurons still showed tonic firing but lacked low frequency oscillatory bursting (Astori et al. 2011; Ghoshal et al. 2020). This weakened the overall synchronized thalamic network oscillations, which underlie sleep spindles. By comparison, low-threshold bursting in the adjacent thalamocortical relay neurons remained intact in $Ca_V3.3$ knockout mice, whereas it was abolished in $Ca_V3.1$ knockouts (Kim et al. 2001). NRT neurons also express $Ca_V3.2$ and the additional knockout of this channel aggravates the thalamic phenotype (Pellegrini et al. 2016). However, $Ca_V3.2$ cannot sustain the rhythmic bursting of NRT cells in the absence of $Ca_V3.3$. Thus, $Ca_V3.3$ initiates burst firing in NRT neurons, and $Ca_V3.1$ in the relay neurons, and both are critical for the rhythmic firing of the thalamocortical network underlying the regulation of the sleep cycles and the spike-wave discharge in generalized absence epilepsy (Fig. 2).

3.4 Further Functions in the Nervous System

Apart from its key function as rhythm generator, $Ca_V3.3$ in NRT neurons is also needed for the formation of NMDA-induced long-term potentiation at thalamocortical synapses (Astori and Lüthi 2013). As spindle waves have been implicated in associative synaptic plasticity and memory consolidation, impaired intellectual abilities may either directly result from loss of $Ca_V3.3$ function or indirectly from altered synaptic connections or the degeneration of NRT neurons due to $Ca_V3.3$ gain-of-function defects (Rosanova and Ulrich 2005; Fogel and Smith 2011; Manoach et al. 2016). Also consistent with its importance for regulation of sleep and attention, $Ca_V3.3$ appears to influence the efficacy of anesthesia, as $Ca_V3.3$ knockout mice are more sensitive to the volatile anesthetic isoflurane than wildtype mice (Feseha et al. 2020). Moreover, $Ca_V3.3$ seems to be involved in the neurotoxicity of local anesthetics (Wen et al. 2018). Treatment of dorsal root ganglia (DRG) with the local anesthetic ropivacaine hydrochloride reduces cell viability and increases the expression of $Ca_V3.3$ and the multifunctional kinase CaMKIIc. Ca_V3 channels and CaMKIIc mutually regulate each other and down regulating either one of them, partly alleviated the anesthetic induced cell injury to DRG neurons.

In sum, $Ca_V3.3$ channels serve a highly specific and critical function in thalamic rhythm generation and further are implicated in functions of the central and peripheral nervous system, including memory formation and pain processing. All of these functions may be affected to varying degrees in patients carrying functionally relevant mutations in the CACNA1I gene.

4 $Ca_V3.3$ and Brain Disease

4.1 Schizophrenia

$Ca_V3.3$ is a schizophrenia risk gene. CACNA1I variants were initially associated with schizophrenia in a genome-wide association study (GWAS) of a large Irish patient cohort with psychiatric disorder (Donnelly et al. 2012) and later confirmed in European and Asian patient cohorts (Schizophrenia Working Group of the Psychiatric Genomics Consortium 2014; Xie et al. 2018). At the same time, two rare de novo missense variants in CACNA1I were identified in persons with schizophrenia but otherwise healthy families using whole-exome sequencing (WES) (Ghoshal et al. 2020). In that study, CACNA1I was the only gene affected in more than one of the 105 included schizophrenia patients. Both de novo mutations (T797M, R1346H) are located in the extracellular loops between pore domain helix S5 and the structured sequences constituting the channel mouth and selectivity filter (Fig. 3). T797M is located in repeat II and R1346 resides immediately adjacent to a conserved N-glycosylation site in repeat III.

Biochemical and electrophysiological analysis in HEK cells demonstrated that the gating properties of these $Ca_V3.3$ mutants were not altered (Andrade et al. 2016). However, R1346H, but not T797M, showed reduced glycosylation and cell surface expression, resulting in the reduction of the current density to 50% of wildtype levels. In simulations of the electrical activity of NRT neurons with the NEURON computer model, such reduced current densities eliminated rebound burst firing. Recently, the physiological significance of this altered channel function has been

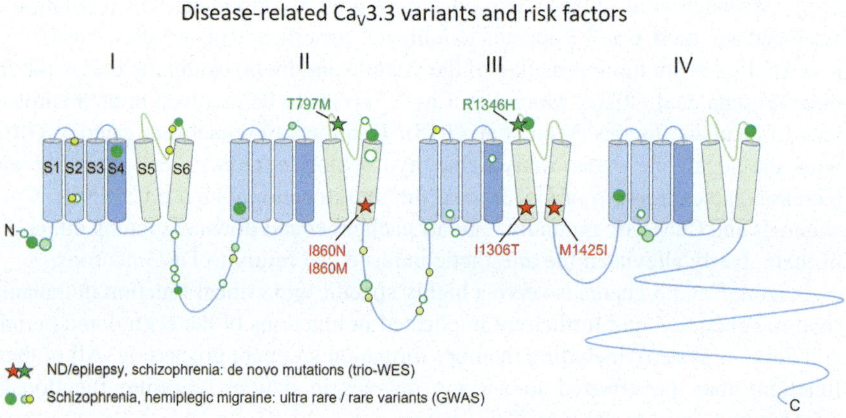

Fig. 3 $Ca_V3.3$ variants associated with psychiatric, neurological, and neurodevelopmental disease. Green: Schizophrenia risk. Red: Neurodevelopmental disease and epilepsy. Yellow: Familial hemiplegic migraine. Asterisks: de novo mutations. Circles: ultra-rare (large) and rare (small) variants; open circles: variants found in schizophrenia and in control group. (Ghoshal et al. 2020; El Ghaleb et al. 2021; Baez-Nieto et al. 2022; Maksemous et al. 2022)

substantiated in a knock-in mouse model expressing Ca$_V$3.3 R1346H (Ghoshal et al. 2020). Electrophysiological recordings from NRT neurons in brain slices demonstrated that T-type calcium currents were significantly reduced in the R1346H knock-in mice compared to controls. This resulted in the reduction of the occurrence and frequency of rebound burst firing between −65 and −50 mV. Finally, electro-encephalogram recordings of R1346H knock-in mice demonstrated that the occurrence and morphology of sleep spindles during non-REM sleep was deficient. Thus, Ca$_V$3.3 R1346H is a loss-of-function mutation that by itself might contribute to abnormal sleep and schizophrenia. Together, these functional studies on the cellular and animal level establish a plausible link between one de novo missense mutation in Ca$_V$3.3 associated with schizophrenia and sleep spindle deficits observed in patients with schizophrenia.

However, how the second de novo mutation found in the trio-WES study (Ghoshal et al. 2020) might be functionally linked to schizophrenia remains elusive, as it did not show altered expression or function in HEK cells and has not been studied in a mouse system. Considering that the occurrence of psychiatric conditions like schizophrenia are multi-factorial, it is not entirely surprising that mutations associated with disease do not show altered current properties when studied in isolation in heterologous expression systems. It might still do so in its native neuronal environment, or may require additional risk factors to become evident.

A systematic functional characterization of rare missense variants of CACNA1I derived from a large Swedish schizophrenia cohort showed that Ca$_V$3.3 variants resulting in reduced peak current density and such resulting in altered voltage-dependence of activation significantly segregated across unaffected and affected groups, respectively (Baez-Nieto et al. 2022). Surprisingly however, more ultra-rare variants with reduced current density were detected in the control group than in the patient group. In light of the reduced currents in the Ca$_V$3.3 R1346H de novo missense variant this is unexpected (Andrade et al. 2016; Ghoshal et al. 2020). Furthermore, the study revealed that ultra-rare Ca$_V$3.3 variants with positive or negative shifts of the voltage-dependence of activation were similarly found in both, the schizophrenia and in the non-affected cohort (Baez-Nieto et al. 2022). For the schizophrenia group this would suggests that not only loss-of-function missense mutations may contribute to schizophrenia but also gain-of-function mutations, and that mutations changing the voltage-dependence of activation might do so as well. By and large, these results caution that mutations leading to reduced currents or altered gating properties in calcium channel variants studied in heterologous cells not necessarily lead to disease. In the large number of rare channel variants reveled in GWAS studies inevitable some mutations associated with schizophrenia will show altered current properties while others will not. However, similarly divergent electriphysiological properties may be found in the non-affected group.

4.2 Other Psychiatric and Neurological Diseases

In addition to the association with schizophrenia, recent GWAS studies of impulsive personality traits and drug experimentation reported an association of the CACNA1I locus with negative urgency; the tendency to act rashly when distressed, frequently associated with problematic substance use (Sanchez-Roige et al. 2019). In a WES study of hemiplegic migraine rare variants of CACNA1I showed a statistically significant increase in the migraine group compared to gnomAD (Fig. 3) (Maksemous et al. 2022). Functional characterization of five of these variants demonstrated modest alterations of current density and gating properties in some but not in other variants. Two amino acid substitutions into histidine showed altered pH sensitivity of the current properties. Clearly, more detailed genetic analysis and functional characterizations will be necessary to establish an association of CACNA1I to these psychiatric and neurological conditions.

4.3 Neurodevelopmental Disease and Epilepsy

Elevated T-type calcium currents in thalamocortical neurons have been implicated in the etiology of absence epilepsy. In three mouse models of generalized epilepsy with mutations in high-voltage activated calcium channel genes ($Ca_V2.1$, β_4, and stargazing/γ_2) the peak T-type calcium currents were augmented and the voltage-dependence of steady-state inactivation was shifted toward depolarizing potentials (Zhang et al. 2002). Therefore, gain of T-type channel function has been suggested to represent a common disease mechanism of these epilepsy mouse models. In human patients, missense variants of the CACNA1H ($Ca_V3.2$) gene have been associated with epilepsy. However, functional analysis revealed gain-of-function as well as loss-of-function defects (Vitko et al. 2005). Upon expression in hippocampal neurons the missense mutation $Ca_V3.2$ C456S caused an increase in spontaneous firing and a reduction of the threshold for rebound burst firing, consistent with a role of $Ca_V3.2$ in causing epilepsy (Eckle et al. 2014). $Ca_V3.2$ and $Ca_V3.3$ are both highly expressed in the thalamus and the knockout of $Ca_V3.3$ in mice abolished low frequency oscillatory burst firing and sleep spindles (Astori et al. 2011). This suggested a role of $Ca_V3.3$ in natural NREM sleep and possibly for spike-wave discharge in generalized absence epilepsy.

Recently heterozygous gain-of-function mutations in CACNA1I were identified in three unrelated patients and one family, presenting variable degrees of congenital neurodevelopmental disorder and epileptic seizures (El Ghaleb et al. 2021). Initially the de novo missense variant c.2579T4A in the CACNA1I gene, predicting the amino acid substitution I860N, was identified using trio whole-exome sequencing (WES) and validated by Sanger sequencing as the only potentially disease-causing variant in a patient with severe developmental retardation and seizures. Subsequently two more unrelated patients with amino acid substitutions I1306T and M1425I, respectively, and a family with three affected patients carrying amino acid substitution I860M in $Ca_V3.3$ were identified. I860N and I1306T are confirmed de novo mutations, whereas for the patient with M1425I only the mother was available for sequencing, who proved unaffected. All patients showed neurodevelopmental disease. In the patients with the de novo mutations, the neurodevelopmental retardation and intellectual impairment were severe with little to no motor development and no development of speech. All showed muscular hypotonia and epileptic seizures. In the affected individuals of the family, the intellectual impairment was much milder and only the mother developed seizures at advanced age. Together the genetic analysis of these patients classified CACNA1I as a disease gene for neurodevelopmental disease and epilepsy.

Electrophysiological analysis of $Ca_V3.3$ carrying the mutations in heterologous mammalian cells demonstrated that the current density was not affected, except in M1425I, in which it was doubled. In contrast, all amino acid substitutions resulted in substantially slowed activation, deactivation, and inactivation and in different degrees of hyperpolarizing shifts of the voltage-dependence of activation and inactivation compared to wildtype $Ca_V3.3$ (El Ghaleb et al. 2021). The magnitude of these effects varied between the mutations, but was strongest in I680N and weakest in I860M. Simulating the effects of the altered gating properties on neuronal excitability in a computer model of thalamic reticular nucleus neurons demonstrated that the mutations lowered the threshold and increased the duration and frequency of action potential firing. Expression of wildtype $Ca_V3.3$, or the I860N and I860M mutants in chromaffin cells shifted the firing mode from low-threshold spikes and burst firing in wildtype to slow oscillations in the mutants.

Together these functional analyses indicated a gain-of-function phenotype with two possible pathways for disease generation. Several of the observed changes of gating properties would result in increased calcium influx in cells expressing the $Ca_V3.3$ mutants. Particularly the slowed current deactivation results in larger tail currents and prolonged calcium influx in the wake of action potentials. On the other hand, the observed hyperpolarizing shifts in the voltage-dependence of activation and inactivation resulted in a left-shift of the window current to voltages near the neuronal resting potential. Consequently, neurons expressing these $Ca_V3.3$ mutants would experience a persisting calcium leak current even at rest. Either one of these defects by itself, and certainly their combined effects, result in an increased calcium load, which might affect the healthy development and function of these neurons and even could cause cell death. This can explain the severe neurodevelopmental defects, including frontal brain atrophy in the severely affected patient with the $Ca_V3.3$ I860N variant. The development of epileptic seizures could secondarily result from these developmental defects. Alternatively, the observed effects of some of the mutations on neuronal excitability could directly cause hyperexcitability of NRT neurons, the altered activity of thalamocortical networks, and ultimately epileptic seizures. This distinction is critical for the treatment of patients. Whereas counteracting the aberrant calcium influx by blocking $Ca_V3.3$ (or T-type currents in general) might slow disease progression but would not reverse the neurodevelopmental defects, treatment with T-type channel blockers might directly prevent the cause of seizures resulting from increased neuronal excitability. In fact, treatment of the patient carrying the $Ca_V3.3$ I860N mutation with the non-selective T-type channel blocker ethosuximide alleviated the seizures.

5 Molecular Patho-Mechanism

The disease phenotype of the patients carrying different $Ca_V3.3$ mutations ranged from borderline intellectual function to severe neurodevelopmental disease. Similarly, the defects of the gating properties and the resulting effects on action potential firing ranged from light to severe for the different CACNA1I variants. Generally, the severity of the clinical features and the strength of the electrophysiological defects correlated with one another, corroborating the causality between the alterations of the channel properties and the disease.

Notably, the patients with the two distinct amino acid substitutions of I860 were at the two extremes of the observed pathological range (El Ghaleb et al. 2021). While the patient carrying the $Ca_V3.3$ I860N mutation developed devastating neurodevelopmental disorder, the three related patients carrying the $Ca_V3.3$ I860M mutation were only mildly affected. Correspondingly, $Ca_V3.3$ I860M showed the least severe channel gating defects. This clearly demonstrates that, even though I860 is fully conserved among the 10 mammalian calcium channel isoforms, not the position or the properties of the mutated amino acid I860 were decisive for developing disease, but the properties of the residues that I860 was substituted for were critical for the outcome. This possibility needs to be considered when judging the pathogenicity of genetic variants.

In an attempt to explain the different functional and pathological outcomes by the alterations of channel structure, we analyzed the possible interactions of the amino acid substitutions in homology structure models of $Ca_V3.3$ in the activated/inactivated and resting states (El Ghaleb et al. 2021). All three affected residues are located in parts of the channel forming the activation gate. I860 and M1425 are positioned at the cytoplasmic end of the pore-forming S6 helices of the second and third repeat, respectively, and I1306 at the cytoplasmic end of the adjacent S5 helix of the third repeat (Fig. 1b and 3). This domain is a hot spot for disease-related mutations in calcium and sodium channels (for review, see Lory et al. 2020). In the activated as well as in the resting state of the $Ca_V3.3$ channel, the affected hydrophobic amino acids are embedded in a hydrophobic surrounding, forming multiple weak hydrophobic interactions with one another. Functionally this makes sense, because the gate must be able to move quickly when the channel opens and closes. In the mutant channels, however, the amino acids replacing I860 and I1306 now formed H-bonds with multiple binding partners in adjacent sequences, thus stabilizing the gate in the resting and the activated states. It seems highly plausible that such ectopic stabilizing interactions would slow the state transitions of the channel gate. This fits well with the measured slowing of both activation and deactivation kinetics. I860N forms two H-bonds, I1306T one H-bond, and I860M one sulfur H-bond. Therefore, the expected strength of the interactions formed by these amino acid substitutions matches the severity of the electrophysiological defects and of the disease.

Only the M1425I mutation does not follow this pattern. In this case, a hydrophobic amino acid is replaced by another hydrophobic amino acid. Either this residue in the S5 helix has different structural and dynamic requirements than those in the adjacent S6 helices, or its van der Waals interactions with I1306 in IIIS6 in the resting state exert sufficient stabilizing actions to impede opening and closing of the gate. Intriguingly, the putative interaction partner of the I1425M variant is I1306, which itself is one of the affected residues (see above). Therefore, the stabilization of I1306 by its substitution for a threonine in I1306T, or by the interaction with the adjacent methionine in I1425M might have similar effects on the mobility of the IIIS6 helix in the channel gate. Clearly, more thorough structure–function analyses of disease mutations in the channel gate will be necessary to substantiate this pathogenic mechanism and to possibly expose the general principles by which amino acid substitutions in the gate of calcium channels alter the channel properties and cause disease.

6 Ca_V3 Pharmacology

Because of their critical contributions to pain processing and neuronal excitability, T-type calcium channels are choice targets for drug development efforts for treatments of these conditions. Unfortunately, clinical use of T-type channel blockers has been hampered by their low specificity, both within the group of Ca_V3 channels and relative to Ca_V1 and Ca_V2 channels (for review, see (Zamponi et al. 2015; Weiss and Zamponi 2019)).

T-type channels can be blocked by nickel and zinc. The efficacy of these inorganic ions is about 10-fold higher for $Ca_V3.2$ than for $Ca_V3.1$ and $Ca_V3.3$. Under specific conditions, zinc can act as agonist of $Ca_V3.3$, but not of $Ca_V3.2$, as it differentially affects the deactivation kinetics of these two T-type channels. Peptide toxins blocking T-type calcium channels have been isolated from scorpion and spider venoms. Again, these show low specificity, mostly blocking $Ca_V3.2$ with higher affinity than $Ca_V3.1$ and $Ca_V3.3$. Both, inorganic ions and peptide toxins are of limited usefulness as therapeutic agents, but readily applied as experimental tools. Interestingly, also natural compounds like lavender and rosemary inhibit Ca_V3 channels, but also in this case the lowest inhibitory effect was found in $Ca_V3.3$ as compared to $Ca_V3.1$ and $Ca_V3.2$ (Alaoui et al. 2017).

Several different classes of small organic compounds blocking T-type channels have been identified. Ethosuximide is a state-dependent, low-affinity blocker of all T-type channels used as anti-epileptic drug. However, it failed clinical trials for neuropathic pain because of multiple adverse side effects. Mibefradil, a non-specific T-type channel blocker, effectively reverses the pain threshold in rat models for neuropathic pain. However, it also blocks L-type calcium channels and was withdrawn from clinical use, because of its interactions with cytochrome P450 and other drugs (Nam 2018). T-type channels are also inhibited by endocannabinoids and synthetic cannabinoid receptor ligands (Chemin et al. 2001b; Bladen et al. 2015). Z944 and ACT-709478 represent two promising synthetic compounds blocking all three T-type channels with greater than 100-fold selectivity over other ion channels (Snutch and Zamponi 2018). They lack off-target effects and are currently studied in phase II clinical trials for the treatment of pain and generalized epilepsy, respectively.

Another potent and specific synthetic inhibitor for T-type channels is the state-dependent drug TTA-A2 (Uebele et al. 2009). Notably, the effect of TTA-A2 on $Ca_V3.3$ shares many features with the inhibitory effects of endogenous lipids, in that these effects exist only at depolarized potentials at which most $Ca_V3.3$ channels are inactivated. Accordingly, a pharmacological interaction between TTA-A2 and endogenous lipids has been suggested and their sites of action on the $Ca_V3.3$ protein seem to overlap (Cazade et al. 2014). Recently, the structure of $Ca_V3.3$ has been solved in complex with the three classes of T-type channel blockers, the antihypertensive drug mibefradil, the antispasmodic drug otilonium bromide, and the antipsychotic drug pimozide (He et al. 2022). All three structurally different compounds bind to the central cavity in the conduction pore, but are mediated by distinct interactions within the binding pocket.

The recent development of specific TAT-based cell penetrating peptides might be a promising approach in the search for selective $Ca_V3.3$ inhibitors. Chronical application of Ca_V3 expressing HEK cells with TAT-C3P, a cell-permeable peptide containing the carboxy proximal region of $Ca_V3.3$, selectively inhibited $Ca_V3.3$, without affecting the other two T-type channels (Cmarko and Weiss 2020). However, this form of drug application may be impractical in vivo or for clinical use and therefore might only offer a new experimental tool to study specific $Ca_V3.3$ function.

Whereas T-channel blockers are available for clinical use and ongoing drug developing efforts hold the prospect of producing compounds with higher specificity for T-type channels and reduced adverse effects, isoform-specific $Ca_V3.3$ blockers are not available. If anything, several of the drugs show a preference for $Ca_V3.2$. Nonetheless, for the treatment of seizures associated with the epileptogenic CACNA1I variants described above, non-selective T-type channel blockers might be just as useful. Because in the thalamus $Ca_V3.3$ cooperates with $Ca_V3.2$ in producing the rhythmic firing of the thalamocortical network, hyperexcitability caused by the gain of $Ca_V3.3$ function might be effectively counteracted by blocking both, $Ca_V3.3$ and $Ca_V3.2$. Accordingly, the therapeutic improvement observed in a patient carrying the CACNA1I I860N variant (El Ghaleb et al. 2021) likely arises from the combined action of ethosuximide on both T-type channel isoforms.

7 Discussion

The newly found $Ca_V3.3$ de novo missense mutations associated with neurodevelopmental disease and epilepsy are located in the channel gate at the cytoplasmic end of the S6 helices and the adjacent S5 helix. This location is reminiscence of a large and steadily growing number of disease associated variants in several other low- and high-voltage activated calcium channels, sometimes subsumed as S6 channelopathies (recently reviewed in (Lory et al. 2020; Striessnig 2021)). The channel gate certainly is a hot spot for calcium channelopathy mutations. Thus, a location in this structure is a predictor of the pathogenicity and the mode of action of calcium channel missense mutations. However, the channel gate not only entails a part of the S6 helices, but also the cytoplasmic end of the S5 helix and the S4–S5 linker, and all these structures contain channelopathy mutations resulting in similar gating defects. Therefore, a more accurate term subsuming these structurally confined and functionally similar channelopathy mutations would be *gate-modifying channelopathy (GMC)* mutations.

Many of these mutations are gain-of-function mutations resulting in increased calcium influx, due to delayed inactivation and/or shifted voltage-dependence of activation and inactivation. Thus, calcium overload leading to aberrant cell differentiation or function, or even loss of cells due to cell death appears to be a common disease mechanism in these GMC channelopathies. The specific and often distinct disease phenotypes are probably determined by the specific expression pattern of the affected calcium channel. For example, slowed inactivation in the widely expressed $Ca_V1.2$ causes Timothy syndrome, a multi-system disorder including long-QT cardiac arrhythmia, immune deficiency, intermittent hypoglycemia, syndactyly, autism, and cognitive abnormalities (Splawski et al. 2005). On the other hand, increased calcium influx through the missense mutants of the much more narrowly expressed $Ca_V3.3$ causes neurodevelopmental disease (El Ghaleb et al. 2021).

For low-voltage activated channels, like $Ca_V3.3$, as well as $Ca_V1.3$, which also is activated at rather negative membrane potentials, a second disease mechanism becomes relevant (Fig. 4). Because both of these channels are involved in

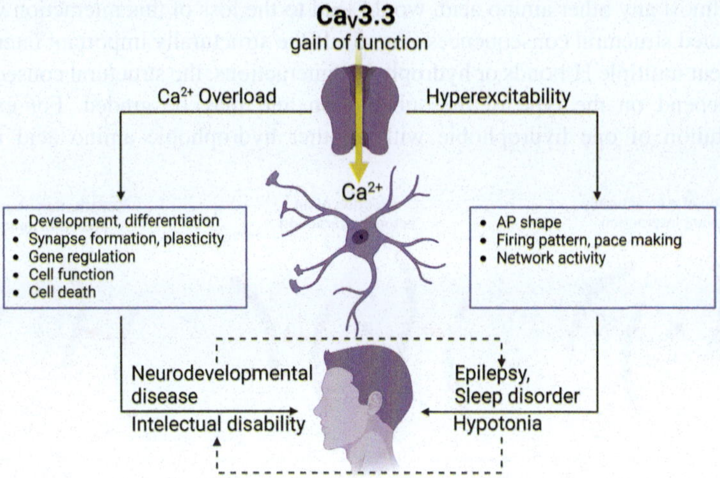

Fig. 4 Two parallel putative pathogenic pathways for $Ca_V3.3$ gain-of-function mutations. Gain-of-function missense mutations may cause calcium overload and aberrant excitability. Calcium overload can alter the development and function of neurons, or cause cell death. This in turn may cause neurodevelopmental disease and intellectual disability. Altered electrical properties may affect the threshold, frequency, and shape of action potentials. This in turn may cause neurological disorders and seizures. Epilepsy may also result secondarily from neurodevelopmental defects. Conversely, epilepsy and sleep disorder may affect intellectual capabilities. *Figure created with* BioRender.com

excitability and rhythm generation, gain-of-function mutations altering their gating properties may result in increased excitability and aberrant firing patterns. Accordingly, the two calcium channels show an overlapping spectrum of disease phenotypes, including neurodevelopmental disease with or without intellectual disability, on the one hand, and neurological and psychiatric symptoms like hypotonia, epilepsy, and autism spectrum disorder, respectively, on the other hand (Pinggera and Striessnig 2016; Pinggera et al. 2017; Hofer et al. 2020; El Ghaleb et al. 2021).

Interestingly, of the $Ca_V3.3$ variants associated with neurodevelopmental disorder two different substitutions of the same residue gave rise to distinct electrophysiological effects and also caused substantially different severity of disease (El Ghaleb et al. 2021). A similar discrepancy of the functional phenotype of two distinct substitutions in the same position has been reported for $Ca_V1.3$ (Hofer et al. 2020). How can this be? Clearly, on the cellular, network, or organismal level multiple factors can contribute to a disease phenotype, and thus the specific environment and the individual status of such factors might either alleviate or aggravate the phenotype. But also on the molecular level amino acid substitutions can produce different outcomes. This may result from two principally different modes of actions. Commonly disease-causing substitutions are viewed as the consequence of the loss of a specific molecular property. For example, a particular residue, especially if highly conserved, may form molecular interactions (like ionic bonds, Pi-Pi stacking) that are important to stabilize a given protein structure. Substitutions of this residue with almost any other amino acid, would lead to the loss of this interaction with the associated structural consequences (Fig. 5). If the structurally important interactions represent multiple H-bonds or hydrophobic interactions, the structural consequences will depend on the type of the substitution and may be graded. For example, substitution of one hydrophobic with another hydrophobic amino acid may be

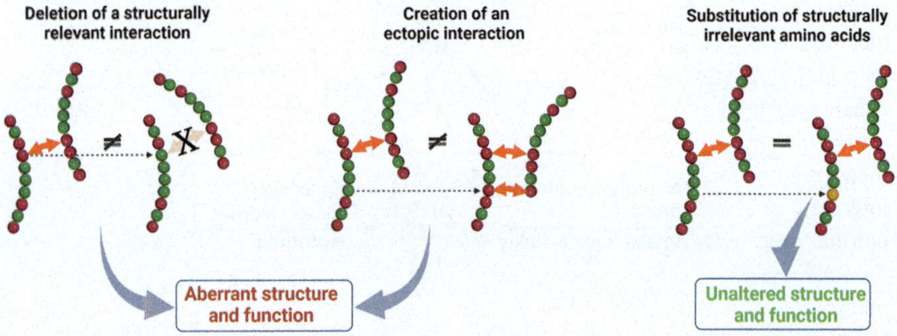

Fig. 5 Possible outcomes of different types of amino acid substitutions. Left: Substitution of a structurally important amino acid with one with different properties results in altered protein structure and function. Center: Introducing a new property, like stabilizing interactions, into a previous structurally irrelevant position will likewise change structure and function. Right: Substitutions with structurally irrelevant amino acids or such with similar properties will not alter structure and function. Red, conserved aa; green, non-conserved aa; yellow, aa with same property as original. Figure created with BioRender.com

inconsequential, whereas substitution of the same residue with a polar amino acid may be detrimental. Furthermore, also substitutions of non-conserved, structurally irrelevant amino acids may cause a functional phenotype, when the new amino acid causes structural clashes or forms ectopic interactions that perturb the protein structure or the dynamic properties necessary for state transitions. Naturally, the functional consequences of such substitutions depend more heavily on the nature of the new amino acid rather than on that of the original. Certain substitutions in a given position may cause disease whereas others in the same position will not.

In the case of $Ca_V3.3$ I860N/M the affected isoleucine is a highly conserved hydrophobic residue embedded in a hydrophobic environment of the channel gate. Substitution with a polar asparagine supports the formation of two hydrogen bonds, thus immobilizing an otherwise mobile part of the IIS6 helix. In contrast, substitution with the also hydrophobic methionine might only form a weak sulfur-H-bond, resulting in a more moderate effect on the mobility of the cytoplasmic end of S6 during channel gating. These differential structural effects likely explain the differing magnitudes of the effects on the gating properties and the different severity of the patients´ phenotypes. In general, these considerations show that prediction of the pathogenicity of a gene defect must not be solely based on the location and conservation of the amino acid replaced, but the effects of the specific substitution (donor and acceptor amino acid) to the channel's structure and stage transitions need to be considered as well. Along these lines, the rapidly increasing number of channel structures and the growing potential of structure modeling will provide the mechanistic insights into channel functions necessary to accurately predict the pathogenic potential of channelopathy missense mutations.

Both de novo mutations associated with neurodevelopmental disease/epilepsy and de novo mutations associated with schizophrenia contain some that do and others that do not lead to functional deficits in channel properties when expressed in heterologous cells. In those cases where channel properties are altered, this evidence substantiates their possible disease-causing nature and may indicate the pathogenic mechanisms. In contrast, lack of physiological effects may call into question the pathogenic nature of the amino acid substitution in cases where alternative disease-causing factors may be present. However, such apparent discrepancies may also arise from limitations in the experimental setup. Outside from their natural environment, channels expressed in heterologous expression systems may lack important interaction partners and modulators, or experience substantially different posttranslational modifications compared to channels in the natural context in neurons. Any missense variant affecting such interactions or modifications may be missed in heterologous expression systems. Therefore, it will be important to further promote functional analyses of putative disease variants in differentiated cell systems like neuronal cultures, patient-derived iPS cells and organoids, and knock-in animal models, and at the same time exploit the potential of in silico models, like NEURON, to simulate the activity of the mutated calcium channels in specific cell types.

References

Alaoui CE, Chemin J, Fechtali T, Lory P (2017) Modulation of T-type Ca 2 + channels by lavender and rosemary extracts. PLoS One 12:1–21

Alza L, Casas-Benito A, Herreros J, Cantí C (2022) Targeting T-type channels in cancer: what is on and what is off? Drug Discov Today 27:743–758

Andrade A, Hope J, Allen A, Yorgan V, Lipscombe D, Pan JQ (2016) A rare schizophrenia risk variant of CACNA1I disrupts CaV3.3 channel activity. Sci Rep 6:1–13

Arteaga-Tlecuitl R, Sanchez AL, Ramirez-Cordero BE, Rosendo-Pineda MJ, Vaca L, Gomora JC (2018) Increase of CaV3 channel activity induced by HVA β1b – subunit is not mediated by a physical interaction. BMC Res Notes 11:1–9

Astori S, Lüthi A (2013) Synaptic plasticity at intrathalamic connections via CaV3.3 T-type Ca2+ channels and GluN2B-containing NMDA receptors. J Neurosci 33:624–630

Astori S, Wimmer RD, Prosser HM, Corti C, Corsi M, Liaudet N, Volterra A, Franken P, Adelman JP, Lüthi A (2011) The CaV3.3 calcium channel is the major sleep spindle pacemaker in thalamus. Proc Natl Acad Sci U S A 108:13823–13828

Baez-Nieto D, Allen A, Akers-Campbell S, Yang L, Budnik N, Pupo A, Shin Y-C, Genovese G, Liao M, Pérez-Palma E, Heyne H, Lal D, Lipscombe D, Pan JQ (2022) Analysing an allelic series of rare missense variants of CACNA1I in a Swedish schizophrenia cohort. Brain 145: 1839–1853

Bladen C, McDaniel SW, Gadotti VM, Petrov RR, Berger ND, Diaz P, Zamponi GW (2015) Characterization of novel cannabinoid based T-type calcium channel blockers with analgesic effects. ACS Chem Nerosci 6:277–287

Blesneac I, Chemin J, Bidaud I, Huc-Brandt S, Vandermoere F, Lory P (2015) Phosphorylation of the Cav3.2 T-type calcium channel directly regulates its gating properties. Proc Natl Acad Sci U S A 112:13705–13710

Bourinet E, Francois A, Laffray S (2016) T-type calcium channels in neuropathic pain. Pain 157: S15–S22

Broicher T, Kanyshkova T, Meuth P, Pape HC, Budde T (2008) Correlation of T-channel coding gene expression, IT, and the low threshold Ca2+ spike in the thalamus of a rat model of absence epilepsy. Mol Cell Neurosci 39:384–399

Cain SM, Snutch TP (2010) Contributions of T-type calcium channel isoforms to neuronal firing. Channels 4:475–482

Carbone E, Calorio C, Vandael DHF (2014) T-type channel-mediated neurotransmitter release. Pflugers Arch Eur J Physiol 466:677–687

Cazade M, Nuss CE, Bidaud I, Renger JJ, Uebele VN, Lory P, Chemin J (2014) Cross-modulation and molecular interaction at the cav3.3 protein between the endogenous lipids and the T-type calcium channel antagonist TTA-A2. Mol Pharmacol 85:218–225

Chemin J, Monteil A, Dubel S, Nargeot J, Lory P (2001a) The α1I T-type calcium channel exhibits faster gating properties when overexpressed in neuroblastoma/glioma NG 108-15 cells. Eur J Neurosci 14:1678–1686

Chemin J, Monteil A, Perez-Reyes E, Nargeot J, Lory P (2001b) Direct inhibition of T-type calcium channels by the endogenous cannabinoid anandamide. EMBO J 20:7033–7040

Chemin J, Taiakina V, Monteil A, Piazza M, Guan W, Stephens RF, Kitmitto A, Pang ZP, Dolphin AC, Perez-Reyes E, Dieckmann T, Guillemette JG, Spafford JD (2017) Calmodulin regulates Cav3 T-type channels at their gating brake. J Biol Chem 292:20010–20031

Chevalier M, Lory P, Mironneau C, Macrez N (2006) T-type Cav3.3 calcium channels produce spontaneous low-threshold action potentials and intracellular calcium oscillations. Eur J Neurosci 23:2321–2329

Cmarko L, Weiss N (2020) Selective inhibition of neuronal Cav3.3 T-type calcium channels by TAT-based channel peptide. Mol Brain 13:4–6

Crunelli V, Tóth TI, Cope DW, Blethyn K, Hughes SW (2005) The 'window' T-type calcium current in brain dynamics of different behavioural states. J Physiol 562:121–129

Cueni L, Canepari M, Luján R, Emmenegger Y, Watanabe M, Bond CT, Franken P, Adelman JP, Lüthi A (2008) T-type Ca2+ channels, SK2 channels and SERCAs gate sleep-related oscillations in thalamic dendrites. Nat Neurosci 11:683–692

Donnelly P, Barroso I, Blackwell J, ISGC and the WTCCC 2 et al (2012) Genome-wide association study implicates HLA-C*01:02 as a risk factor at the major histocompatibility complex locus in schizophrenia. Biol Psychiatry 72:620–628

Dreyfus FM, Tscherter A, Errington AC, Renger JJ, Shin HS, Uebele VN, Crunelli V, Lambert RC, Leresche N (2010) Selective T-type calcium channel block in thalamic neurons reveals channel redundancy and physiological impact of ITwindow. J Neurosci 30:99–109

Dubel SJ, Altier C, Chaumont S, Lory P, Bourinet E, Nargeot J (2004) Plasma membrane expression of T-type calcium channel α1 subunits is modulated by high voltage-activated auxiliary subunits. J Biol Chem 279:29263–29269

Eckle V-S, Shcheglovitov A, Vitko I, Dey D, Yap CC, Winckler B, Perez-Reyes E (2014) Mechanisms by which a *CACNA1H* mutation in epilepsy patients increases seizure susceptibility. J Physiol 592:795–809

El Ghaleb Y, Schneeberger PE, Fernández-Quintero ML, Geisler SM, Pelizzari S, Polstra AM, Van HJM, Denecke J, Campiglio M, Liedl KR, Stevens CA, Person RE, Rentas S, Marsh ED, Conlin LK, Tuluc P, Kutsche K, Flucher BE (2021) CACNA1I gain-of-function mutations differentially affect channel gating and cause neurodevelopmental disorders. Brain 144:2092–2106

Feseha S, Stamenic TT, Wallace D, Tamag C, Yang L, Pan JQ, Todorovic SM (2020) Global genetic deletion of CaV3.3 channels facilitates anaesthetic induction and enhances isoflurane-sparing effects of T-type calcium channel blockers. Sci Rep 10:21510

Fogel SM, Smith CT (2011) The function of the sleep spindle: a physiological index of intelligence and a mechanism for sleep-dependent memory consolidation. Neurosci Biobehav Rev 35:1154–1165

Ghoshal A et al (2020) Effects of a patient-derived de novo coding alteration of CACNA1I in mice connect a schizophrenia risk gene with sleep spindle deficits. Transl Psychiatry 10:1–12

Gomora JC, Murbartián J, Arias JM, Lee J-H, Perez-Reyes E (2002) Cloning and expression of the human T-type channel Cav3.3: insights into prepulse facilitation. Biophys J 83:229–241

Hamid J, Peloquin JB, Monteil A, Zamponi GW (2006) Determinants of the differential gating properties of Cav3.1 and Cav3.3 T-type channels: a role of domain IV? Neuroscience 143:717–728

Hansen PBL (2015) Functional importance of T-type voltage-gated calcium channels in the cardiovascular and renal system: news from the world of knockout mice. Am J Physiol Regul Integr Comp Physiol 308:R227–R237

Harding EK, Zamponi GW (2022) Central and peripheral contributions of T-type calcium channels in pain. Mol Brain 15:1–13

He L, Yu Z, Geng Z, Huang Z, Zhang C, Dong Y, Gao Y, Wang Y, Chen Q, Sun L, Ma X, Huang B, Wang X, Zhao Y (2022) Structure, gating, and pharmacology of human CaV3.3 channel. Nat Commun 13:1–9

Hering S, Berjukow S, Aczél S, Timin EN (1998) Ca2+ channel block and inactivation: common molecular determinants. Trends Pharmacol Sci 19:439–443

Hofer NT, Tuluc P, Ortner NJ, Nikonishyna YV, Fernándes-quintero ML, Liedl KR, Flucher BE, Cox H, Striessnig J (2020) Biophysical classification of a CACNA1D de novo mutation as a high-risk mutation for a severe neurodevelopmental disorder. Mol Autism 11:1–18

Huguenard JR, McCormick DA (2007) Thalamic synchrony and dynamic regulation of global forebrain oscillations. Trends Neurosci 30:350–356

Jurkovicova-Tarabova B, Cmarko L, Rehak R, Zamponi GW, Lacinova L, Weiss N (2019) Identification of a molecular gating determinant within the carboxy terminal region of Ca v 3.3 T-type channels. Mol Brain 12:1–11

Kang HW, Park JY, Lee JH (2008) Distinct contributions of different structural regions to the current kinetics of the Cav3.3 T-type Ca2+ channel. Biochim Biophys Acta Biomembr 1778: 2740–2748

Kim D, Song I, Keum S, Lee T, Jeong MJ, Kim SS, McEnery MW, Shin HS (2001) Lack of the burst firing of thalamocortical relay neurons and resistance to absence seizures in mice lacking α1G T-type Ca2+ channels. Neuron 31:35–45

Klöckner U, Lee JH, Cribbs LL, Daud A, Hescheler J, Pereverzev A, Perez-Reyes E, Schneider T (1999) Comparison of the Ca2+ currents induced by expression of three cloned α1 subunits, α1G, α1H and α1I, of low-voltage-activated T-type Ca2+ channels. Eur J Neurosci 11:4171–4178

Latour I, Hamid J, Beedle AM, Zamponi GW, Macvicar BA (2003) Expression of voltage-gated Ca2+ channel subtypes in cultured astrocytes. Glia 41:347–353

Lee JH, Daud AN, Cribbs LL, Lacerda AE, Pereverzev A, Klöckner U, Schneider T, Perez-Reyes E (1999) Cloning and expression of a novel member of the low voltage-activated T- type calcium channel family. J Neurosci 19:1912–1921

Lee SE, Lee J, Latchoumane C, Lee B, Oh SJ, Saud ZA, Park C, Sun N, Cheong E, Chen CC, Choi EJ, Lee CJ, Shin HS (2014) Rebound burst firing in the reticular thalamus is not essential for pharmacological absence seizures in mice. Proc Natl Acad Sci U S A 111:11828–11833

Lee N, Jeong S, Kim K-C, Kim J-A, Park J-Y, Kang H-W, Perez-Reyes E, Lee J-H (2017) Ca2+ regulation of Cav3.3 T-type Ca2+ channel is mediated by calmodulin. Mol Pharmacol 92:347–357

Leresche N, Lambert RC (2017) T-type calcium channels in synaptic plasticity. Channels 11:121–139

Li J, Stevens L, Klugbauer N, Wray D (2004) Roles of molecular regions in determining differences between voltage dependence of activation of CaV3.1 and CaV1.2 calcium channels. J Biol Chem 279:26858–26867

Liu X-B, Murray KD, Jones EG (2011) Low-threshold calcium channel subunit Ca(v) 3.3 is specifically localized in GABAergic neurons of rodent thalamus and cerebral cortex. J Comp Neurol 519:1181–1195

Lory P, Nicole S, Monteil A (2020) Neuronal Cav3 channelopathies: recent progress and perspectives. Pflugers Arch 472:831–844

Maksemous N, Blayney CD, Sutherland HG, Smith RA, Lea RA, Tran KN, Ibrahim O, Mcarthur JR, Haupt LM, Cader MZ, Finol-urdaneta RK, Adams DJ, Griffiths LR, Lory P, Adams DJ (2022) Investigation of CACNA1I Cav3. 3 dysfunction in hemiplegic migraine. Front Mol Neurosci 15:892820

Manoach DS, Pan JQ, Purcell SM, Stickgold R (2016) Reduced sleep spindles in schizophrenia: a treatable endophenotype that links risk genes to impaired cognition? Biol Psychiatry 80:599–608

Marksteiner R, Schurr P, Berjukow S, Margreiter E, Perez-Reyes E, Hering S (2001) Inactivation determinants in segment IIIS6 of CaV3.1. J Physiol 537:27–34

Mckay BE, Mcrory JE, Molineux ML, Hamid J, Snutch TP, Zamponi GW, Turner RW (2006) CaV3 T-type calcium channel isoforms differentially distribute to somatic and dendritic compartments in rat central neurons. Eur J Neurosci 24:2581–2594

Mittman S, Guo J, Emerick MC, Agnew WS (1999) Structure and alternative splicing of the gene encoding a. Mol Divers 269:121–124

Monteil A, Chemin J, Leuranguer V, Altier C, Mennessier G, Bourinet E, Lory P, Nargeot J (2000) Specific properties of T-type calcium channels generated by the human alpha 1I subunit. J Biol Chem 275:16530–16535

Murbartián J, Arias JM, Perez-Reyes E (2004) Functional impact of alternative splicing of human T-type Cav3.3 calcium channels. J Neurophysiol 92:3399–3407

Nam G (2018) T-type calcium channel blockers: a patent review (2012–2018). Expert Opin Ther Pat 28:883–901

Park JY, Kang HW, Jeong SW, Lee JH (2004) Multiple structural elements contribute to the slow kinetics of the Ca v3.3 T-type channel. J Biol Chem 279:21707–21713

Pellegrini C, Lecci S, Lüthi A, Astori S (2016) Suppression of sleep spindle rhythmogenesis in mice with deletion of CaV3.2 and CaV3.3 T-type Ca2+ channels. Sleep 39:875–885

Perez-Reyes E (2003) Molecular physiology of low-voltage-activated T-type calcium channels. Physiol Rev 83:117–161

Perez-Reyes E (2006) Molecular characterization of T-type calcium channels. Cell Calcium 40:89–96

Perez-Reyes E (2010) Characterization of the gating brake in the I-II loop of CaV3 T-type calcium channels. Channels 4:453–458

Pinggera A, Striessnig J (2016) Cav1.3 (CACNA1D) L-type Ca2+ channel dysfunction in CNS disorders. J Physiol 594:5839–5849

Pinggera A, Mackenroth L, Rump A, Schallner J, Beleggia F, Wollnik B, Striessnig J (2017) New gain-of-function mutation shows CACNA1D as recurrently mutated gene in autism spectrum disorders and epilepsy. Hum Mol Genet 26:2923–2932

Rosanova M, Ulrich D (2005) Behavioral/systems/cognitive pattern-specific associative long-term potentiation induced by a sleep spindle-related spike train. J Neurosci 25:9398–9405

Sanchez-Roige S, Fontanillas P, Elson SL, Gray JC, de Wit H, MacKillop J, Palmer AA (2019) Genome-wide association studies of impulsive personality traits (BIS-11 and UPPS-P) and drug experimentation in up to 22,861 adult research participants identify loci in the CACNA1I and CADM2 genes. J Neurosci 39:2562–2572

Sanchez-Sandoval AL, Carrillo ZH, Velásquez CED, Delgadillo DM, Rivera HM, Gomora JC (2018) Contribution of S4 segments and S4-S5 linkers to the low-voltage activation properties of T-type CaV3.3 channels. PLoS One 13:1–29

Schizophrenia Working Group of the Psychiatric Genomics Consortium (2014) Biological insights from 108 schizophrenia-associated genetic loci. Nature 511:421–427

Snutch TP, Zamponi GW (2018) Recent advances in the development of T-type calcium channel blockers for pain intervention. Br J Pharmacol 175:2375–2383

Splawski I, Timothy KW, Decher N, Kumar P, Sachse FB, Beggs AH, Sanguinetti MC, Keating MT (2005) Severe arrhythmia disorder caused by cardiac L-type calcium channel mutations. Proc Natl Acad Sci U S A 102:8089–8096

Striessnig J (2021) Voltage-gated Ca2+-channel α1-subunit de novo missense mutations: gain or loss of function – implications for potential therapies. Front Synaptic Neurosci 13:1–25

Talavera K, Nilius B (2006) Biophysics and structure – function relationship of T-type Ca 2 + channels. Cell Calcium 40:97–114

Talley EM, Cribbs LL, Lee J-H, Daud A, Perez-Reyes E, Bayliss DA (1999) Differential distribution of three members of a gene family encoding low voltage-activated (T-type) calcium channels. J Neurosci 19:1895–1911

Uebele VN et al (2009) Positive allosteric interaction of structurally diverse T-type calcium channel antagonists. Cell Biochem Biophys 55:81–93

Vitko I, Chen Y, Arias JM, Shen Y, Wu X, Perez-reyes E (2005) Functional characterization and neuronal modeling of the effects of childhood absence epilepsy variants of CACNA1H, a T-type calcium channel. J Neurosci 25:4844–4855

Weiss N, Zamponi GW (2019) T-type channel druggability at a crossroads. ACS Chem Nerosci 10: 1124–1126

Weiss N, Zamponi GW (2020) Genetic T-type calcium channelopathies. J Med Genet 57:1–10

Wen X, Liang H, Li H, Ou W, Wang HB, Liu H, Li S (2018) In vitro neurotoxicity by ropivacaine is reduced by silencing Cav3.3 T-type calcium subunits in neonatal rat sensory neurons. Artif Cells Nanomed Biotechnol 46:1617–1624

Xie Y, Huang D, Wei L, Luo X-J (2018) Further evidence for the genetic association between CACNA1I and schizophrenia. Hereditas 155:16

Xu W, Liu Y, Chen J, Guo Q, Liu K, Wen Z, Zhou Z, Song Z, Zhou J, He L, Yi Q, Shi Y (2018) Genetic risk between the CACNA1I gene and schizophrenia in Chinese Uygur population. Hereditas 155:5

Yu J, Shi Y, Zhao K, Yang G, Yu L, Li Y, Andersson EM, Ämmälä C, Yang SN, Berggren PO (2020) Enhanced expression of β cell CaV3.1 channels impairs insulin release and glucose homeostasis. Proc Natl Acad Sci U S A 117:448–453

Yunker AMR, Sharp AH, Sundarraj S, Ranganathan V, Copeland TD, McEnery MW (2003) Immunological characterization of T-type voltage-dependent calcium channel Cav3.1 (alpha1G) and Cav3.3 (alpha1I) isoforms reveal differences in their localization, expression, and neural development. Neuroscience 117:321–335

Zamponi GW, Striessnig J, Koschak A, Dolphin AC (2015) The physiology, pathology, and pharmacology of voltage-gated calcium channels and their future therapeutic potential. Pharmacol Rev 67:821–870

Zhang Y, Mori M, Burgess DL, Noebels JL (2002) Mutations in high-voltage-activated calcium channel genes stimulate low-voltage-activated currents in mouse thalamic relay neurons. J Neurosci 22:6362–6371

Zhao Y, Huang G, Wu Q, Wu K, Li R, Lei J, Pan X, Yan N (2019) Cryo-EM structures of apo and antagonist-bound human CaV3.1. Nature 576:492–497

Pathophysiological Roles of Auxiliary Calcium Channel $\alpha_2\delta$ Subunits

Manuel Hessenberger, Sabrin Haddad, and Gerald J. Obermair

Contents

M. Hessenberger · G. J. Obermair (✉)
Division Physiology, Department Pharmacology, Physiology and Microbiology, Karl Landsteiner University of Health Sciences, Krems, Austria
e-mail: Manuel.Hessenberger@kl.ac.at; Gerald.Obermair@kl.ac.at

S. Haddad
Division Physiology, Department Pharmacology, Physiology and Microbiology, Karl Landsteiner University of Health Sciences, Krems, Austria

Institute of Physiology, Medical University of Innsbruck, Innsbruck, Austria
e-mail: Sabrin.Haddad@kl.ac.at

J. Striessnig (ed.), *Voltage-gated Ca^{2+} Channels: Pharmacology, Modulation and their Role in Human Disease*, Handbook of Experimental Pharmacology 279,
https://doi.org/10.1007/164_2022_630

Abstract

$\alpha_2\delta$ proteins serve as auxiliary subunits of voltage-gated calcium channels, which are essential components of excitable cells such as skeletal and heart muscles, nerve cells of the brain and the peripheral nervous system, as well as endocrine cells. Over the recent years, $\alpha_2\delta$ proteins have been identified as critical regulators of synaptic functions, including the formation and differentiation of synapses. These functions require signalling mechanisms which are partly independent of calcium channels. Hence, in light of these features it is not surprising that the genes encoding for the four $\alpha_2\delta$ isoforms have recently been linked to neurological and neurodevelopmental disorders including epilepsy, autism spectrum disorders, schizophrenia, and depressive and bipolar disorders. Despite the increasing number of identified disease-associated mutations, the underlying pathophysiological mechanisms are only beginning to emerge. However, a thorough understanding of the pathophysiological role of $\alpha_2\delta$ proteins ideally serves two purposes: first, it will contribute to our understanding of general pathological mechanisms in synaptic disorders. Second, it may support the future development of novel and specific treatments for brain disorders. In this context, it is noteworthy that the antiepileptic and anti-allodynic drugs gabapentin and pregabalin both act via binding to $\alpha_2\delta$ proteins and are among the top sold drugs for treating neuropathic pain. In this book chapter, we will discuss recent developments in our understanding of the functions of $\alpha_2\delta$ proteins, both as calcium channel subunits and as independent regulatory entities. Furthermore, we present and summarize recently identified and likely pathogenic mutations in the genes encoding $\alpha_2\delta$ proteins and discuss potential underlying pathophysiological consequences at the molecular and structural level.

Keywords

Brain disorders · CACNA2D1–4 · Ca_V · Gabapentin · Pregabalin · Synapse formation · Synaptic transmission · Synaptopathy · Voltage-gated calcium channels

1 Introduction

Voltage-gated calcium (Ca_V) channels play important and crucial roles in a variety of excitable cells. They regulate, for example, skeletal and cardiac muscle contraction, the release of neurotransmitters in nerve cells, hormone secretion and gene regulation. Based on their sequence similarity and their biophysical properties, Ca_V

channels can be divided into three low-voltage-gated ($Ca_V3.1$–3.3) and seven high-voltage-gated ($Ca_V1.1$–1.4, $Ca_V2.1$–2.3) channels. Overall, Ca_V channels are tightly regulated to fine-tune calcium influx and channel dysfunction can lead to severe disorders in all cells and tissues with critical functions depending on calcium channels. These disorders include, but are not limited to, neurological, retinal, cardiac and muscular pathologies such as epilepsy and intellectual disability (Vergult et al. 2015), night blindness (Wycisk et al. 2006a, b) and short QT syndrome (Templin et al. 2011), respectively. The calcium channel complex is assembled around the α_1 subunit, which contains the pore for calcium flux and determines the major biophysical properties of the channel. To fine-tune channel function and to insure its functional plasma membrane expression, the α_1 subunit exists in a hetero-multimeric complex with the auxiliary cytoplasmic β and the entirely extracellular $\alpha_2\delta$ subunits. At least in skeletal muscle, the complex is also associated with a transmembrane γ subunits (reviewed in Obermair et al. 2008; Dolphin 2020; Loh et al. 2020; Ferron et al. 2021; Striessnig 2021; Tuluc et al. 2021; Gandini and Zamponi 2022; Williams et al. 2022).

Seven of the ten pore-forming α_1 subunits ($Ca_V1.1$–1.4, $Ca_V2.1$–2.3) can interact and associate with one out of four $\alpha_2\delta$ subunits (encoded by CACNA2D1-4 genes) and one out of four β subunits (encoded by CACNB1-4 genes), giving rise to a potentially large number of heteromeric channel complexes with a huge diversity of properties and functions. Both β and $\alpha_2\delta$ subunits play major roles in enhancing the functional expression of the channel, as well as influencing their biophysical properties. However, accumulating evidence suggests additional roles for these auxiliary subunits that are not directly dependent on the channel complex. For example, several studies have revealed the involvement of $\alpha_2\delta$ subunits in synaptogenesis (Eroglu et al. 2009; Kurshan et al. 2009) and that β_3 and β_4 may function as transcriptional regulators (Hibino et al. 2003; Zhang et al. 2010). Therefore, it is not surprising that auxiliary subunits also play a pathological role in various diseases (Dolphin 2016; Ablinger et al. 2020), further emphasized by disorders associated with knockout mouse models and mutations. $\alpha_2\delta$ subunits have been linked to neuropathic pain and epilepsy (Neely et al. 2010; Edvardson et al. 2013; Pippucci et al. 2013), mutations in β_4 gene have been found in patients with generalized epilepsy and episodic ataxia (Escayg et al. 2000) and neurodevelopmental disorder (de Bagneaux et al. 2020) and a spontaneous mutation in β_4 gene in the lethargic mouse causes an ataxia, lethargic behaviour and spontaneous focal motor seizures (Burgess et al. 1999). Hence, future therapeutic directions including structural biological approaches should also be included to identify drugs targeting these auxiliary subunits.

$\alpha_2\delta$ proteins became of particular interest as mutations in these proteins are linked to a variety of neurodevelopmental disorders. Up to now, four genes (CACNA2D1-4) have been identified, encoding for four $\alpha_2\delta$ proteins. Worth mentioning, an $\alpha_2\delta$-like protein termed Cachd1 was bioinformatically identified, comprising partly similar effects on calcium channels (Cottrell et al. 2018; Stephens and Cottrell 2019; Ablinger et al. 2022). Besides showing cell and tissue type specific expression, $\alpha_2\delta$ proteins also form multiple transcripts by alternative splicing. Importantly, such splice variants can have considerable functional heterogeneity, even within the same cell type (Lana et al. 2014; Geisler et al. 2019). Furthermore, post translational modifications such as glycosylation and proteolytic cleavage regulate $\alpha_2\delta$ protein maturation and Ca_V channel-dependent and independent trafficking (reviewed in Geisler et al. (2015); Celli et al. (2017); Dolphin (2018); Ablinger et al. (2020); Risher and Eroglu (2020); Tuluc et al. (2021)). In this chapter we focus on the structure and function of the four different $\alpha_2\delta$ subunits, their involvement in channel-dependent and independent functions, as well as their disease associations including new insights into potential pharmacological strategies.

2 Structural Domains and Processing of $\alpha_2\delta$ Subunit

The $\alpha_2\delta$-1 protein (CACNA2D1) was first cloned and discovered in skeletal muscle. Gene sequencing and biochemical studies revealed two highly glycosylated α_2 and δ peptides (Fig. 1), which are encoded by the same gene but transformed into its mature state by proteolytic cleavage and disulphide bond formation (Ellis et al. 1988; De Jongh et al. 1990) (Fig. 2). The other identified subunits (CACNA2D2-4) show similar post translational modifications and glycosylation motifs, compared to $\alpha_2\delta$-1 (Brickley et al. 1995; Gurnett et al. 1996; Qin et al. 2002). Glycosylation is a process by which a sugar moiety is covalently attached to an amino acid and which is

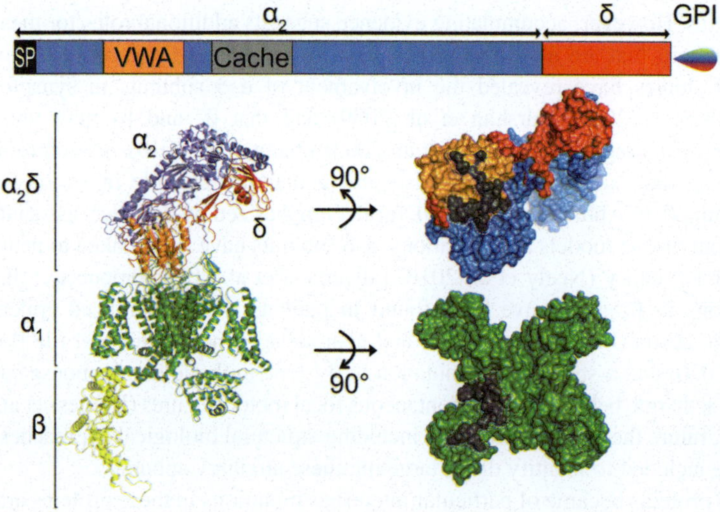

Fig. 1 (Top) Domain architecture of $\alpha_2\delta$ showing the N-terminal signalling peptide (SP) in black, the von Willebrand factor-A (VWA) domain in orange, the Cache domain in grey, the δ-region in red and the GPI-anchor in rainbow. (Bottom) Three-dimensional structure of $Ca_V1.3$ channel complex solved by cryo-EM (PDB code: 7uhg). The domains of $\alpha_2\delta$ were coloured similarly to the domain architecture scheme. The pore-forming subunit is coloured in green and the intracellular β subunit in yellow. The pore-forming α_1 subunit interacts with $\alpha_2\delta$ subunit via specific aa residues, outlined as grey surface

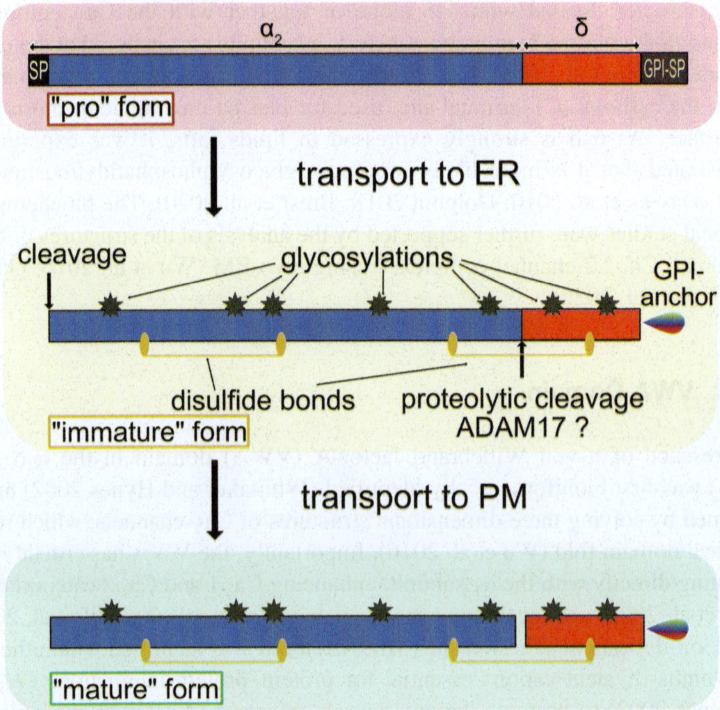

Fig. 2 Maturation of $\alpha_2\delta$ proteins includes several modification steps such as cleavage of the signalling peptide (SP), glycosylation, GPI anchoring, proteolytic cleavage and disulphide bond formation. *GPI-SP* GPI-signalling peptide, *ER* endoplasmic reticulum, *PM* plasma membrane

functionally crucial for protein folding and subsequently cell signalling. Such modifications have been shown to be important for functional maturation of $\alpha_2\delta$ subunits (Davies et al. 2010). The cryogenic electron microscopy (cryo-EM) study of the skeletal muscle $Ca_V1.1$ gained first structural insights into the domain architecture and potential contact sites (Wu et al. 2016), which were later supported by studies on the $Ca_V2.2$ and $Ca_V1.3$ channel complex (Gao et al. 2021; Yao et al. 2022) (Fig. 1). In addition, it was shown that $\alpha_2\delta$-1 is directly binding via its VWA domain to extended extracellular loops of the α_1 subunit, which are defined by specific amino acids (Fig. 1).

2.1 The Amino and Carboxyl Terminus

After the initial cloning of $\alpha_2\delta$-1 an amino (N)-terminal signal sequence was identified (Figs. 1 and 2). Such motifs ensure the trafficking and synthesis into the endoplasmic reticulum (ER), where the immature protein undergoes folding, glycosylation, cleaving off the signal sequence and shuttling to the Golgi apparatus.

Finally, vesicles, loaded with $\alpha_2\delta$ alone or together with the Ca_V complex, are travelling to the plasma membrane, where $\alpha_2\delta$ subunits are anchored at the extracellular side (Canti et al. 2005). First biochemical studies suggested a transmembrane helix at the carboxy (C)-terminal site, used for plasma membrane anchoring at the cell surface. As $\alpha_2\delta$ is strongly expressed in lipids rafts, it was experimentally demonstrated that it is more likely forming a glycosylphosphatidylinositol (GPI)-anchor (Davies et al. 2010; Dolphin 2018; Brest et al. 2020). The biochemical and functional studies were further supported by the analysis of the structures of $Ca_V1.1$, $Ca_V1.3$ and $Ca_V2.2$ channel complexes using cryo-EM (Wu et al. 2016; Gao et al. 2021; Yao et al. 2022).

2.2 VWA Domain

The presence of a von Willebrand factor-A (VWA) domain in the $\alpha_2\delta$ subunit (Fig. 1) was first bioinformatically identified (Whittaker and Hynes 2002) and later confirmed by solving three-dimensional structures of Ca_V channels, which revealed its typical domain fold (Wu et al. 2016). Importantly, the VWA has crucial roles by interacting directly with the α_1 subunit, enhancing Ca_V1 and Ca_V2 calcium currents (Canti et al. 2005) and localizing synaptic calcium channels (Schöpf et al. 2021). A special ion-dependent adhesion site (MIDAS) motif was identified within the VWA, which binds divalent cations essential for protein–protein interactions (Whittaker and Hynes 2002) to promote the anterograde transport of calcium channels (Canti et al. 2005; Hoppa et al. 2012).

2.3 Cache Domain

A calcium channel and chemotaxis receptor (Cache) domain, which is present in $\alpha_2\delta$ proteins, is known to have an important role in small-molecule recognition in a wide range of proteins. First bioinformatic analyses revealed two cache domains (Anantharaman and Aravind 2000) but cryo-EM structures detected four cache domain folds (Wu et al. 2016). The interaction of $\alpha_2\delta$ proteins with extracellular small molecules via the Cache domain may be important in cellular regulatory processes.

2.4 Molecular Structure of $\alpha_2\delta$

The first ever structural information was obtained by a cryo-EM analysis of mouse skeletal muscle $Ca_V1.1$. This structure allowed the prediction of individual domains, post translational modifications, and supports membrane attachment via a C-terminally located GPI-anchor. Furthermore, it confirmed that the VWA domain of $\alpha_2\delta$-1 has physical contact with several loops of the skeletal muscle $Ca_V1.1$ α_1 subunit. The binding of the VWA domain to the loop of the voltage sensing domain of α_1 could explain the mechanism underlying calcium current modulation by $\alpha_2\delta$ subunits (Wu et al. 2016). Still, until today similarly precise insights into the structural features of the isoforms $\alpha_2\delta$-2 to -4 are missing.

2.5 $\alpha_2\delta$ Processing

The four different $\alpha_2\delta$ subunits are encoded by four individual genes CACNA2D1-4 (Ellis et al. 1988; Gao et al. 2000; Qin et al. 2002; Klugbauer et al. 2003). However, these proteins can vary in their molecular weight according to alternative splicing processes, which results in multiple and functionally diverse transcripts (Dolphin 2012; Lana et al. 2014; Geisler et al. 2019). Independent of this, all $\alpha_2\delta$ variants encode an N-terminal signalling peptide, followed by α_2 and at the C-terminal located δ sequence (Fig. 2). Consequently, the pro-form protein is synthesized into the ER, where it undergoes multiple modifications such as protein folding, glycosylation, GPI anchoring, formation of disulphide bonds between the α_2 and δ subunit and cleaving off the signalling sequence (De Jongh et al. 1990; Brickley et al. 1995; Calderon-Rivera et al. 2012). The immature $\alpha_2\delta$ proteins are shuttled to the Golgi apparatus to be finally processed into their mature form by proteolytic cleavage (Kadurin et al. 2016; Ferron et al. 2018). It is assumed that the protease ADAM17 is important for the "separation" of α_2 and δ subunit, which are later linked via disulphide bonds (Kadurin et al. 2022). Even though, $\alpha_2\delta$ subunits are able to be endocytosed without the α_1 channel, it became certain that only the mature form of $\alpha_2\delta$ provides an activation step for Ca_V channel function (Kadurin et al. 2016; Ferron et al. 2018).

2.6 Splicing

With the process of splicing, multiple protein variants can be created from a single gene. CACNA2D1 and CACNA2D2 contain 39 exons including three known alternative splicing regions (termed regions A, B and C) (Angelotti and Hofmann 1996). Certain splice variants were detected in a cell-type specific manner. For example, the $\alpha_2\delta$-1 splice variant, which is lacking region C, could only be found in skeletal muscle, whereas the heart contains different splice variants (Angelotti and Hofmann 1996; Lana et al. 2014). In contrast, fragment C of Cacna2d2 has been confirmed to be essential for the trans-synaptic recruitment of postsynaptic gamma-aminobutyric acid ($GABA_A$)-receptors ($GABA_AR$) and axonal wiring (Geisler et al. 2019). Overall, such protein processing mechanisms seem to be essential for certain physiological functions. At the experimental level the functional variability arising due to alternative splicing makes it even harder to elucidate the specific molecular mechanism in distinct biological processes.

3 Physiological Role of $\alpha_2\delta$ Proteins

After first biochemical and bioinformatic analyses of $\alpha_2\delta$-1, it was believed that these subunits are stably associated with the calcium channel complex to modulate calcium influx (Sharp et al. 1987). Over the years it became evident that $\alpha_2\delta$ proteins have calcium channel-dependent (Obermair et al. 2005; Meyer et al. 2019) but also independent functions like their involvement in synapse formation (Eroglu et al. 2009; Kurshan et al. 2009; Chen et al. 2018; Geisler et al. 2019; Schöpf et al. 2021). The expression of individual $\alpha_2\delta$ subunits in distinct cell types and tissues provides an overall basis for their specific functions (Table 1). Heterologously expressed calcium channels, which are lacking an $\alpha_2\delta$ subunit show a significant reduction in calcium currents (Gurnett et al. 1996, 1997; Klugbauer et al. 1999; Yamaguchi et al. 2000) when compared to channels co-expressed with an $\alpha_2\delta$ subunit. Mechanistically this can be explained by a reduced functional membrane expression of the channel complex (Cassidy et al. 2014; Dolphin 2018). Interestingly, co-expression of $\alpha_2\delta$ subunits was found to increase Ca_V1- and Ca_V2-mediated currents in an α_1 isoform-dependent manner (Obermair et al. 2005; Meyer et al. 2019) (reviewed in Klugbauer et al. (1999); Obermair et al. (2008); Dolphin (2012)). In contrast, knockdown and knockout of $\alpha_2\delta$ subunits resulted in reduced densities of Ca_V channels at the plasma membrane (Nieto-Rostro et al. 2018; Schöpf et al. 2021) or even a complete loss and mis-localization (reviewed in Geisler et al. 2015). Overall, the large extracellular proportion of $\alpha_2\delta$ proteins may also interact with other pre- and postsynaptic proteins to regulate calcium channel activation and kinetics (Chen et al. 2018; Li et al. 2021). An interaction between $\alpha_2\delta$-1 and the GluA1 and GluA2 subunits of the α-amino-3-hydroxy-5-methyl-4-isoxazolepropionic acid (AMPA)-receptors has been described in the hypothalamus, which was shown to be involved in regulating the AMPA receptor calcium permeability (Zhou et al. 2022) (see also below).

Table 1 Overall tissue-specific expression and neuronal distribution of $\alpha_2\delta$ isoforms

Subunit	Tissue-specific expression	Neuronal distribution
$\alpha_2\delta$-1	Muscle (skeletal, smooth, cardiac) Kidney Pancreas	Dorsal root ganglia neurons, glutamatergic neurons (hippocampus, cortex, thalamus and cerebellum)
$\alpha_2\delta$-2	Muscle (skeletal, cardiac) Lung	Cerebellar Purkinje cells Parvalbumin positive interneurons (cortex, hippocampus and thalamus)
$\alpha_2\delta$-3	Muscle (skeletal, cardiac)	Neurons in the caudate putamen and auditory system Cortex and hippocampus
$\alpha_2\delta$-4	Retinal photoreceptor cells adrenal and pituitary glands	Very low expression in CNS

Data are based on multiple studies (Gao et al. 2000; Barclay et al. 2001; Gong et al. 2001; Qin et al. 2002; Cole et al. 2005; Obermair et al. 2005; Schlick et al. 2010; Pirone et al. 2014; Mastrolia et al. 2017)

Furthermore, $\alpha_2\delta$-2 of cochlear inner hair cells is important to align presynaptic L-type $Ca_V1.3$ calcium channels with postsynaptic AMPA receptors (Fell et al. 2016). Also, $\alpha_2\delta$-1 was found to interact with the ionotropic glutamate receptor NMDA (NMDAR) (Chen et al. 2018). This points to an $\alpha_2\delta$ specific function in ion channel clustering. However, as the interaction with NMDAR involves the C-terminus, which is cleaved off during GPI anchoring, the physiological relevance of this proposed interaction has to be further elucidated. Moreover, $\alpha_2\delta$ proteins are found to fulfil important roles in synapse formation and differentiation (Eroglu et al. 2009; Schöpf et al. 2021). Furthermore, as $\alpha_2\delta$-1 was shown to bind insulin granules in pancreatic β-cells and to tether these granules to calcium channels in the plasma membrane, these subunits also play crucial roles in glucose metabolism (reviewed in Tuluc et al. (2021)).

3.1 Channel-Dependent Function

As already stated above, Ca_V channels have to be tightly regulated to control important cellular functions like muscle contraction and neuronal signalling, among others. Heterologous co-expression of $\alpha_2\delta$ is able to enhance current densities and modulate the activation and inactivation kinetics of individually co-expressed α_1 channel (Obermair et al. 2008; Dolphin 2012). Particularly in nerve cells, where calcium influx triggers presynaptic neurotransmitter release and regulates postsynaptic excitability and signal integration, modulatory functions of auxiliary proteins may be critical. Using whole-cell recordings of mouse Purkinje cells, $\alpha_2\delta$-2 was shown to regulate the coupling between postsynaptic calcium entry and calcium-dependent effector systems (Beeson et al. 2022). Another very important function of $\alpha_2\delta$ subunits is to maintain plasma membrane expression of Ca_V channels, which was confirmed in the case of $Ca_V2.2$ (Cassidy et al. 2014; Nieto-Rostro et al. 2018). Consequently, triple knockout/knockdown of the three major neuronal $\alpha_2\delta$ isoforms resulted in strongly reduced presynaptic clustering of $Ca_V2.1$ and $Ca_V2.2$ (Schöpf et al. 2021). Interestingly, even Ca_V channels with mutations in their selectivity filter were shown to be efficiently recruited by $\alpha_2\delta$-1 (Meyer et al. 2019). However, $\alpha_2\delta$ has a dual role in regulating Ca_V membrane expression and activation/inactivation kinetics. These functions can be clearly distinguished as knockdown of $\alpha_2\delta$-1 in differentiated cell models only affected the current kinetics of $Ca_V1.1$ channels without changing its membrane expression, modulated current kinetics and slightly reduced membrane expression of $Ca_V1.2$ channels, and strongly reduced the current density of $Ca_V2.1$ and $Ca_V2.2$ channels without affecting current kinetics (Tuluc et al. 2007; Obermair et al. 2008). Very recently, a study identified $\alpha_2\delta$-4, which is only weakly expressed in the brain, as a regulator of calcium channel function but also being crucial in synapse formation and differentiation. This points to a dual mechanism in channel-dependent and independent functions, fulfilled by a single $\alpha_2\delta$ isoform (Ablinger et al. 2022). It is worth mentioning that the auxiliary cytoplasmic β subunits are not only essential for Ca_V channel trafficking but also show β-dependent current modulation in different cell systems, expressing individual $\alpha_2\delta$ subunits (Yasuda et al. 2004) (reviewed in Buraei and Yang (2010)).

3.2 Channel Independent Function

Besides the well-studied channel-dependent functions of $\alpha_2\delta$ proteins, studies over the last 15 years point towards several distinct functions independent of calcium channel complexes.

Even though $\alpha_2\delta$ subunits are predominantly found in presynaptic structures, they fulfil important roles at the postsynaptic site such as serving as a receptor for the astrocyte-secreted and synaptogenic thrombospondins (TSP) (Eroglu et al. 2009). Interestingly, $\alpha_2\delta$-1 interacts with TSP-1 to induce excitatory synapse formation, followed by recruitment of NMDAR, shown in retinal ganglion cells (Risher et al. 2018). Also, TSP-4 was shown to weakly bind $\alpha_2\delta$-1 (Lana et al. 2016; El-Awaad et al. 2019). Fundamental roles in synapse formation were recently demonstrated by triple knockout/knockdown of all brain $\alpha_2\delta$ isoforms in cultured hippocampal neurons. These dramatic changes in pre- and postsynaptic differentiation could be rescued by the expression of each individual $\alpha_2\delta$ isoforms, pointing to their critical role in basic neuronal functions (Schöpf et al. 2021). However, presynaptic expression of a splice variant of $\alpha_2\delta$-2 (ΔExon 23) is sufficient to regulate postsynaptic $GABA_AR$ abundance in inhibitory synapses (Geisler et al. 2019). Also, $\alpha_2\delta$-3 was shown to regulate presynaptic differentiation in vertebrates, as demonstrated by investigating the consequence of $\alpha_2\delta$-3 knockout on auditory nerve fibres (Pirone et al. 2014). The retinally expressed $\alpha_2\delta$-4 subunit (Wycisk et al. 2006a; Knoflach et al. 2013; Lee et al. 2015) is essential in regulating membrane expression of α_1 subunits and synaptic transmission of rods and cones (Kerov et al. 2018).

4 $\alpha_2\delta$ Mutations and Related Diseases

Mutations affecting $\alpha_2\delta$ subunits on the genetic or protein level have recently been associated with severe diseases including epilepsy, autism spectrum disorders (ASD), schizophrenia (SCZ), bipolar disorders (BPD) and many more (Surmeier 2007; Edvardson et al. 2013; Pietrobon 2013; Rajakulendran and Hanna 2016; Striessnig 2016; Zamponi 2016; Celli et al. 2017; Andrade et al. 2019; Ablinger et al. 2020). A variety of different genomic analyses tools such as single nucleotide variant (SNV) analysis, single nucleotide polymorphism (SNP) genotyping and copy number variation (CNV) analysis detected potential genetic variations, which do not directly affect $\alpha_2\delta$ at the protein level but rather at the genetic level. In the next chapter we will summarize known disease mutations and distinguish between mutations at the genetic (CACNA2D1-4) and protein ($\alpha_2\delta$-1 to $\alpha_2\delta$-4) level. All potential pathologically relevant mutations are mapped to the $\alpha_2\delta$-1 structure and consequential homology models of $\alpha_2\delta$-2 to $\alpha_2\delta$-4 (Fig. 3). Furthermore, the disease-associated mutations are listed in Table 2.

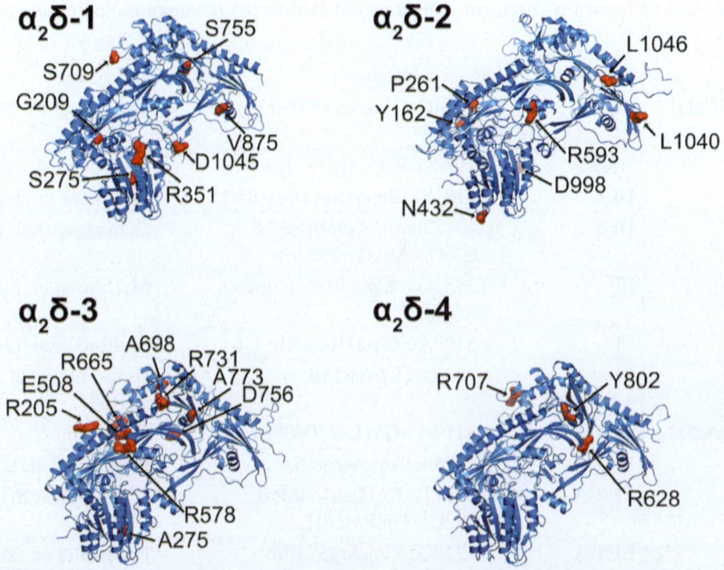

Fig. 3 Cryo-EM structure of $\alpha_2\delta$-1 and homology models of $\alpha_2\delta$-2 to $\alpha_2\delta$-4, calculated by Alfafold, are coloured in blue. Residues which are associated with specific diseases are labeled and shown as red spheres (for details, see text)

4.1 Mutations of $\alpha_2\delta$ at the Protein Level

Most of the disease-related mutations were recently reviewed (Ablinger et al. 2020). Generally, missense, nonsense and silent mutations can affect the folding, degradation, localization or the function of the protein. Consequently, such alterations could directly influence Ca_V channel trafficking, current density and activation and inactivation kinetics, and specific synaptic functions. In contrast, CNV could affect the $\alpha_2\delta$ expression level, which may lead to overall loss- or gain-of-function phenotypes.

All $\alpha_2\delta$ subunits show diverse and overlapping, but also very specific expression patterns throughout different cell types (see Table 1). Therefore, certain diseases are linked to mutations of $\alpha_2\delta$ subunits, which are expressed in a tissue-specific manner. For example, $\alpha_2\delta$-2 mutations are more likely to induce epileptic encephalopathy due to the predominant expression of $\alpha_2\delta$-2 in the cerebellum, whereas mutations of $\alpha_2\delta$-4, which is the predominant isoform in retinal photoreceptor cells, may lead to diseases affecting vision.

4.1.1 $\alpha_2\delta$-1

Disease mutations in $\alpha_2\delta$-1 have been identified mainly in patients with epileptic disorders. Genetic analyses of patients with epileptic encephalopathy (EE) showed two large genomic deletions with 8.2 and 3.9 Mb in size, respectively (Mefford et al. 2011). Additionally, two families have been identified with affected twins having

Table 2 List of human $\alpha_2\delta$ protein mutations reported in the literature including the associated diseases

Gene	Phenotype	Sequence changes	Reference
CACNA2D1	SQTS	c.2264G>C/p.Ser755Thr (corresp. Ser767)	Templin et al. (2011)
	ASD	c.1052G>C/p.Arg351Thr	Iossifov et al. (2014)
	DEE	c.818_821dup/p.Ser275Asnfs*13	Dahimene et al. (2022)
	DEE	c.13_23dup/p.Leu9Alafs*5 c.626G>A/p.Gly209Asp	Dahimene et al. (2022)
	EE	c.2382G>A/p.S709N (corresp. Ser696)	Mefford et al. (2011)
	EE	c.3390A>C/p.D1045A1	Mefford et al. (2011)
	West syndrome	c.2625delT/p.V875fs	Hino-Fukuyo et al. (2015)
CACNA2D2	EE/CA	c.3119A>G/p.Leu1040Pro	Edvardson et al. (2013)
	EE/CA	c.1295delA/p.Asn432fs	Pippucci et al. (2013)
	EE/CA	c.3137T>C/p.Leu1046Pro c.782C>T/p.Pro261Leu	Butler et al. (2018)
	EE/CA	c.1778G>C/p.Arg593Pro	Punetha et al. (2019)
	EE/CA	c.485_486delAT/p.Tyr162Ter	Punetha et al. (2019)
	EE/CA	c.2992G>A/p.Asp998Asn	Valence et al. (2019)
CACNA2D3	ASD	c.2057-2A>G splice site	Iossifov et al. (2014)
	ASD	c.823G>A/p.Ala275Thr	Wang et al. (2016)
	ASD	c.1993C>T/p.Arg665Cys	De Rubeis et al. (2014)
	ASD	c.2092G>A/p.Ala698Thr	De Rubeis et al. (2014)
	ASD	c.1733G>A/p.Arg578Gln	De Rubeis et al. (2014)
	ASD	c.1522G>T/p.Glu508Ter	De Rubeis et al. (2014)
	ASD	c.2191C>G/p.Arg731Gly	De Rubeis et al. (2014)
	ASD	c.2266G>A/p.Asp756Asn	De Rubeis et al. (2014)
	ASD	c.613C>G/p.Arg205Gly	De Rubeis et al. (2014)
	ASD	c.2318C>T/p.Ala773Val	Wang et al. (2016)
	ASD	c.2749G>A/p.Ala917Thr*	Guo et al. (2018)
	ASD	Gene disruptions	Redin et al. (2017)
CACNA2D4	NB	c.2406C>A/p.Tyr 802Ter	Wycisk et al. (2006b)
	BD	Deletion of exons 17–26	Van Den Bossche et al. (2012)
	CRD	c.2120G>A/p.Arg707His	Huang et al. (2015)
	RD	c.1882C>T/p.Arg628Ter	Ba-Abbad et al. (2016)
	RD	Deletion of exons 17–26	Ba-Abbad et al. (2016)

SQTS short QT syndrome, *ASD* autism spectrum disorder, *DEE* developmental and epileptic encephalopathy, *EE/CA* epileptic encephalopathy and cerebellar atrophy, *NB* night blindness, *BD* Belgian late onset bipolar disorder I, *CRD* cone-rod dystrophy, *RD* retinal dysfunction

*This residue is not resolved within the structure and therefore not included in Fig. 3

a p.Ser709Asn or p.Asp1045Ala mutation. Furthermore, mutation p.Val875fs seems to induce age-dependent epilepsy (West syndrome) (Hino-Fukuyo et al. 2015) and p. Arg351Thr mutation ASDs (Iossifov et al. 2014). In addition, biallelic CACNA2D1 loss-of-function variants were found in two unrelated individuals with developmental and epileptic encephalopathy, which have been recently functionally characterized (Dahimene et al. 2022). Not only mutations may potentially contribute to specific diseases. There are two case reports in the literature reporting auto-antibodies against $\alpha_2\delta$-1 proteins potentially linked to amyotrophic lateral sclerosis associated with type 2 diabetes and autoimmune encephalitis (Shi et al. 2019; Lee et al. 2021). Studies on Cacna2d1 knockout mice focusing on neuropathic pain following nerve injury showed reduced sensitivity to mechanical stimulation (Patel et al. 2013).

4.1.2 $\alpha_2\delta$-2

Mutations in the $\alpha_2\delta$-2 protein are predominantly associated with epileptic disorders and cerebellar ataxia, which could be explained by its almost exclusive expression in the cerebellum. For example, p.Leu1040Pro (Edvardson et al. 2013) and p. Asp991Asn missense mutations are linked to EE and cerebellar ataxia (CA) (Valence et al. 2019). Two families with EE were identified, harbouring p.Arg593Pro missense or p.Tyr162Ter nonsense mutation (Punetha et al. 2019). Also, patients with a p.Asn432fs frameshift mutation were found, who are suffering from EE (Pippucci et al. 2013). Additionally, Cacna2d2 knockout mice were used to confirm the specific phenotypes of cerebellar ataxia and epilepsy (Barclay et al. 2001; Brodbeck et al. 2002; Brill et al. 2004; Ivanov et al. 2004; Donato et al. 2006; Geisler et al. 2021).

4.1.3 $\alpha_2\delta$-3

Mutations in $\alpha_2\delta$-3 have been primarily associated with ASDs. Hence, a long list of $\alpha_2\delta$-3 mutations exists, including p.Arg578Gln, p.Arg665Cys, p.Ala698Thr, p. Arg731Gly, p.Asp756Asn, p.Arg205Gly and p.Glu508Ter, which were identified in patients with ASDs (De Rubeis et al. 2014). Further mutations were identified in the Chinese ASD cohort (p.Ala773Val and p.Ala275Thr) (Wang et al. 2016) and from the autism clinical and genetic resources in China (p.Ala917Thr and p. Ala698Val) (Guo et al. 2018).

4.1.4 $\alpha_2\delta$-4

As $\alpha_2\delta$-4 is predominantly expressed in the retina, it is not surprising that the majority of mutations are linked to retinal disorders and only a few brain-related phenotypes are known until today. Hence, the nonsense mutation p.Tyr802Ter induces night blindness (Wycisk et al. 2006a) and the p.Arg707His mutation causes cone-rod dystrophy (Ba-Abbad et al. 2016). Regarding potential brain-related disorders, the deletion of amino acids 574–850 was reported in patients with bipolar disorder (BPD) (Van Den Bossche et al. 2012) and only recently a specific role of $\alpha_2\delta$-4 in sensorimotor gating and motor function was discovered in knockout mice (Klomp et al. 2022).

4.2 Genetic Variations in CACNA2D Genes

The non-coding genome sequence contains various functional elements such as promoters, enhancers and silencers, all of which are involved in the regulation of gene expression. Hence, genetic variations including SNVs and CNVs in the non-coding regions of the human genome may cause gene dysregulation. In addition, most of the identified genetic variations in genome-wide association studies (GWAS), which increased the risk of complex diseases such as autism, have been mapped to non-coding regions.

In this chapter we will summarize human CACNA2D disease associations (for details, see also Ablinger et al. (2020)). However, the interpretation of these non-coding variations is difficult and their potential pathophysiological consequences are still poorly understood.

4.2.1 CACNA2D1

Various psychiatric disorders such as major depressive disorders (MDDs), BPD and SCZ are associated with mutations of CACNA2D1, encoding the $\alpha_2\delta$-1 isoform. Two identified SNPs (rs258668 and rs258677) are related to neuroticism disorders (Okbay et al. 2016) and another SNP (rs2367911) has a high risk for BPD (Winham et al. 2014). Furthermore, a schizophrenia patient within a Swedish population shows a frameshift variant (Purcell et al. 2014). Interestingly, six specific SNPs were identified in epileptic patients, out of which only one is located in the coding sequence (exon 11) (Klassen et al. 2011). Large and even entire chromosomal deletions were found in patients suffering from epilepsy and intellectual disability (Vergult et al. 2015). In addition, CACNA2D1 mutations are reported to induce cardiac dysfunction (Templin et al. 2011) and Brugada syndrome (Burashnikov et al. 2010).

4.2.2 CACNA2D2

Several SNPs have been identified in CACNA2D2, which have been linked to multiple psychiatric disorders, such as schizophrenia (Purcell et al. 2014) and BPD. The SNP rs11917269 induces a premature stop codon or a disrupted splice donor site, which is found in schizophrenia patients (Stahl et al. 2019).

4.2.3 CACNA2D3

Similar to the protein mutations (see above) also genetic mutations within CACNA2D3 are linked to ASD (Iossifov et al. 2012). CACNA2D3 was suggested also to be associated with pain transduction, as SNP rs6777055 has been shown to reduce thermal pain response (Neely et al. 2010) making it a potentially attractive drug target. Over the last years, CACNA2D3 was, similar to other isoforms, linked to neurodevelopmental and neuropsychiatric disorders such as SCZ, BPD, MDD and ASD. The chromosomal region 3p21.1_1, which includes the CACNA2D3 gene, has been identified as a risk region for SCZ and BPD (Liu et al. 2018). In addition, an intronic SNP (rs75407252) within this region has been linked to MDD (Dunn et al. 2016). Furthermore, several additional SNPs are directly connected to ASD, SCZ and epilepsy (Klassen et al. 2011; Iossifov et al. 2012; Meda et al. 2014; Moons et al. 2016; Zhang et al. 2018; Smedler et al. 2019). In one case of ASD, affected patients harbour a SNP (rs3773540), which is located at a splice site (Skafidas et al. 2014). It is worth mentioning that several inherited variations in CACNA2D3 were identified with so far unknown effects (De Rubeis et al. 2014). Interestingly, there is also a link between this gene and hearing (Pirone et al. 2014), which sofar could only be confirmed in a knockout mouse models but not in humans.

4.2.4 CACNA2D4

Even though this gene is almost exclusively expressed in the retina, very low levels have been detected in hippocampus (Schlick et al. 2010; van Loo et al. 2019). Consequently, mutations and certain SNPs may not only be linked to retina degeneration but they might also have a role in neurological disorders such as SNP rs1024582, which is associated with SCZ and BPD (Consortium 2013). Other specific SNPs like rs4765847 are connected to SCZ (Meda et al. 2014), whereas SNP rs2041140 and rs2041141 are linked to BPD (Prabhu and Pe'er 2012). A splice donor variant from a patient with MDD was identified in a post-mortem brain sample (Tombacz et al. 2017) and a homozygous deletion (12p13.33) was found in an ASD patient (Smith et al. 2012). Interestingly, 13 specific SNPs were only found in epilepsy patients (Klassen et al. 2011).

5 $\alpha_2\delta$s Disease Mechanism

$\alpha_2\delta$ proteins are important modulators of calcium channels. As described previously, $\alpha_2\delta$ subunits have calcium channel-dependent but also independent functions such as synapse formation and differentiation. Therefore, $\alpha_2\delta$ proteins can directly or corporately interact with the channel complex by affecting Ca_V channels to induce severe diseases. Genetic modifications in the coding sequence of CACNA2D genes can lead to non-functional proteins, complete or partial protein degradation and also to mis-localization of affected subunit. In contrast, the consequence of mutations, which affect $\alpha_2\delta$ at the genetic level, are not easily predictable as these modifications may only affect the non-coding DNA sequences. However, diseases like MDD ($\alpha_2\delta$-1, $\alpha_2\delta$-2 and $\alpha_2\delta$-3), SCZ ($\alpha_2\delta$-1, $\alpha_2\delta$-2, $\alpha_2\delta$-3 and $\alpha_2\delta$-4), BPD ($\alpha_2\delta$-1, $\alpha_2\delta$-2 and $\alpha_2\delta$-4) and possibly also nicotine dependence ($\alpha_2\delta$-3) (Yin et al. 2017) are often debilitating disorders and it is hence crucial to understand the pathophysiological mechanisms. Principally, specific mutations and deletions within $\alpha_2\delta$ subunits can exert their pathophysiological mechanisms either by altering cellular calcium handling (calcium channel-dependent disease mechanisms) or by interfering with specific functions (e.g., synapse formation; calcium channel independent disease mechanisms). For example, the voltage sensitivity of the channel is influenced by $\alpha_2\delta$-1, therefore, a mutation in $\alpha_2\delta$ may alter the channel activation (Savalli et al. 2016). Of note, $\alpha_2\delta$ subunits have been shown to increase the plasma membrane expression of $Ca_V2.2$ (Cassidy et al. 2014), hence mutations in $\alpha_2\delta$ show inability of $\alpha_2\delta$ to increase the trafficking of Ca_V2 channels (Dahimene et al. 2022).

Similar results have been reported on Ca_V1 channels, which show affected activation and inactivation kinetics due to $\alpha_2\delta$-1 knockdown (Obermair et al. 2005; Meyer et al. 2019). Interestingly, interactions of $\alpha_2\delta$ subunits with other proteins can also influence the channel functions. For example, the low-density lipoprotein receptor-related protein 1 (LRP1) influences $Ca_V2.2$ trafficking via $\alpha_2\delta$-1 (Kadurin et al. 2017) and the synaptic cell-adhesion molecule α-neurexin together with $\alpha_2\delta$-1 controls $Ca_V2.1$-dependent calcium influx (Brockhaus et al. 2018). Overall, all $\alpha_2\delta$ subunits seem to affect channel surface expression (reviewed in Klugbauer et al. (2003); Davies et al. (2007); Geisler et al. (2015); Macabuag and Dolphin (2015)). Besides mutations, splice variants can also influence specific interactions, such as the presence of exon 23 in $\alpha_2\delta$-2, which blocks the trans-synaptic recruitment of postsynaptic $GABA_A$ receptors (Geisler et al. 2021). Hence, splice site mutations in $\alpha_2\delta$ may result in abnormal protein expression, which could alter synaptic and trans-synaptic functions regulated by $\alpha_2\delta$ proteins.

Many potential disease-related point mutations were found in $\alpha_2\delta$-1 to $\alpha_2\delta$-4 at the protein level, but the detailed molecular mechanisms underlying the disease phenotypes remain elusive. However, potential functional consequences of these mutations can be predicted from the known functional domain structure of the protein and the cryo-EM structure: mutations may affect protein expression levels, glycosylation, or affect calcium channel surface expression and calcium current properties. For example, the VWA domain is directly interacting with the α_1 subunit (Fig. 1), mutations in this region could reduce the binding affinity and therefore negatively influence the membrane expression of the affected Ca_V channel. Indeed, mutations within the VWA domain of $\alpha_2\delta$-2 render the protein unable to enhance $Ca_V1.2$, $Ca_V2.1$, and $Ca_V2.2$ currents (Canti et al. 2005). Similarly, the frameshift mutation Asp432fs in $\alpha_2\delta$-2 (Fig. 3) could negatively affect the binding between the VWA domain and the α_1 subunit. The cryo-EM structures of $Ca_V1.1$, $Ca_V1.3$, and $Ca_V2.2$ channel complexes revealed a highly conserved asparagine within the VWA domain of $\alpha_2\delta$-1, which interacts specifically with the S5-S6 loop of $Ca_V1.1$, $Ca_V1.3$, and $Ca_V2.2$ (Gao et al. 2021). Moreover, the single point mutation p. Leu1040Pro in CACNA2D2 (Fig. 3) reduced current densities and suppressed inactivation, which may alter the excitatory-inhibitory balance in the brain and hence cause epilepsy (Edvardson et al. 2013). Another point mutation in the VWA domain of $\alpha_2\delta$-1 (p.Arg351Thr) found in ASD patients leads to a reduced plasma membrane expression of $\alpha_2\delta$-1 (Iossifov et al. 2014). Overall, both mutations might negatively influence the expression of pre- and postsynaptic calcium channels which may also induce developmental adaptations and together alter the excitatory-inhibitory balance. To better understand disease-related mutations within the $\alpha_2\delta$ subunits it will be necessary to obtain high resolution structures of each individual isoforms with and without the calcium channel complex.

6 $\alpha_2\delta$ Proteins as Drug Targets

Cellular calcium handling has to be tightly regulated, as an imbalance can lead to disorders such as depression, osteoporosis, cardiac arrhythmia and many more. The pharmacological market provides a wide selection of calcium antagonists, many of which are blocking or modulating calcium influx through Ca_V channels. Compounds of the dihydropyridine class are used predominantly to reduce systemic vascular resistance and arterial pressure, whereas non-dihydropyridine drugs are selective for myocardial (phenylalkylamine) or for vascular calcium channels (benzothiazepine). Hence, most listed compounds have a direct impact on the cardio-vascular system. Interestingly, two small molecules, gabapentin and pregabalin from the gabapentinoid class of drugs, were identified to specifically bind $\alpha_2\delta$ proteins (predominantly $\alpha_2\delta$-1). Furthermore, in 2018, a Japanese group described a newly synthesized compound, mirogabalin, targeting $\alpha_2\delta$ proteins with similar binding specificities to treat pain, associated with diabetic peripheral neuropathy and postherpetic neuralgia (Komatsu et al. 2021). The gabapentinoid binding site was identified upstream of the VWA domain comprising an RRR motif (Brown and Gee 1998; Wang et al. 1999), which is only present in $\alpha_2\delta$-1 and -2, likely explaining why these drugs do not bind to the other $\alpha_2\delta$ subunits (Gong et al. 2001; Marais et al. 2001). However, many studies showed that the therapeutic effect are mediated by binding to $\alpha_2\delta$-1 and not to $\alpha_2\delta$-2 (Lotarski et al. 2014). Gabapentin was initially identified as an antiepileptic drug (Crawford et al. 1987) but is nowadays a widely accepted treatment for neuropathic pain conditions (Wiffen et al. 2005). The underlying mechanism of action of these drugs is still insufficiently understood. $\alpha_2\delta$-1 expression levels in spinal cord are increased strongly following nerve injury (Luo et al. 2001; Li et al. 2006). Chronic pregabalin treatment of sensory nerve-injured rats showed to inhibit the trafficking of $\alpha_2\delta$-1 to presynaptic terminals in the dorsal horn associated with neuropathic pain in several animal models (Bauer et al. 2009). In vitro experiments indeed indicate that gabapentin reduces the cell surface expression of $\alpha_2\delta$-1 and $\alpha_2\delta$-2 (Field et al. 1999; Li et al. 2011). In fact, gabapentin was shown to specifically disrupt rab11-dependent $\alpha_2\delta$-2 recycling from endosomal compartments, which leads to a reduction of calcium currents (Tran-Van-Minh and Dolphin 2010). Importantly, gabapentin also affects Ca_V channels, as long incubation of cells resulted in a reduction in calcium currents and membrane expression of Ca_V2 α_1 subunits. In contrast to long-term incubation (24–48 h), short-term incubation resulted in only a slight reduction in calcium currents (Hendrich et al. 2008; Tran-Van-Minh and Dolphin 2010). As gabapentin shows its maximum effect on synaptic transmission and cell surface expression of $\alpha_2\delta$-1 and -2 and $Ca_V2.2$ only after long-term incubation, this may point to a mechanism influencing channel surface expression levels or altering interaction with unknown proteins and channels. However, it is speculated that at the spinal level, also acute treatment may target channel cycling pathways (Patel and Dickenson 2016). Apparently, upregulated $\alpha_2\delta$-1 can also interact via its C-terminus with both GluA1 and GluA2 subunits of AMPA receptors, inhibiting the GluA1/GluA2 heteromeric assembly leading to an increased prevalence of GluA2-lacking Ca^{2+}-permeable

AMPA receptors on the cell surface (Li et al. 2021). Additionally, it has been shown that $\alpha_2\delta$-1 also interacts with NMDA receptors via its C-terminus (Chen et al. 2018). Application of gabapentin restored the heteromeric assembly and surface expression of GluA1/GluA2 receptors and reduced the trafficking of NMDARs to the cell surface. Similar effects were observed after intrathecal injection of an $\alpha_2\delta$-1 C-terminus-interfering peptide (Chen et al. 2018; Deng et al. 2019). However, the physiological relevance of these interactions must be further investigated, as the mentioned interactions involve the distal C-terminus of $\alpha_2\delta$-1, which is expected to be post-translationally cleaved during the addition of the GPI-anchor.

Gabapentin was found to block the interaction between TSP and $\alpha_2\delta$-1 in vitro thereby disrupting synaptogenesis (Eroglu et al. 2009), which may be relevant in neuropathic pain and epilepsy. $\alpha_2\delta$-2 was shown to suppress axon growth and regeneration in adult mouse sensory cells (Tedeschi et al. 2016; Goins et al. 2021). Due to this fact, gabapentin and pregabalin were successfully used to promote the regeneration of ascending sensory axons (Sun et al. 2020). Even though gabapentin and pregabalin have side effects causing dizziness and problems walking, they have positive effects on chronic pain including neck and back pain in older adults (Fu and Perloff 2022). Very recently, the cryo-EM structure of gabapentin-bound $\alpha_2\delta$-1 was analyzed (Chen et al. 2022). These data will be essential to further analyse the binding specificity and to understand the mode of action in modulating calcium currents and synaptic transmission. To sum up, gabapentin and pregabalin are important drugs, which are widely used to improve life quality of patients suffering from neuropathic pain conditions after nerve injury induced by trauma, surgery or by hyperglycaemia of diabetic patients.

7 Summary and Outlook

After several years of intense studies on the physiological functions of $\alpha_2\delta$ proteins there is no doubt about their critical roles in regulating calcium channels and in controlling synapse formation and wiring. However, due to their pre- and postsynaptic localization in neurons, several questions need further clarifications: Which mechanisms are rather general and which are related to their specific subcellular neuronal localization? Generally, $\alpha_2\delta$ proteins are important regulators of calcium channel trafficking to the plasma membrane and modulators of the activation kinetics of the channel. The emerging picture on localization-specific functions is more heterogeneous. For example, postsynaptic $\alpha_2\delta$-1 can act as receptors for thrombospondins, glia-secreted molecules promoting excitatory synaptogenesis, indicating a general postsynaptic mechanism in synaptogenesis. On the other hand, experiments in cultured hippocampal neurons demonstrated that triple knockout/knockdown of presynaptic $\alpha_2\delta$ proteins leads to a severe defect in both presynaptic differentiation and function. The lack of presynaptic $\alpha_2\delta$ proteins also strongly affected postsynaptic receptor distribution and structure, however, the absence of $\alpha_2\delta$ proteins from the postsynaptic compartment, although affecting synapse densities (Bikbaev et al. 2020), did not alter glutamatergic synapse composition. Together this implicates presynaptic $\alpha_2\delta$s as key players in pre- and postsynaptic synapse formation by a trans-synaptic mechanism. The severe defect in synapse formation, induced by $\alpha_2\delta$ triple knockout/knockdown, can be rescued by the expression of any of the $\alpha_2\delta$ isoforms and even the $\alpha_2\delta$-like protein Cachd1 (Ablinger et al. 2022). While this general and critical role in synapse formation and differentiation can apparently be redundantly mediated by all $\alpha_2\delta$ isoforms, isoform-specific effects on calcium signalling seem to be more subtle (Ablinger et al. 2022). Together, these findings point towards critical redundant and modulatory heterogeneous functions of individual $\alpha_2\delta$ isoforms in defined locations at pre- or postsynaptic sites. Consequently, studying the specific functions of individual $\alpha_2\delta$ isoforms in regulating calcium channels on the one hand, and synapse formation and differentiation on the other, both at pre- and postsynaptic sites of excitatory and inhibitory neurons, will be important for understanding the detailed physiological roles of $\alpha_2\delta$ proteins. The experimental setup for in-depth in vivo and in vitro studies is experimentally challenging and requires a combination of biochemical, cell- and neurobiological, as well as bioinformatic approaches. Further complicating the experimental challenges is the fact that the neuronal expression patterns of $\alpha_2\delta$ isoforms are partly overlapping and partly diverse and they also vary between individual cell types and during embryonic development. Additionally, the distribution of $\alpha_2\delta$ isoforms within different cell types is still unknown including its antero- and retrograde transport-mechanism. Of note in this context is that the interplay between $\alpha_2\delta$-1 and $\alpha_2\delta$-3 during early development has been shown to be essential to drive the formation of the neuronal network (Bikbaev et al. 2020). Furthermore, $\alpha_2\delta$-4, which was initially suggested to be exclusively expressed in the retina, turned out to be expressed at very low levels in the brain, nevertheless critical for development and disease. Therefore, it will be important to experimentally map the mRNA and protein levels of all $\alpha_2\delta$ isoforms during development and at high resolution.

Besides these open questions, gabapentin and pregabalin, which specifically target $\alpha_2\delta$-1 and $\alpha_2\delta$-2 within a defined binding cavity, are widely prescribed drugs to treat mainly neuropathic pain and also epilepsy. Interestingly, despite the wide use of these drugs, the molecular mechanism in modulating $\alpha_2\delta$-dependent calcium channels in short- or long-term application is insufficiently understood. Up to now, we are lacking detailed structural information of $\alpha_2\delta$ proteins bound to any of the antiepileptic drugs or potential endogenous ligands such as leucine and isoleucine (Brown et al. 1998; Dooley et al. 2007). Additionally, the binding affinities and other mechanisms defining isoform specificity between different $\alpha_2\delta$ proteins and the pore-forming α_1 subunits have not yet been elucidated. However, such a molecular and biochemical characterization is essential to understand the regulatory mechanism and differences of $\alpha_2\delta$-dependent calcium channel modulation.

To sum up, the implementation of heterologous expression systems, creating knockout and mutant mouse models, and employing OMICs technologies, made it possible to gain insights into the roles of $\alpha_2\delta$ isoforms in modulating calcium channels, the formation of neuronal networks and synaptic transmission. Ultimately, to understand the precise molecular physiological and pathophysiological mechanisms of the individual $\alpha_2\delta$ isoforms it will be necessary to develop novel experimental tools in order to implement reliable and specific experimental read-outs and controls.

References

Ablinger C, Geisler SM, Stanika RI et al (2020) Neuronal alpha2delta proteins and brain disorders. Pflugers Arch 472:845–863. https://doi.org/10.1007/s00424-020-02420-2

Ablinger C, Eibl C, Geisler SM et al (2022) α2δ-4 and Cachd1 proteins are regulators of presynaptic functions. Int J Mol Sci 23. https://doi.org/10.3390/ijms23179885

Anantharaman V, Aravind L (2000) Cache-a signalling domain common to animal Ca channel subunits and a class of prokaryotic chemotaxis receptors. Trends Biochem Sci 25:535–537

Andrade A, Brennecke A, Mallat S et al (2019) Genetic associations between voltage-gated calcium channels and psychiatric disorders. Int J Mol Sci 20. https://doi.org/10.3390/ijms20143537

Angelotti T, Hofmann F (1996) Tissue-specific expression of splice variants of the mouse voltage-gated calcium channel a2/d subunit. FEBS Lett 397:331–337

Ba-Abbad R, Arno G, Carss K et al (2016) Mutations in CACNA2D4 cause distinctive retinal dysfunction in humans. Ophthalmology 123:668–671.e2. https://doi.org/10.1016/j.ophtha.2015.09.045

Barclay J, Balaguero N, Mione M et al (2001) Ducky mouse phenotype of epilepsy and ataxia is associated with mutations in the Cacna2d2 gene and decreased calcium channel current in cerebellar Purkinje cells. J Neurosci 21:6095–6104

Bauer CS, Nieto-Rostro M, Rahman W et al (2009) The increased trafficking of the calcium channel subunit a2d-1 to presynaptic terminals in neuropathic pain is inhibited by the a2d ligand pregabalin. J Neurosci 29:4076–4088

Beeson KA, Westbrook GL, Schnell E (2022) alpha2delta-2 is required for depolarization-induced suppression of excitation in Purkinje cells. J Physiol 600:111–122. https://doi.org/10.1113/JP282438

Bikbaev A, Ciuraszkiewicz-Wojciech A, Heck J et al (2020) Auxiliary alpha2delta1 and alpha2delta3 subunits of calcium channels drive excitatory and inhibitory neuronal network development. J Neurosci 40:4824–4841. https://doi.org/10.1523/JNEUROSCI.1707-19.2020

Brest P, Refae S, Mograbi B et al (2020) Host polymorphisms may impact SARS-CoV-2 infectivity. Trends Genet 36:813–815. https://doi.org/10.1016/j.tig.2020.08.003

Brickley K, Campbell V, Berrow N et al (1995) Use of site-directed antibodies to probe the topography of the a2 subunit of voltage-gated Ca2+ channels. FEBS Lett 364:129–133

Brill J, Klocke R, Paul D et al (2004) entla, a novel epileptic and ataxic Cacna2d2 mutant of the mouse. J Biol Chem 279:7322–7330

Brockhaus J, Schreitmuller M, Repetto D et al (2018) alpha-neurexins together with alpha2delta-1 auxiliary subunits regulate Ca(2+) influx through Cav2.1 channels. J Neurosci 38:8277–8294. https://doi.org/10.1523/JNEUROSCI.0511-18.2018

Brodbeck J, Davies A, Courtney J-M et al (2002) The ducky mutation in Cacna2d2 results in altered Purkinje cell morphology and is associated with the expression of a truncated a2d-2 protein with abnormal function. J Biol Chem 277:7684–7693

Brown JP, Gee NS (1998) Cloning and deletion mutagenesis of the a2d calcium channel subunit from porcine cerebral cortex. J Biol Chem 273:25458–25465

Brown JP, Dissanayake VU, Briggs AR et al (1998) Isolation of the [3H]gabapentin-binding protein/alpha 2 delta Ca2+ channel subunit from porcine brain: development of a radioligand binding assay for alpha 2 delta subunits using [3H]leucine. Anal Biochem 255:236–243

Buraei Z, Yang J (2010) The beta subunit of voltage-gated Ca2+ channels. Physiol Rev 90:1461–1506

Burashnikov E, Pfeiffer R, Barajas-Martinez H et al (2010) Mutations in the cardiac L-type calcium channel associated with inherited J-wave syndromes and sudden cardiac death. Heart Rhythm 7: 1872–1882

Burgess DL, Biddlecome GH, McDonough SI et al (1999) β subunit reshuffling modifies N- and P/Q-type Ca2+ channel subunit compositions in lethargic mouse brain. Mol Cell Neurosci 13: 293–311. https://doi.org/10.1006/mcne.1999.0748

Butler KM, Holt PJ, Milla SS et al (2018) Epileptic encephalopathy and cerebellar atrophy resulting from compound heterozygous CACNA2D2 variants. Case Rep Genet 2018:6308283. https://doi.org/10.1155/2018/6308283

Calderon-Rivera A, Andrade A, Hernandez-Hernandez O et al (2012) Identification of a disulfide bridge essential for structure and function of the voltage-gated Ca(2+) channel alpha(2)delta-1 auxiliary subunit. Cell Calcium 51:22–30

Canti C, Nieto-Rostro M, Foucault I et al (2005) The metal-ion-dependent adhesion site in the Von Willebrand factor-A domain of alpha2delta subunits is key to trafficking voltage-gated Ca2+ channels. Proc Natl Acad Sci U S A 102:11230–11235

Cassidy JS, Ferron L, Kadurin I et al (2014) Functional exofacially tagged N-type calcium channels elucidate the interaction with auxiliary alpha2delta-1 subunits. Proc Natl Acad Sci U S A 111: 8979–8984. https://doi.org/10.1073/pnas.1403731111

Celli R, Santolini I, Guiducci M et al (2017) The alpha2delta subunit and absence epilepsy: beyond calcium channels? Curr Neuropharmacol 15:918–925. https://doi.org/10.2174/1570159X15666170309105451

Chen J, Li L, Chen SR et al (2018) The alpha2delta-1-NMDA receptor complex is critically involved in neuropathic pain development and gabapentin therapeutic actions. Cell Rep 22: 2307–2321. https://doi.org/10.1016/j.celrep.2018.02.021

Chen Z, Mondal A, Minor DL Jr (2022) Structural basis for CaVα2δ:gabapentin binding. bioRxiv. https://doi.org/10.1101/2022.12.05.519047

Cole RL, Lechner SM, Williams ME et al (2005) Differential distribution of voltage-gated calcium channel alpha-2 delta (α2δ) subunit mRNA-containing cells in the rat central nervous system and the dorsal root ganglia. J Comp Neurol 491:246–269. https://doi.org/10.1002/cne.20693

Consortium C group of PG (2013) Identification of risk loci with shared effects on five major psychiatric disorders: a genome-wide analysis. Lancet 381:1371–1379. https://doi.org/10.1016/S0140-6736(12)62129-1

Cottrell GS, Soubrane CH, Hounshell JA et al (2018) CACHD1 is an alpha2delta-like protein that modulates CaV3 voltage-gated calcium channel activity. J Neurosci. https://doi.org/10.1523/JNEUROSCI.3572-15.2018

Crawford P, Ghadiali E, Lane R et al (1987) Gabapentin as an antiepileptic drug in man. J Neurol Neurosurg Psychiatry 50:682–686

Dahimene S, von Elsner L, Holling T et al (2022) Biallelic CACNA2D1 loss-of-function variants cause early-onset developmental epileptic encephalopathy. Brain 145:2721–2729. https://doi.org/10.1093/brain/awac081

Davies A, Hendrich J, Van Minh AT et al (2007) Functional biology of the alpha(2)delta subunits of voltage-gated calcium channels. Trends Pharmacol Sci 28:220–228. https://doi.org/10.1016/j.tips.2007.03.005

Davies A, Kadurin I, Alvarez-Laviada A et al (2010) The α2δ subunits of voltage-gated calcium channels form GPI-anchored proteins, a posttranslational modification essential for function. Proc Natl Acad Sci U S A 107:1654–1659. https://doi.org/10.1073/pnas.0908735107

de Bagneaux PC, von Elsner L, Bierhals T et al (2020) A homozygous missense variant in CACNB4 encoding the auxiliary calcium channel beta4 subunit causes a severe neurodevelopmental disorder and impairs channel and non-channel functions. PLoS Genet 16(3):e1008625

De Jongh KS, Warner C, Catterall WA (1990) Subunits of purified calcium channels. a2 and d are encoded by the same gene. J Biol Chem 265:14738–14741

De Rubeis S, He X, Goldberg AP et al (2014) Synaptic, transcriptional and chromatin genes disrupted in autism. Nature 515:209–215. https://doi.org/10.1038/nature13772

Deng M, Chen SR, Chen H, Pan HL (2019) alpha2delta-1-bound N-methyl-D-aspartate receptors mediate morphine-induced hyperalgesia and analgesic tolerance by potentiating glutamatergic input in rodents. Anesthesiology 130:804–819. https://doi.org/10.1097/ALN.0000000000002648

Dolphin AC (2012) Calcium channel auxiliary α 2 δ and β subunits: trafficking and one step beyond. Nat Rev Neurosci 13:542–555. https://doi.org/10.1038/nrn3311

Dolphin AC (2016) Voltage-gated calcium channels and their auxiliary subunits: physiology and pathophysiology and pharmacology. J Physiol 594:5369–5390. https://doi.org/10.1113/JP272262

Dolphin AC (2018) Voltage-gated calcium channel alpha 2delta subunits: an assessment of proposed novel roles. F1000Res 7:1830. https://doi.org/10.12688/f1000research.16104.1

Dolphin AC (2020) Functions of presynaptic voltage-gated calcium channels. Function 2:1–10. https://doi.org/10.1093/function/zqaa027

Donato R, Page KM, Koch D et al (2006) The ducky2J mutation in Cacna2d2 results in reduced spontaneous Purkinje cell activity and altered gene expression. J Neurosci 26:12576–12586

Dooley DJ, Taylor CP, Donevan S, Feltner D (2007) Ca2+ channel α2δ ligands: novel modulators of neurotransmission. Trends Pharmacol Sci 28:75–82. https://doi.org/10.1016/j.tips.2006.12.006

Dunn EC, Wiste A, Radmanesh F et al (2016) Genome-wide association study (Gwas) and genome-wide by environment interaction study (Gweis) of depressive symptoms in African American and Hispanic/Latina women. Depress Anxiety 33:265–280. https://doi.org/10.1002/da.22484

Edvardson S, Oz S, Abulhijaa FA et al (2013) Early infantile epileptic encephalopathy associated with a high voltage gated calcium channelopathy. J Med Genet 50:118–123. https://doi.org/10.1136/jmedgenet-2012-101223

El-Awaad E, Pryymachuk G, Fried C et al (2019) Direct, gabapentin-insensitive interaction of a soluble form of the calcium channel subunit alpha2delta-1 with thrombospondin-4. Sci Rep 9:16272. https://doi.org/10.1038/s41598-019-52655-y

Ellis SB, Williams ME, Ways NR et al (1988) Sequence and expression of mRNAs encoding the a1 and a2 subunits of a DHP-sensitive calcium channel. Science 241:1661–1664

Eroglu C, Allen NJ, Susman MW et al (2009) Gabapentin receptor alpha2delta-1 is a neuronal thrombospondin receptor responsible for excitatory CNS synaptogenesis. Cell 139:380–392

Escayg A, De Waard M, Lee DD et al (2000) Coding and noncoding variation of the human calcium-channel β4-subunit gene CACNB4 patients with idiopathic generalized epilepsy and episodic ataxia. Am J Hum Genet 66:1531–1539. https://doi.org/10.1086/302909

Fell B, Eckrich S, Blum K et al (2016) alpha2delta2 controls the function and trans-synaptic coupling of Cav1.3 channels in mouse inner hair cells and is essential for normal hearing. J Neurosci 36:11024–11036. https://doi.org/10.1523/JNEUROSCI.3468-14.2016

Ferron L, Kadurin I, Dolphin AC (2018) Proteolytic maturation of alpha2delta controls the probability of synaptic vesicular release. elife 7:e37507. https://doi.org/10.7554/eLife.37507

Ferron L, Koshti S, Zamponi GW (2021) The life cycle of voltage-gated Ca2+ channels in neurons: an update on the trafficking of neuronal calcium channels. Neuronal Signal 5:1–17. https://doi.org/10.1042/ns20200095

Field MJ, McCleary S, Hughes J, Singh L (1999) Gabapentin and pregabalin, but not morphine and amitriptyline, block both static and dynamic components of mechanical allodynia induced by streptozocin in the rat. Pain 80:391–398

Fu JL, Perloff MD (2022) Pharmacotherapy for spine-related pain in older adults. Drugs Aging 39:523–550. https://doi.org/10.1007/s40266-022-00946-x

Gandini MA, Zamponi GW (2022) Voltage-gated calcium channel nanodomains: molecular composition and function. FEBS J 3:614–633

Gao B, Sekido Y, Maximov A et al (2000) Functional properties of a new voltage-dependent calcium channel auxiliary subunit gene (CACNA2D2). J Biol Chem 275:12237–12242. https://doi.org/10.1074/jbc.275.16.12237

Gao S, Yao X, Yan N (2021) Structure of human Cav2.2 channel blocked by the painkiller ziconotide. Nature. https://doi.org/10.1038/s41586-021-03699-6

Geisler SM, Schöpf CL, Obermair GJ (2015) Emerging evidence for specific neuronal functions of auxiliary calcium channel α2δ subunits. Gen Physiol Biophys 34:105–118

Geisler S, Schopf CL, Stanika R et al (2019) Presynaptic alpha2delta-2 calcium channel subunits regulate postsynaptic GABAA receptor abundance and axonal wiring. J Neurosci 39:2581–2605. https://doi.org/10.1523/JNEUROSCI.2234-18.2019

Geisler SM, Benedetti A, Schopf CL et al (2021) Phenotypic characterization and brain structure analysis of calcium channel subunit alpha2delta-2 mutant (ducky) and alpha2delta double knockout mice. Front Synaptic Neurosci 13:634412. https://doi.org/10.3389/fnsyn.2021.634412

Goins A, Patel K, Alles SRA (2021) The gabapentinoid drugs and their abuse potential. Pharmacol Ther 227:107926. https://doi.org/10.1016/j.pharmthera.2021.107926

Gong HC, Hang J, Kohler W et al (2001) Tissue-specific expression and gabapentin-binding properties of calcium channel alpha2delta subunit subtypes. J Membr Biol 184:35–43

Guo H, Wang T, Wu H et al (2018) Inherited and multiple de novo mutations in autism/developmental delay risk genes suggest a multifactorial model. Mol Autism 9:64. https://doi.org/10.1186/s13229-018-0247-z

Gurnett CA, De Waard M, Campbell KP (1996) Dual function of the voltage-dependent Ca2+ channel a2d subunit in current stimulation and subunit interaction. Neuron 16:431–440

Gurnett CA, Felix R, Campbell KP (1997) Extracellular interaction of the voltage-dependent Ca2+ channel a2d and a1 subunits. J Biol Chem 272:18508–18512

Hendrich J, Tran-Van-Minh A, Heblich F et al (2008) Pharmacological disruption of calcium channel trafficking by the a2d ligand gabapentin. Proc Natl Acad Sci U S A 105:3628–3633

Hibino H, Pironkova R, Onwumere O et al (2003) Direct interaction with a nuclear protein and regulation of gene silencing by a variant of the Ca2+−channel β4 subunit. Proc Natl Acad Sci U S A 100:307–312. https://doi.org/10.1073/pnas.0136791100

Hino-Fukuyo N, Kikuchi A, Arai-Ichinoi N et al (2015) Genomic analysis identifies candidate pathogenic variants in 9 of 18 patients with unexplained West syndrome. Hum Genet 134:649–658. https://doi.org/10.1007/s00439-015-1553-6

Hoppa MB, Lana B, Margas W et al (2012) alpha2delta expression sets presynaptic calcium channel abundance and release probability. Nature 486:122–125. https://doi.org/10.1038/nature11033

Huang XF, Huang F, Wu KC et al (2015) Genotype-phenotype correlation and mutation spectrum in a large cohort of patients with inherited retinal dystrophy revealed by next-generation sequencing. Genet Med 17:271–278. https://doi.org/10.1038/gim.2014.138

Iossifov I, Ronemus M, Levy D et al (2012) De novo gene disruptions in children on the autistic spectrum. Neuron 74:285–299

Iossifov I, O'Roak BJ, Sanders SJ et al (2014) The contribution of de novo coding mutations to autism spectrum disorder. Nature 515:216–221. https://doi.org/10.1038/nature13908

Ivanov SV, Ward JM, Tessarollo L et al (2004) Cerebellar ataxia, seizures, premature death, and cardiac abnormalities in mice with targeted disruption of the Cacna2d2 gene. Am J Pathol 165:1007–1018

Kadurin I, Ferron L, Rothwell SW et al (2016) Proteolytic maturation of α2δ represents a checkpoint for activation and neuronal trafficking of latent calcium channels. elife 5:e21143

Kadurin I, Rothwell SW, Lana B et al (2017) LRP1 influences trafficking of N-type calcium channels via interaction with the auxiliary alpha2delta-1 subunit. Sci Rep 7:43802. https://doi.org/10.1038/srep43802

Kadurin I, Dahimene S, Page KM et al (2022) ADAM17 mediates proteolytic maturation of voltage-gated calcium channel auxiliary α2δ subunits, and enables calcium current enhancement. Function 3:1–13. https://doi.org/10.1093/function/zqac013

Kerov V, Laird JG, Joiner ML et al (2018) alpha2delta-4 is required for the molecular and structural organization of rod and cone photoreceptor synapses. J Neurosci 38:6145–6160. https://doi.org/10.1523/JNEUROSCI.3818-16.2018

Klassen T, Davis C, Goldman A et al (2011) Exome sequencing of ion channel genes reveals complex profiles confounding personal risk assessment in epilepsy. Cell 145:1036–1048. https://doi.org/10.1016/j.cell.2011.05.025

Klomp A, Omichi R, Iwasa Y et al (2022) The voltage-gated Ca2+ channel subunit alpha2delta-4 regulates locomotor behavior and sensorimotor gating in mice. PLoS One 17:e0263197. https://doi.org/10.1371/journal.pone.0263197

Klugbauer N, Lacinova L, Marais E et al (1999) Molecular diversity of the calcium channel a2-d subunit. J Neurosci 19:684–691

Klugbauer N, Marais E, Hofmann F (2003) Calcium channel alpha2delta subunits: differential expression, function, and drug binding. J Bioenerg Biomembr 35:639–647

Knoflach D, Kerov V, Sartori SB et al (2013) Cav1.4 IT mouse as model for vision impairment in human congenital stationary night blindness type 2. Channels (Austin) 7:503–513. https://doi.org/10.4161/chan.26368

Komatsu S, Nakamura S, Nonaka T et al (2021) Analgesic characteristics of a newly developed alpha2delta ligand, mirogabalin, on inflammatory pain. Mol Pain 17:17448069211052168. https://doi.org/10.1177/17448069211052167

Kurshan PT, Oztan A, Schwarz TL (2009) Presynaptic alpha(2)delta-3 is required for synaptic morphogenesis independent of its Ca(2+)-channel functions. Nat Neurosci 12:1415–1423

Lana B, Schlick B, Martin S et al (2014) Differential up-regulation in DRG neurons of an alphadelta-1 splice variant with a lower affinity for gabapentin after peripheral sensory nerve injury. Pain 155:522–533. https://doi.org/10.1016/j.pain.2013.12.001

Lana B, Page KM, Kadurin I et al (2016) Thrombospondin-4 reduces binding affinity of [3H]-gabapentin to calcium-channel α2δ-1-subunit but does not interact with α2δ-1 on the cell-surface when co-expressed. Sci Rep 6:24531

Lee A, Wang S, Williams B et al (2015) Characterization of Cav1.4 complexes (alpha11.4, beta2, and alpha2delta4) in HEK293T cells and in the retina. J Biol Chem 290:1505–1521. https://doi.org/10.1074/jbc.M114.607465

Lee ST, Lee BJ, Bae JY et al (2021) CaV alpha2delta autoimmune encephalitis: a novel antibody and its characteristics. Ann Neurol 89:740–752. https://doi.org/10.1002/ana.26017

Li CY, Zhang XL, Matthews EA et al (2006) Calcium channel alpha(2)delta(1) subunit mediates spinal hyperexcitability in pain modulation. Pain 125:20–34

Li Z, Taylor CP, Weber M et al (2011) Pregabalin is a potent and selective ligand for alpha(2)delta-1 and alpha(2)delta-2 calcium channel subunits. Eur J Pharmacol 667:80–90

Li L, Chen SR, Zhou MH et al (2021) alpha2delta-1 switches the phenotype of synaptic AMPA receptors by physically disrupting heteromeric subunit assembly. Cell Rep 36:109396. https://doi.org/10.1016/j.celrep.2021.109396

Liu J, Chen J, Perrone-Bizzozero NI et al (2018) Regional enrichment analyses on genetic profiles for schizophrenia and bipolar disorder. Schizophr Res 192:240–246. https://doi.org/10.1016/j.schres.2017.04.033

Loh KWZ, Liang MC, Soong TW, Hu Z (2020) Regulation of cardiovascular calcium channel activity by post-translational modifications or interacting proteins. Pflugers Arch Eur J Physiol 472:653–667. https://doi.org/10.1007/s00424-020-02398-x

Lotarski S, Hain H, Peterson J et al (2014) Anticonvulsant activity of pregabalin in the maximal electroshock-induced seizure assay in alphadelta (R217A) and alphadelta (R279A) mouse mutants. Epilepsy Res 108:833–842. https://doi.org/10.1016/j.eplepsyres.2014.03.002

Luo ZD, Chaplan SR, Higuera ES et al (2001) Upregulation of dorsal root ganglion a2d calcium channel subunit and its correlation with allodynia in spinal nerve-injured rats. J Neurosci 21: 1868–1875
Macabuag N, Dolphin AC (2015) Alternative splicing in CaV2.2 regulates neuronal trafficking via adaptor protein complex-1 adaptor protein binding motifs. J Neurosci 35:14636–14652
Marais E, Klugbauer N, Hofmann F (2001) Calcium channel alpha(2)delta subunits – structure and gabapentin binding. Mol Pharmacol 59:1243–1248
Mastrolia V, Flucher SM, Obermair GJ et al (2017) Loss of alpha2delta-1 calcium channel subunit function increases the susceptibility for diabetes. Diabetes 66:897–907. https://doi.org/10.2337/db16-0336
Meda SA, Ruano G, Windemuth A et al (2014) Multivariate analysis reveals genetic associations of the resting default mode network in psychotic bipolar disorder and schizophrenia. Proc Natl Acad Sci U S A 111:E2066–E2075. https://doi.org/10.1073/pnas.1313093111
Mefford HC, Yendle SC, Hsu C et al (2011) Rare copy number variants are an important cause of epileptic encephalopathies. Ann Neurol 70:974–985. https://doi.org/10.1002/ana.22645
Meyer JO, Dahimene S, Page KM et al (2019) Disruption of the key Ca2+ binding site in the selectivity filter of neuronal voltage-gated calcium channels inhibits channel trafficking. Cell Rep 29:22–33
Moons T, De Hert M, Gellens E et al (2016) Genetic evaluation of schizophrenia using the illumina humanexome chip. PLoS One 11:e0150464. https://doi.org/10.1371/journal.pone.0150464
Neely GG, Hess A, Costigan M et al (2010) A genome-wide Drosophila screen for heat nociception identifies alpha2delta3 as an evolutionarily conserved pain gene. Cell 143:628–638
Nieto-Rostro M, Ramgoolam K, Pratt WS et al (2018) Ablation of alpha2delta-1 inhibits cell-surface trafficking of endogenous N-type calcium channels in the pain pathway in vivo. Proc Natl Acad Sci U S A 115:E12043–E12052. https://doi.org/10.1073/pnas.1811212115
Obermair GJ, Kugler G, Baumgartner S et al (2005) The Ca2+ channel alpha2delta-1 subunit determines Ca2+ current kinetics in skeletal muscle but not targeting of alpha1S or excitation-contraction coupling. J Biol Chem 280:2229–2237
Obermair GJ, Tuluc P, Flucher BE (2008) Auxiliary Ca(2+) channel subunits: lessons learned from muscle. Curr Opin Pharmacol 8:311–318. https://doi.org/10.1016/j.coph.2008.01.008
Okbay A, Baselmans BM, De Neve JE et al (2016) Genetic variants associated with subjective well-being, depressive symptoms, and neuroticism identified through genome-wide analyses. Nat Genet 48:624–633. https://doi.org/10.1038/ng.3552
Patel R, Dickenson AH (2016) Mechanisms of the gabapentinoids and α2δ-1 calcium channel subunit in neuropathic pain. Pharmacol Res Perspect 4:1–13. https://doi.org/10.1002/prp2.205
Patel R, Bauer CS, Nieto-Rostro M et al (2013) alpha2delta-1 gene deletion affects somatosensory neuron function and delays mechanical hypersensitivity in response to peripheral nerve damage. J Neurosci 33:16412–16426. https://doi.org/10.1523/JNEUROSCI.1026-13.2013
Pietrobon D (2013) Calcium channels and migraine. Biochim Biophys Acta 1828:1655–1665. https://doi.org/10.1016/j.bbamem.2012.11.012
Pippucci T, Parmeggiani A, Palombo F et al (2013) A novel null homozygous mutation confirms CACNA2D2 as a gene mutated in epileptic encephalopathy. PLoS One 8:e82154. https://doi.org/10.1371/journal.pone.0082154
Pirone A, Kurt S, Zuccotti A et al (2014) alpha2delta3 is essential for normal structure and function of auditory nerve synapses and is a novel candidate for auditory processing disorders. J Neurosci 34:434–445. https://doi.org/10.1523/JNEUROSCI.3085-13.2014
Prabhu S, Pe'er I (2012) Ultrafast genome-wide scan for SNP-SNP interactions in common complex disease. Genome Res 22:2230–2240. https://doi.org/10.1101/gr.137885.112
Punetha J, Karaca E, Gezdirici A et al (2019) Biallelic CACNA2D2 variants in epileptic encephalopathy and cerebellar atrophy. Ann Clin Transl Neurol 6:1395–1406. https://doi.org/10.1002/acn3.50824
Purcell SM, Moran JL, Fromer M et al (2014) A polygenic burden of rare disruptive mutations in schizophrenia. Nature 506:185–190. https://doi.org/10.1038/nature12975

Qin N, Yagel S, Momplaisir ML et al (2002) Molecular cloning and characterization of the human voltage-gated calcium channel a2d-4 subunit. Mol Pharmacol 62:485–496

Rajakulendran S, Hanna MG (2016) The role of calcium channels in epilepsy. Cold Spring Harb Perspect Med 6:a022723. https://doi.org/10.1101/cshperspect.a022723

Redin C, Brand H, Collins RL et al (2017) The genomic landscape of balanced cytogenetic abnormalities associated with human congenital anomalies. Nat Genet 49:36–45. https://doi.org/10.1038/ng.3720

Risher WC, Eroglu C (2020) Emerging roles for α2δ subunits in calcium channel function and synaptic connectivity. Curr Opin Neurobiol 63:162–169. https://doi.org/10.1016/j.conb.2020.04.007

Risher WC, Kim N, Koh S et al (2018) Thrombospondin receptor alpha2delta-1 promotes synaptogenesis and spinogenesis via postsynaptic Rac1. J Cell Biol. https://doi.org/10.1083/jcb.201802057

Savalli N, Pantazis A, Sigg D et al (2016) The alpha2delta-1 subunit remodels CaV1.2 voltage sensors and allows Ca2+ influx at physiological membrane potentials. J Gen Physiol 148:147–159. https://doi.org/10.1085/jgp.201611586

Schlick B, Flucher BE, Obermair GJ (2010) Voltage-activated calcium channel expression profiles in mouse brain and cultured hippocampal neurons. Neuroscience 167:786–798

Schöpf CL, Ablinger C, Geisler SM et al (2021) Presynaptic alpha2delta subunits are key organizers of glutamatergic synapses. Proc Natl Acad Sci U S A 118. https://doi.org/10.1073/pnas.1920827118

Sharp AH, Imagawa T, Leung AT, Campbell KP (1987) Identification and characterization of the dihydropyridine-binding subunit of the skeletal muscle dihydropyridine receptor. J Biol Chem 262:12309–12315

Shi Y, Park KS, Kim SH et al (2019) IgGs from patients with amyotrophic lateral sclerosis and diabetes target CaValpha2delta1 subunits impairing islet cell function and survival. Proc Natl Acad Sci U S A 116:26816–26822. https://doi.org/10.1073/pnas.1911956116

Skafidas E, Testa R, Zantomio D et al (2014) Predicting the diagnosis of autism spectrum disorder using gene pathway analysis. Mol Psychiatry 19:504–510. https://doi.org/10.1038/mp.2012.126

Smedler E, Bergen SE, Song J, Landen M (2019) Genes, biomarkers, and clinical features associated with the course of bipolar disorder. Eur Neuropsychopharmacol 29:1152–1160. https://doi.org/10.1016/j.euroneuro.2019.07.132

Smith M, Flodman PL, Gargus JJ et al (2012) Mitochondrial and ion channel gene alterations in autism. Biochim Biophys Acta 1817:1796–1802. https://doi.org/10.1016/j.bbabio.2012.04.004

Stahl EA, Breen G, Forstner AJ et al (2019) Genome-wide association study identifies 30 loci associated with bipolar disorder. Nat Genet 51:793–803. https://doi.org/10.1038/s41588-019-0397-8

Stephens GJ, Cottrell GS (2019) CACHD1: a new activity-modifying protein for voltage-gated calcium channels. Channels 13:120–123. https://doi.org/10.1080/19336950.2019.1600968

Striessnig J (2016) Voltage-gated calcium channels – from basic mechanisms to disease. J Physiol 594:5817–5821. https://doi.org/10.1113/JP272619

Striessnig J (2021) Voltage-gated Ca2+−channel α1-subunit de novo missense mutations: gain or loss of function – implications for potential therapies. Front Synaptic Neurosci 13:1–25. https://doi.org/10.3389/fnsyn.2021.634760

Sun W, Larson MJ, Kiyoshi CM et al (2020) Gabapentinoid treatment promotes corticospinal plasticity and regeneration following murine spinal cord injury. J Clin Invest 130:345–358. https://doi.org/10.1172/JCI130391

Surmeier DJ (2007) Calcium, ageing, and neuronal vulnerability in Parkinson's disease. Lancet Neurol 6:933–938. https://doi.org/10.1016/S1474-4422(07)70246-6

Tedeschi A, Dupraz S, Laskowski CJ et al (2016) The calcium channel subunit Alpha2delta2 suppresses axon regeneration in the adult CNS. Neuron 92:419–434. https://doi.org/10.1016/j.neuron.2016.09.026

Templin C, Ghadri JR, Rougier JS et al (2011) Identification of a novel loss-of-function calcium channel gene mutation in short QT syndrome (SQTS6). Eur Heart J 32:1077–1088

Tombacz D, Maroti Z, Kalmar T et al (2017) High-coverage whole-exome sequencing identifies candidate genes for suicide in victims with major depressive disorder. Sci Rep 7:7106. https://doi.org/10.1038/s41598-017-06522-3

Tran-Van-Minh A, Dolphin AC (2010) The alpha2delta ligand gabapentin inhibits the Rab11-dependent recycling of the calcium channel subunit alpha2delta-2. J Neurosci 30:12856–12867. https://doi.org/10.1523/JNEUROSCI.2700-10.2010

Tuluc P, Kern G, Obermair GJ, Flucher BE (2007) Computer modeling of siRNA knockdown effects indicates an essential role of the Ca2+ channel alpha2delta-1 subunit in cardiac excitation-contraction coupling. Proc Natl Acad Sci U S A 104:11091–11096. https://doi.org/10.1073/pnas.0700577104

Tuluc P, Theiner T, Jacobo-Piqueras N, Geisler SM (2021) Role of high voltage-gated ca2+ channel subunits in pancreatic β-cell insulin release. From structure to function. Cell 10. https://doi.org/10.3390/cells10082004

Valence S, Cochet E, Rougeot C et al (2019) Exome sequencing in congenital ataxia identifies two new candidate genes and highlights a pathophysiological link between some congenital ataxias and early infantile epileptic encephalopathies. Genet Med 21:553–563. https://doi.org/10.1038/s41436-018-0089-2

Van Den Bossche MJ, Strazisar M, De Bruyne S et al (2012) Identification of a CACNA2D4 deletion in late onset bipolar disorder patients and implications for the involvement of voltage-dependent calcium channels in psychiatric disorders. Am J Med Genet B Neuropsychiatr Genet 159B:465–475. https://doi.org/10.1002/ajmg.b.32053

van Loo KMJ, Rummel CK, Pitsch J et al (2019) Calcium channel subunit alpha2delta4 is regulated by early growth response 1 and facilitates epileptogenesis. J Neurosci 39:3175–3187. https://doi.org/10.1523/JNEUROSCI.1731-18.2019

Vergult S, Dheedene A, Meurs A et al (2015) Genomic aberrations of the CACNA2D1 gene in three patients with epilepsy and intellectual disability. Eur J Hum Genet 23:628–632. https://doi.org/10.1038/ejhg.2014.141

Wang MH, Offord J, Oxender DL, Su TZ (1999) Structural requirement of the calcium-channel subunit a2d for gabapentin binding. Biochem J 342:313–320

Wang T, Jones RT, Whippen JM, Davis GW (2016) alpha2delta-3 is required for rapid transsynaptic homeostatic signaling. Cell Rep 16:2875–2888. https://doi.org/10.1016/j.celrep.2016.08.030

Whittaker CA, Hynes RO (2002) Distribution and evolution of von Willebrand/integrin A domains: widely dispersed domains with roles in cell adhesion and elsewhere. Mol Biol Cell 13:3369–3387

Wiffen PJ, McQuay HJ, Edwards JE, Moore RA (2005) Gabapentin for acute and chronic pain. Cochrane Database Syst Rev:CD005452. https://doi.org/10.1002/14651858.CD005452

Williams B, Maddox W, Lee A (2022) Calcium channels in retinal function and disease. Annu Rev Vis Sci 8

Winham SJ, Cuellar-Barboza AB, McElroy SL et al (2014) Bipolar disorder with comorbid binge eating history: a genome-wide association study implicates APOB. J Affect Disord 165:151–158. https://doi.org/10.1016/j.jad.2014.04.026

Wu J, Yan Z, Li Z et al (2016) Structure of the voltage-gated calcium channel Cav1.1 at 3.6 A resolution. Nature 537:191–196. https://doi.org/10.1038/nature19321

Wycisk KA, Budde B, Feil S et al (2006a) Structural and functional abnormalities of retinal ribbon synapses due to Cacna2d4 mutation. Invest Ophthalmol Vis Sci 47:3523–3530

Wycisk KA, Zeitz C, Feil S et al (2006b) Mutation in the auxiliary calcium-channel subunit CACNA2D4 causes autosomal recessive cone dystrophy. Am J Hum Genet 79:973–977

Yamaguchi H, Okuda M, Mikala G et al (2000) Cloning of the b2a subunit of the voltage-dependent calcium channel from human heart: cooperative effect of a2/d and b2a on the membrane expression of the a1c subunit. Biochem Biophys Res Commun 267:156–163

Yao X, Gao S, Yan N (2022) Structural basis for pore blockade of human voltage-gated calcium channel Cav1.3 by motion sickness drug cinnarizine. Cell Res:3–5. https://doi.org/10.1038/s41422-022-00663-5

Yasuda T, Chen L, Barr W et al (2004) Auxiliary subunit regulation of high-voltage activated calcium channels expressed in mammalian cells. Eur J Neurosci 20:1–13. https://doi.org/10.1111/j.1460-9568.2004.03434.x

Yin X, Bizon C, Tilson J et al (2017) Genome-wide meta-analysis identifies a novel susceptibility signal at CACNA2D3 for nicotine dependence. Am J Med Genet B Neuropsychiatr Genet 174:557–567. https://doi.org/10.1002/ajmg.b.32540

Zamponi GW (2016) Targeting voltage-gated calcium channels in neurological and psychiatric diseases. Nat Rev Drug Discov 15:19–34. https://doi.org/10.1038/nrd.2015.5

Zhang Y, Yamada Y, Fan M et al (2010) The β subunit of voltage-gated Ca2+ channels interacts with and regulates the activity of a novel isoform of Pax6. J Biol Chem 285:2527–2536. https://doi.org/10.1074/jbc.M109.022236

Zhang T, Zhu L, Ni T et al (2018) Voltage-gated calcium channel activity and complex related genes and schizophrenia: a systematic investigation based on Han Chinese population. J Psychiatr Res 106:99–105. https://doi.org/10.1016/j.jpsychires.2018.09.020

Zhou JJ, Shao JY, Chen SR et al (2022) alpha2delta-1 protein promotes synaptic expression of Ca (2+) permeable-AMPA receptors by inhibiting GluA1/GluA2 heteromeric assembly in the hypothalamus in hypertension. J Neurochem 161:40–52. https://doi.org/10.1111/jnc.15573

The manufacturer's authorised representative in the EU is Springer Nature Customer Service Centre GmbH, Europaplatz 3, 69115 Heidelberg, Germany. If you have any concerns regarding our products, please contact ProductSafety@springernature.com

Printed and bound by CPI Group (UK) Ltd, Croydon, CR0 4YY
01/10/2026
02225905-0001